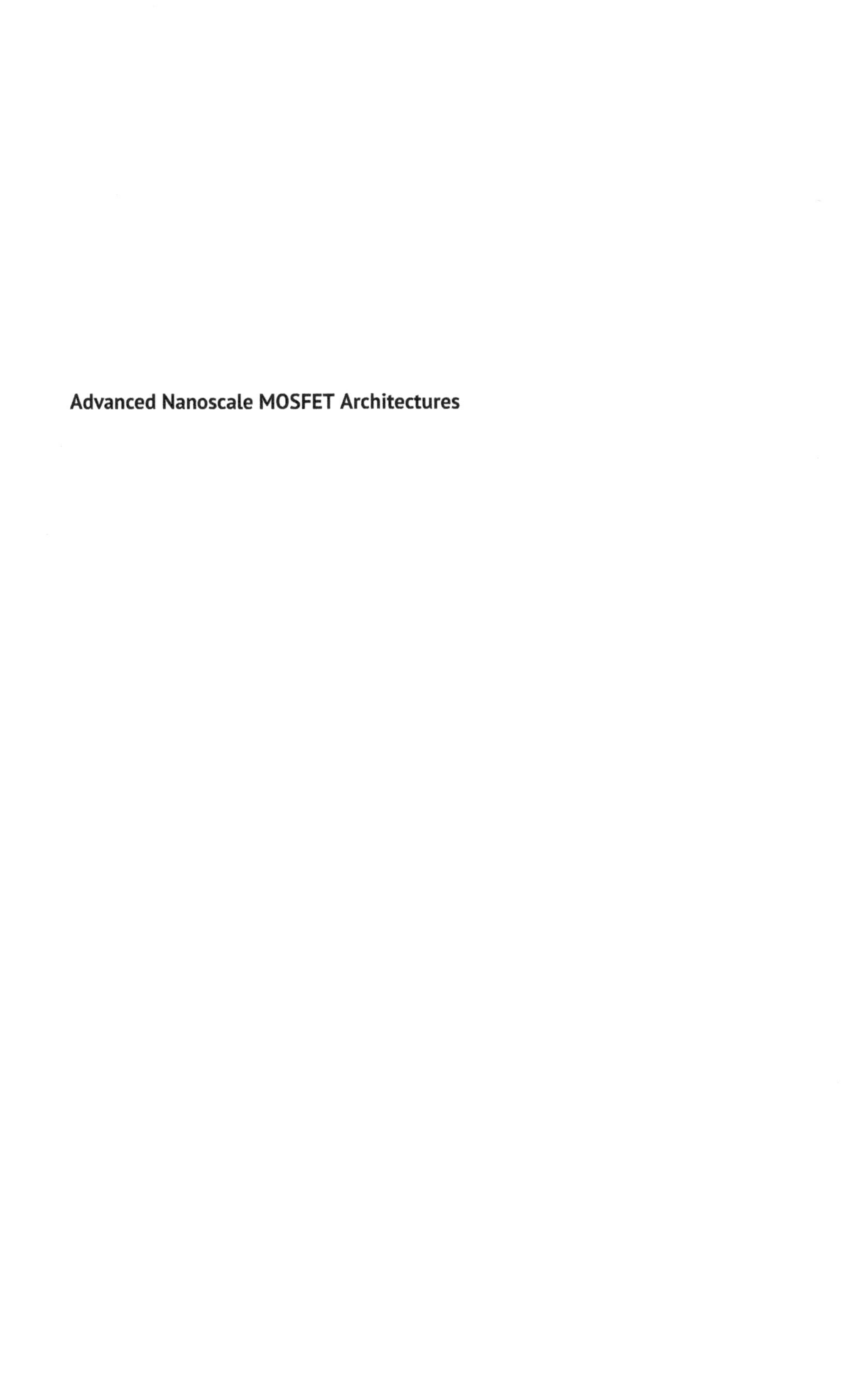

Advanced Nanoscale MOSFET Architectures

Advanced Nanoscale MOSFET Architectures

Current Trends and Future Perspectives

Edited by

Kalyan Biswas
MCKV Institute of Engineering
West Bengal
India

Angsuman Sarkar
Kalyani Govt. Engineering College
West Bengal
India

Published by John Wiley & Sons, Inc., Hoboken, New Jersey.
Published simultaneously in Canada.

For general information on our other products and services or for technical support, please contact our Customer Care Department within the United States at (800) 762-2974, outside the United States at (317) 572-3993 or fax (317) 572-4002.

Wiley also publishes its books in a variety of electronic formats. Some content that appears in print may not be available in electronic formats. For more information about Wiley products, visit our web site at www.wiley.com.

Library of Congress Cataloging-in-Publication Data applied for:

Hardback ISBN: 9781394188949

Cover Design: Wiley
Cover Image: © Darren Robb/Getty Images

Set in 9.5/12.5pt STIXTwoText by Straive, Chennai, India

Contents

About the Editors

Dr. Kalyan Biswas is an a Assistant Professor in the Department of Electronics and Communication Engineering, MCKV Institute of Engineering, Liluah, Howrah, India. He obtained his B Tech and M Tech degrees from the Department of Applied Physics, University of Calcutta, and received his PhD (Engg.) from Jadavpur University. He is a Senior Member of IEEE since 2013 and is currently serving as the Secretary of IEEE SSCS Kolkata Chapter. He worked in different research and industry positions in Japan and Singapore before joining MCKV Institute of Engineering. Along with his teaching, he is involved in research work in the fields of nanoscale electronic devices, MEMS-based sensors, fibre Bragg gratings, electronics packaging, etc. He has more than 50 publications in reputed international journals, conferences and has contributed in many book chapters. He has served as an organizing committee member and reviewer for several international conferences. He is also a reviewer for many international journals.

Angsuman Sarkar is Professor of Electronics and Communication Engineering at Kalyani Government Engineering College, West Bengal, India. He received his M Tech degree and his PhD from Jadavpur University. His current research interests span the study of short-channel effects of sub-100-nm MOSFETs and nano-device modelling. He is a Senior Member of IEEE, Life Member of the Indian Society for Technical Education (ISTE), Associate Life Member of the Institution of Engineers (IE) India, and is currently serving as the Chairman of the IEEE Electron Device Society, Kolkata Chapter. He has authored 6 books, 23 contributed book chapters, 97 journal papers in international refereed journals, and 57 research papers in national and international conferences. He is a member of the board of editors of various journals. He is a reviewer for various international journals. He is currently supervising eight PhD scholars and has already guided seven students successfully as principal supervisor. He has delivered invited talks/tutorial speech/expert talks at various international conferences/technical programs. He has organized IEEE international conferences and several workshops/seminars.

List of Contributors

Zahra Ahangari
Department of Electronic
Yadegar -e- Imam Khomeini (RAH)
Shahre Rey Branch
Islamic Azad University
Tehran
Iran

Ruthramurthy Balachandran
Department of Electronics and Communication Engineering
SOEEC
ASTU
Adama
Ethiopia

Arvind Bisht
Department of Electronics Engineering
National Institute of Technology Uttarakhand
Srinagar Garhwal
Uttarakhand
India

Kalyan Biswas
ECE Department
MCKV Institute of Engineering
Liluah
Howrah
West Bengal
India

Rishu Chaujar
Department of Applied Physics
Delhi Technological University
New Delhi
India

Taraprasanna Dash
Department of ECE
Siksha 'O' Anusandhan (Deemed to be University)
Bhubaneswar
India

Debashis De
Department of Computer Science and Engineering
Maulana Abul Kalam Azad University of Technology
Kolkata
India

and

Department of Physics
University of Western Australia
Perth
Western Australia
Australia

Palasri Dhar
Electronics and Communication Engineering Department
Guru Nanak Institute of Technology
Maulana Abul Kalam Azad University of Technology
Kolkata
India

Yoshitaka Fujimoto
Graduate School of Engineering
Kyushu University
Fukuoka
Japan

Jhansirani Jena
Department of ECE
Siksha 'O' Anusandhan (Deemed to be University)
Bhubaneswar
India

Annada S. Lenka
Department of Electrical Engineering
Nano-Electronics Lab
NIT
Rourkela
India

Chinmay K. Maiti
SouraNiloy
Kolkata
India

Bansi Dhar Malhotra
Department of Biotechnology
Delhi Technological University
New Delhi
India

Girish S. Mishra
EECE
School of Technology
GITAM
Bengaluru
India

Nagarajan Mohankumar
Symbiosis Institute of Technology
Nagpur Campus
Symbiosis International (Deemed University)
Pune
India

Eleena Mohapatra
Department of ECE
RV College of Engineering
Visvesvaraya Technological University
Bengaluru
India

Koyel Mukherjee
Centre of Advanced Study
Institute of Radio Physics and Electronics
University of Calcutta
Kolkata
India

Pankaj K. Pal
Department of Electronics Engineering
National Institute of Technology Uttarakhand
Srinagar Garhwal
Uttarakhand
India

Soumya Pandit
Centre of Advanced Study
Institute of Radio Physics and Electronics
University of Calcutta
Kolkata
India

Yash Pathak
Department of Applied Physics
Delhi Technological University
New Delhi
India

Soumik Poddar
Electronics and Communication Engineering Department
Guru Nanak Institute of Technology
Maulana Abul Kalam Azad University of Technology
Kolkata
India

Yogendra P. Pundir
Department of Electronics and Communication Engineering
HNB Garhwal (A Central) University
Srinagar Garhwal
Uttarakhand
India

and

Department of Electronics Engineering
National Institute of Technology Uttarakhand
Srinagar Garhwal
Uttarakhand
India

Debarati D. Roy
Department of Electronics and Communication Engineering
B. P. Poddar Institute of Management and Technology
Kolkata
West Bengal
India

and

Department of Computer Science and Engineering
Maulana Abul Kalam Azad University of Technology
Kolkata
India

Pradipta Roy
Department of Computer Application
Dr. B. C. Roy Academy of Professional Courses
Durgapur
West Bengal
India

Sunipa Roy
Electronics and Communication Engineering Department
Guru Nanak Institute of Technology
Maulana Abul Kalam Azad University of Technology
Kolkata
India

Prasanna K. Sahu
Department of Electrical Engineering
Nano-Electronics Lab
NIT
Rourkela
India

Angsuman Sarkar
ECE Department
Kalyani Government Engineering College
Kalyani
Nadia
West Bengal
India

Savitesh M. Sharma
Chinmaya Vishwa Vidyapeeth
Ernakulam
Kerala
India

Avtar Singh
Department of Electronics and Communication Engineering
SOEEC
ASTU
Adama
Ethiopia

Asutosh Srivastava
School of Computer & Systems Sciences
Jawaharlal Nehru University
New Delhi
India

Kajal Verma
Department of Applied Physics
Delhi Technological University
New Delhi
India

Preface

The field of metal–oxide–semiconductor field-effect transistor (MOSFET) devices has observed swift growth in the last decade. In recent years, scientists' views on the use of technology have increased. Nanotechnology is a technology that has the potential to significantly impact almost all areas of human activity, raising great hopes for finding solutions to the major needs of society. The fields of application of research in nanoscience include aerospace, defense, national security, electronics, biology, and medicine. In recent years, human knowledge has made great progress through both theoretical analysis and experimental findings in the area of nanoscience and nanoscale devices.

Nanoelectronic devices are the basis of today's powerful computers and are attracting many new applications, including electronic switching, sensing, and other computational applications. However, our purpose is not to discuss specific tools or applications. Rather, it is to illustrate the concept that has emerged in the last two years to understand the flow of electricity at the atomic scale. This is important not only for the creation of new nanoscale materials but also for the insights it provides into some long-standing questions in transport and quantum physics.

Reasonable attention has been given to editing this book to promote knowledge exchange and collaboration among different stakeholders in the field of nanoscale materials. Nano-devices include new and broad fields of activity such as physics, chemistry, biology, and materials engineering focusing on the nanoscale. To understand how these devices work, it is crucial to understand the structure, properties, and quantum behavior of these devices.

Modern life is revolutionized by the advancements of complementary metal–oxide–semiconductor (CMOS) technology. Performance of MOSFET has been improved continuously at a dramatic rate via gate length scaling since its invention. In order to serve the next-generation high-performance requirements with lower operating power, remorseless scaling of CMOS technology has now reached the atomic scale dimensions. Conventional MOSFET scaling not only involves the reduction of device size but also requires the reduction in the transistor supply voltage (V_{DD}). With the reduction of V_{DD}, the threshold voltage (V_{th}) must be scaled down simultaneously in order to attain reasonable ON-state current, reduce delay, and maintain sufficient gate overdrive voltage. As a consequence of scaling of device following Moore's law, the channel length of the MOSFET is reducing every year, causing short channel effects (SCEs). Different strategies have been considered to surmount SCEs using different device architectures and material compositions.

In this book, the problems associated with the emerging nanoscale MOSFET devices and their trends are highlighted. This book is focused on the evaluation of the present development of nanoscale electronic devices and the future projection of device technologies. Basic device physics and MOSFET operation are presented at the beginning. A widespread discussion on basics of MOSFETs and potential difficulties related to scaling and its remedies is presented. Next, discussion on the impact of high-*k* gate dielectrics in next-generation transistors is included. The effects of trap charges on dielectric defects for multiple gate devices, strain engineering for advanced devices like FinFETs, gate all around nanosheet transistors, etc., have been discussed in different chapters. TCAD analysis is a very important methodology for device performance analysis. TCAD simulation is discussed for negative capacitance field-effect transistors (FETs) and their linearity performance. Quantum-mechanical tunnelling effect for electrically doped nano-devices is also included in the scope of this book. The principles and operations of tunnel FETs, graphene-based FETs, and related issues are discussed. Applications of GaN devices are considered for optoelectronics applications. Performance analysis of nanosheet transistors and low-power circuit design using advanced MOSFETs is also discussed. Finally, an FET-based biosensor with negative capacitance is included.

Readers can feel pleasure in learning about nanoscale devices in real-world applications. Throughout this book, one can discover the amazing developments of nanoelectronics, its challenges, and its future prospects. We hope that this book will appear as a one volume reference for postgraduate students, prospective researchers, and professionals requiring knowledge for design of integrated circuits using nanoscale devices.

November 4th, 2024

Kalyan Biswas
Angsuman Sarkar
Kolkata, West Bengal, India

Acknowledgments

We would like to take this golden opportunity to express our gratitude to all those who have helped us complete this book. First and foremost, we would like to convey our gratitude to all the contributors to this book for contributing chapters and all the necessary information throughout this project.

We would like to express our gratitude to Prof. (Dr.) Chandan Kumar Sarkar, a retired professor at Jadavpur University, who supported us throughout the entire work. Lots of useful discussions with him and his advice on device physics and device simulations made us stay confident and helped us to finalize this book project.

Special thanks to the management of MCKV Institute of Engineering and Kalyani Government Engineering College for their necessary support.

We would like to thank our family members for always cheering us up and helping us a lot unconditionally in all the ways that they can. Finally, we would like to express our gratitude to our colleagues for their love, encouragement, and generous support all the time.

Kalyan Biswas
Angsuman Sarkar

1

Emerging MOSFET Technologies

Kalyan Biswas[1] and Angsuman Sarkar[2]

[1]*ECE Department, MCKV Institute of Engineering, Liluah, Howrah, West Bengal, India*
[2]*ECE Department, Kalyani Govt. Engineering College, Kalyani, Nadia, West Bengal, India*

1.1 Introduction: Transistor Action

The human life of the modern generation has been revolutionized by the progress of complementary metal–oxide–semiconductor (CMOS) technology. Metal–oxide–semiconductor field-effect transistor (MOSFET) is one of the most noteworthy inventions of the twentieth century. One important milestone in the progress of semiconductor integrated circuits was the famous – Moore's law [1]. Following Moore's law, the performance of MOSFET has been improved continuously at an intense rate through gate length scaling. To serve the next-generation high-performance requirements with lower operating power, unrelenting scaling of CMOS technology has now reached the atomic scale dimensions. The trend will continue with emerging areas of applications such as the internet of things (IoT), e-mobility, artificial intelligence, and 5G. The cutting-edge innovation in MOSFET technologies is the most important and at the heart of these emerging technologies. A schematic diagram of the Conventional Bulk MOSFET Structure is shown in Figure 1.1.

1.2 MOSFET Scaling

This downscaling of dimensions of the device is critical to integrate the greater number of devices in integrated circuits (ICs). As a consequence of the Moore's law, every year channel length of the MOSFET sinks, causing short channel effect (SCEs). SCEs are affecting power consumption of the circuits [2–9]. The transistor scaling target has been made reachable because of the advanced lithographic

Advanced Nanoscale MOSFET Architectures: Current Trends and Future Perspectives,
First Edition. Edited by Kalyan Biswas and Angsuman Sarkar.

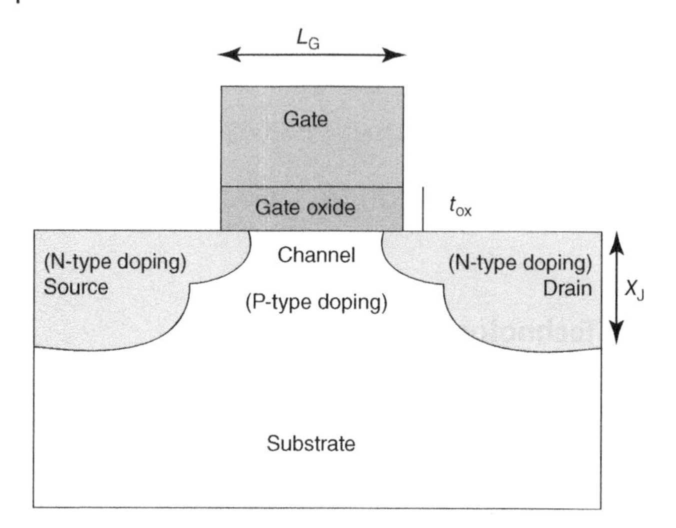

Figure 1.1 Schematic diagram of the Conventional Bulk MOSFET Structure.

capability to make shorter/thinner channels. In the early stage, scaling was possible with conventional structures and material technology, but it is understood that conventional scaling technology cannot continue forever. Therefore, investigation of non-classical device structures became necessary.

1.3 Challenges in Scaling the MOSFET

Scaling of MOSFETS is not an easy task but faces lots of challenges. Normally, six different short-channel effects can be distinguished such as "Sub-Threshold Slope," DIBL and threshold voltage roll-off, velocity saturation, hot carrier effects, and direct source to drain tunneling [10–12].

As the SCEs set hurdles to device operation and degrade device performance, these effects should be removed or minimized, so that a device with a shorter physical channel length can preserve the required device characteristics. Researchers tried to overcome these problems by reducing the gate oxide thickness and the depth of source/drain junction while reducing the gate length in conventional bulk MOSFETs. But these scales reached the physical limit of dimension. As a remedy, gate dielectric materials with higher permittivity were used. The use of these high-k materials as gate oxide allowed for achieving smaller equivalent oxide thickness with a thicker physical dimension. But shrinking of MOSFET to the sub-10 nm scale is challenging and new technologies were necessary. As per ITRS forecasts and published literature, it is understood that the main research is going on in two different directions: possible modification of the planar architecture and use of non-planner 3D structure [13–17] to push for its physical limits, or a new way of making transistors, such as devices based on III–V group materials, use of nanomaterials and nanotechnologies like silicon nanowires, carbon nanotubes (CNTs) or graphene, single electron transistors, and also some other emerging devices such as quantum cellular automata and spin-based electronics.

1.4 Emerging MOSFET Architectures

For decades, traditional scaling techniques based on sinking its physical dimensions have largely dominated the development path of MOSFETs. However, this traditional scaling technique is not valid for emerging nanoscale devices. As device scaling enters beyond the 22 nm node, various significant changes in terms of device architecture and materials in the traditional MOSFET would be required for the competent operation of the device and to extend Moore's law [18–21]. To surmount SCEs, researchers are employing different strategies for nanoscale devices. The main approaches are (i) by employing different structures such as multigate MOSFETs (ii) advanced device physics approaches, such as junctionless MOSFET, tunnel FET (TFET), and (iii) different channel materials having higher carrier mobility such as III–V-based materials, strained silicon, CNTs, Graphene, etc. for continuing the progress in nanoscale.

1.4.1 Tunnel FET

To reduce power consumption in MOSFETs without degrading device performance, operating voltage (V_{dd}) and threshold voltage (V_{th}) of the device need to be scaled down. If V_{th} is reduced keeping sub threshold swing (SS) of MOSFET unchanged, the power consumption increases. The TFET, which is based on the principle of band-to-band quantum tunneling, is one of the most favorable devices, having a steep slope for applications in low-power circuits. The device structure of a TFET differs from that of the conventional MOSFET as a type of doping in the source region and drain region of TFET are of opposite types. A schematic diagram of single-gate n-type TFET is shown in Figure 1.2. A positive voltage in the gate and reverse bias between the source and drain is required to switch the n-type device ON. It is a semiconductor device based on the band-to-band tunneling principle of electrons rather than thermal emission. TFETs operate by tunneling through the S/D barrier rather than diffusion over the barrier [22–31]. The device switches between ON-state as well as OFF-state at lower voltages than the V_{dd} of the MOSFET, making it a suitable choice for low-power consumption applications in the era of emerging nanoscale devices. This type of device can

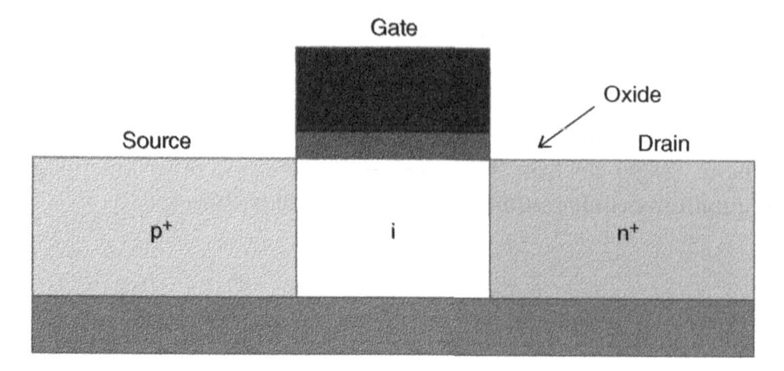

Figure 1.2 Schematic diagram of tunnel FET.

provide extremely low OFF-current and steeper sub-threshold slope than conventional MOSFET. Tunneling occurs for an electron between the valence band of the semiconductor to the conduction band through a potential barrier without having enough energy required for this transition, and this phenomenon can only be explained by quantum mechanical physics. The output characteristics of a TFET are dependent on the parameters such as the doping, the gate work function, etc. Therefore, these parameters can be modified to obtain the desired output characteristics of a TFET. However, from the fabrication point of view, TFET faces a few challenges such as the fabrication of an ultra-thin body required for robust electrostatics, formation of abrupt junction, III–V/high-*k* interface with low trap density, etc.

Two-dimensional crystal semiconductors are being investigated as the materials of the channels for field effect transistors (FETs). The main advantages of such 2D-transistors consist of outstanding electrostatic control of the gate terminal because of the considerably higher surface-to-volume ratio, pristine surfaces to confirm better interface quality with the insulators, and greater electrical conductivity owing to the ballistic/quasi-ballistic transport. It also offers tunable electronic properties dependent on the layer and stacking providing further flexibility in transistor design. These distinctive attributes offer the chance to acquaint with 2D materials in the design of TFET, which can concurrently combine the benefits of greater electrostatic integrity and tunneling barrier engineering. As a result, the arena of TFET design based on 2D materials has grown significantly in recent years.

1.4.2 Nanowire FET

In the era of sub-10-nm technology nodes, cylindrical-shaped structures with gates all around were proposed to provide better gate controllability on the

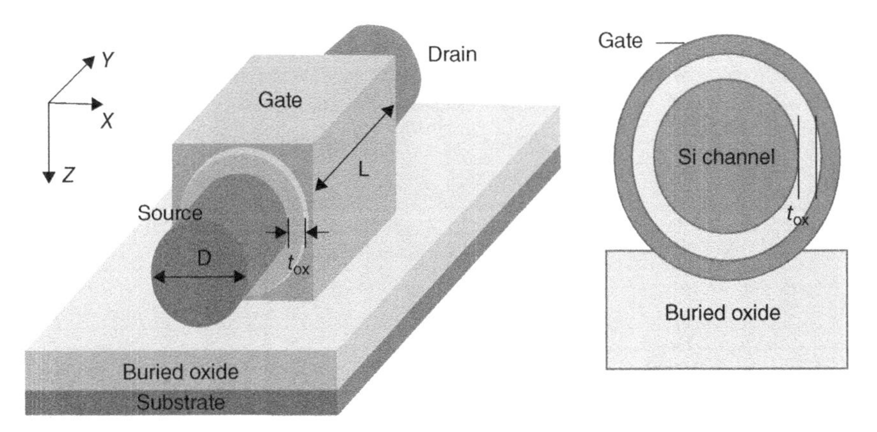

Figure 1.3 Schematic 3D view and a cross-sectional view of a cylindrical FET.

channel and reduce "Short Channel Effects" [32–35]. In this structure, a gate is wrapped around the cylindrical-shaped channel region and termed a silicon nanowire FET (Figure 1.3). Nanowires can be fabricated with single-crystal structures, controllable doping, and diameters as small as several nanometers. Though the silicon nanowire transistors (SNWT) improves device performance, the fluctuations in process parameters rigorously affect the device characteristics. As per the projection of the International Technology Roadmap for Semiconductor (ITRS), the multiple-gate SOI MOSFETs will be able to scale up to sub-10 nm dimensions and are capable candidates for nanoscale devices in the future.

1.4.3 Nanosheet FET

Nanosheet FETs are considered as a transistors of next-generation technology, which have been broadly adopted by the industry to carry on logic scaling beyond 5 nm technology nodes, and beyond FinFETs. Scaling of FinFET beyond 7 nm node results worsened SCEs, forced them to move from tri-date to gate all-around structures. Among different gate all-around structures, wider nanosheets provide higher "ON" current and better electrostatic control [36]. FinFETs were the first architectural change of devices in transistor history and gate-all-around nanosheet FETs are the milestones in the history of transistor devices as they utilize the complete architectural change. To obtain the full advantages of nanosheet FETs, multiple nanosheets should be stacked on one another. The channel thickness during the stacking is fully dependent on the lithographical limit of the fabrication process. Induction of strain to increase hole mobility has also been adopted recently to improve the device's performance (Figure 1.4).

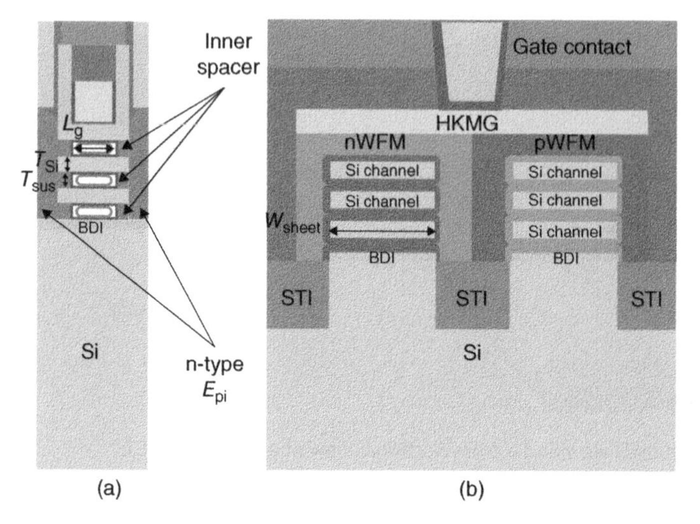

Figure 1.4 Schematic diagram of a gate-all-around nanosheet FET [36]/MDPI/CC by 4.0. Cross section view across a) source-drain region b) gate region.

1.4.4 Negative Capacitance FET

The negative capacitance field effect transistor (NCFET) has become a good solution for extending Moore's Law due to its process compatibility, high on/off current ratio, and low subthreshold swing. In these devices, a layer of ferroelectric material is sandwiched between the gate oxide and gate metal and utilizes the property of polarization inversion of the ferroelectric material under the influence of gate voltage to provide negative capacitance (Figure 1.5).

Additionally, the use of ferroelectric layers, for example, NCs in the gate stack, helps to reduce the sub-threshold slope of the FET to less than the theoretical limit

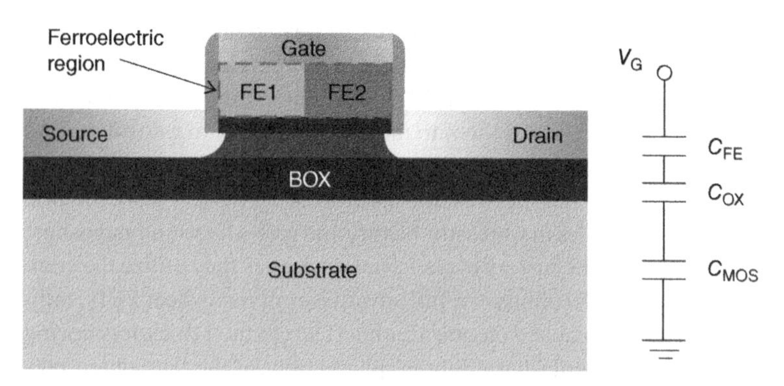

Figure 1.5 (a) Schematic diagram of the DFR-negative capacitance FET, (b) equivalent capacitance model of the device [37]/MDPI/CC by 4.0.

of 60 mV/decade [37]. Various additives such as Al (HAO), Zr (HZO), and Si (HSO) in hafnium-based ferroelectric materials have also been considered to improve the performance of NCFETs.

1.4.5 Graphene FET

CNTs are planar graphite sheets known as graphene that are wrapped into tube shapes. CNTs have outstanding electrical characteristics and they can be fabricated with very small dimensions, as small as 4–8 Å in diameter. The encouraging electrical properties of a CNT depend on its diameter and the wrapping angle of the graphene. Theory shows that the structure of CNTs may be expressed by a chiral vector linked with two integers (n, m). CNTs can be metallic or semiconducting depending on the difference of values in fundamental tube indices (n, m), and their bandgap is dependent on the diameter. The analysis also indicates that semiconducting CNTs have very high low-field mobility, large current-carrying capability, excellent thermal and mechanical stability, and high thermal conductivity [38–40]. Because of their superior material properties, nanotubes are attractive as future interconnects and show enormous advantages as a channel material of high-performance MOSFETs. Though CNT-based MOSFETs promise great performance lots of processing issues remain such as fabrication of identical nanotubes, control of abrupt doping profiles, etc. A sketch of the graphene FET is shown in Figure 1.6 [39].

1.4.6 III–V Material-based MOSFETS

As the performance improvement of silicon-based MOSFETs reaches its limit of scaling. Interest has been greatly increased in introducing non-silicon materials as

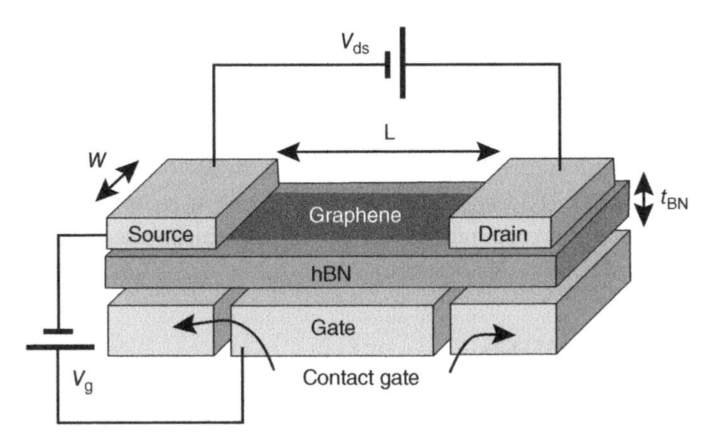

Figure 1.6 A sketch of the graphene FET [39]/MDPI/CC by 4.0.

a channel. III–V-based MOSFETs are considered one of the most efficient devices for high-performance digital logic applications. Currently, III–V MOSFETs are expected to allow higher drive currents and greater flexibility than silicon-based MOSFETs. A wide range of compound semiconductor materials can be obtained using elements from the Periodic Table's columns III and V, like GaAs, InP, and $In_xGa_{(1-x)}As$. The main parameter which defines the important characteristics of these materials is the bandgap energy. The integration of Ge/III–V and Si CMOS platforms is promising in providing low-power integrated circuits in 10 nm technology nodes and beyond [41]. One of the key challenges of the III–V MOSFET technology is thermodynamically stable, high-quality gate dielectrics that passivate the interface states.

1.4.7 HEMT

In recent times, high electron mobility transistor (HEMT) accomplished excessive interest due to its superior electron transport. HEMT devices are facing tremendous challenges and replacing traditional field-effect transistors (FETs) because of their outstanding performance at high frequencies [42]. HEMT technology was first innovated by T. Mimura who was involved in compound semiconductor device development at Fujitsu Laboratories Ltd, Japan [43]. HEMT devices incorporate heterojunctions formed at the junction of two different bandgap materials in which electrons are trapped in quantum wells to avoid scattering by impurities. Thanks to their higher electron mobility and dielectric constant, GaAs having direct bandgap have been used in high-frequency applications and the field of optoelectronic integrated circuits. AlGaAs having nearly similar lattice constant but larger bandgap in comparison to GaAs, are considered the most suitable contender for barrier material and one of the most prevalent choices to be used in HEMTs [44–46]. However, another excellent material that has been widely studied for HEMT devices in recent years is AlGaN/GaN. AlGaN/GaN HEMTs can operate at very high frequencies with high breakdown strength and high saturation electron velocity. GaN also shows very robust piezoelectric polarization that helps to accumulate huge carriers at the interface of AlGaN/GaN. The performance of the MEMS devices depends on many factors such as a combination of material layers, concentration of doping, and different layer thicknesses, which provide flexibility in the device design process.

1.4.8 Strain Engineered MOSFETs

Strained silicon technology based on the improvement of carrier mobility under the influence of axial strain. Proper use of strain in the silicon channel has emerged as a powerful technique for improved MOSFET performance [47].

Usually, epitaxial layer growth of Si on SiGe allows the formation of strained silicon. Carrier mobility can also be altered by changing content of the "Ge" on "SiGe" layer [48]. The application of strain helps to enhance the ON-current considerably without altering the transistor design and helps to meet the projected performance improvement. Strain engineering has evolved from the conventional structure of 2D MOSFET to 3D FinFET structure in silicon-based advanced CMOS technology. Strain engineering is also studied in non-silicon based functional materials like gallium nitride (GaN) and 2D materials [49]. Tremendous growth with innovative ideas of introducing stress into the device channel is realized with each technology node.

1.5 Organization of this Book

This book is focused on the evaluation of the present development of nanoscale electronic devices and the future projection of device technologies. Basic device physics and MOSFET operation are presented in Chapter 2. In this chapter, a widespread discussion on the basics of MOSFETs and potential difficulties related to scaling and its remedies is presented. Chapter 3 discusses the impact of high-k gate dielectrics in next-generation transistors. Chapter 4 deals with the effects of trap charges on dielectric defects for multiple gate devices. Strain engineering for advanced devices like FinFETs, gates all around nanosheet transistors, etc. has been discussed in Chapter 5. TCAD analysis for NCFETs and its linearity performance is presented in Chapter 6. In Chapter 7, quantum-mechanical tunnelling effect for electrically doped nanodevices is discussed. Chapter 8 elaborates on the principles and operations of TFETs. Applications of GaN devices are provided for optoelectronics in Chapter 9. An illustrative idea of graphene-based FETs and related issues are presented in chapter 10. Performance analysis of nanosheet transistors is considered Chapter 11. Low-power circuit design using advanced MOSFETs is discussed in Chapter 12. Chapter 13 presents FET-based biosensors with negative capacitance. An overall conclusion is drawn in Chapter 14.

References

1 Moore, G.E. (1965). Cramming more components onto integrated circuits. *Electronics* 38 (8).

2 Björkqvist, K. and Arnborg, T. (1981). Short channel effects in MOS-transistors. *Physica Scripta* 24 (2): 418–421.

3 Sarkar, A., De, S., Nagarajan, M. et al. (2008). Effect of fringing fields on subthreshold surface potential of channel engineered short channel MOSFETs. *TENCON 2008 – 2008 IEEE Region 10 Conference*. 1–6.

4 Sarkar, A., De, S., and Sarkar, C.K. (2011). *VLSI Design and EDA Tools*. India: Scitech Publications.
5 Pal, A. and Sarkar, A. (2014). Analytical study of dual material surrounding gate MOSFET to suppress short-channel effects (SCEs). *Journal of Engineering Science and Technology* https://doi.org/10.1016/J.JESTCH.2014.06.002.
6 Bari, S., De, D., and Sarkar, A. (2015). Effect of gate engineering in JLSRG MOSFET to suppress SCEs: an analytical study. *Physica E: Low-dimensional Systems and Nanostructures* 67: 143–151.
7 Baral, B., Das, A.K., De, D., and Sarkar, A. (2016). An analytical model of triple-material double-gate metal–oxide–semiconductor field-effect transistor to suppress short-channel effects. *International Journal of Numerical Modelling: Electronic Networks, Devices and Fields* 29 (1): 47–62.
8 Sarkar, A. (2013). Device simulation using silvaco ATLAS tool. In: *Technology Computer Aided Design: Simulation for VLSI MOSFET*, 187.
9 Sarkar, A., De, S., Chanda, M., and Sarkar, C.K. (2016). *Low Power VLSI Design: Fundamentals*. Walter de Gruyter GmbH & Co KG.
10 Young, K.K. (1989). Short-channel effect in fully depleted SOI MOSFETs. *IEEE Transactions on Electron Devices* 36 (2): 399–402.
11 Biswas, K., Sarkar, A., and Sarkar, C.K. (2015). Impact of barrier thickness on analog, RF and linearity performance of nanoscale DG heterostructure MOSFET. *Superlattices and Microstructures* 86: 95–104.
12 Biswas, K., Sarkar, A., and Sarkar, C.K. (2017). Assessment of dielectrics and channel doping impact in nanoscale double gate III–V MOSFET with heavily doped source/drain region. *Materials Focus* 6 (2): 116–120.
13 Colinge, J.P. (2008). *FinFETs and Other Multi-Gate Transistors*. Springer.
14 Biswas, K., Sarkar, A., and Sarkar, C.K. (2016). Impact of Fin width scaling on RF/analog performance of junctionless accumulation-mode bulk FinFET. *ACM Journal on Emerging Technologies in Computing Systems (JETC)* 12 (4): 1–12.
15 Biswas, K., Sarkar, A., and Sarkar, C.K. (2017). Spacer engineering for performance enhancement of junctionless accumulation-mode bulk FinFETs. *IET Circuits, Devices and Systems* 11 (1): 80–88.
16 Biswas, K., Sarkar, A., and Sarkar, C.K. (2018). Fin shape influence on analog and RF performance of junctionless accumulation-mode bulk FinFETs. *Microsystem Technologies* 24: 2317–2324.
17 Ghoshhajra, R., Biswas, K., and Sarkar, A. (2022). Device performance prediction of nanoscale junctionless FinFET using MISO artificial neural network. *Silicon* 1–10.
18 Sarkar, A., De, S., and Sarkar, C.K. (2013). Asymmetric halo and symmetric SHDMG & DHDMGn-MOSFETs characteristic parameter modeling. *Indian Journal Of Nuclear Medicine*, Wiley, USA 26 (1): 41–55.

19 Basak, A. and Sarkar, A. (2020). Drain current modelling of asymmetric junctionless dual material double gate MOSFET with high K gate stack for analog and RF performance. *Silicon* 1–12.

20 Basak, A. and Sarkar, A. (2020). Analog/RF performance of AJDMDG Stack MOSFET. *Solid State Electronics Letters* 2: 117–123.

21 Ghoshhajra, R., Biswas, K., and Sarkar, A. (2021). A review on machine learning approaches for predicting the effect of device parameters on performance of nanoscale MOSFETs. 489–493.

22 Biswal, S.M., Baral, B., De, D., and Sarkar, A. (2019). Simulation and comparative study on analog/RF and linearity performance of III–V semiconductor-based staggered heterojunction and InAs nanowire (nw) Tunnel FET. *Microsystem Technologies* 25: 1855–1861.

23 Chakraborty, A., Singha, D., and Sarkar, A. (2018). Staggered heterojunctions-based tunnel-FET for application as a label-free biosensor. *International Journal Of Nanoparticles* 10 (1–2): 107–116.

24 Sarkar, A. and Sarkar, C.K. (2013). RF and analogue performance investigation of DG tunnel FET. *International Journal of Electronics Letters* 1 (4): 210–217.

25 Chakraborty, A. and Sarkar, A. (2015). Staggered heterojunctions-based nanowire tunneling field-effect transistors for analog/mixed-signal system-on-chip applications. *Nano* 10 (02): 1550027.

26 Chakraborty, A. and Sarkar, A. (2015). Investigation of analog/RF performance of staggered heterojunctions based nanowire tunneling field-effect transistors. *Superlattices and Microstructures* 80: 125–135.

27 Biswal, S.M., Baral, B., De, D., and Sarkar, A. (2016). Study of effect of gate-length downscaling on the analog/RF performance and linearity investigation of InAs-based nanowire tunnel FET. *Superlattices and Microstructures* 91: 319–330.

28 Biswal, S.M., Baral, B., De, D., and Sarkar, A. (2016). Analog/RF performance and linearity investigation of Si-based double gate tunnel FET. *Advances in Industrial Engineering and Management* 5 (1): 1501–1556.

29 Baral, B., Biswal, S.M., De, D., and Sarkar, A. (2017). Effect of gate-length downscaling on the analog/RF and linearity performance of InAs-based nanowire tunnel FET. *International Journal of Numerical Modelling: Electronic Networks, Devices and Fields* 30 (3–4).

30 Deyasi, A., Mukhopadhyay, S., and Sarkar, A. (2020). Novel analytical model for computing subthreshold current in heterostructure p-MOSFET incorporating band-to-band tunneling effect. *Journal of Physics: Conference Series* 1579: 012009.

31 Choi, W.Y., Park, B., Lee, J.D., and Liu, T.K. (2007). Tunnelling field-effect transistors (TFETs) with subthreshold swing (SS) less than 60 mV/dec. *IEEE Electron Device Letters* 28 (8): 743–745.

32 Sarkar, A., De, S., Dey, A., and Sarkar, C.K. (2012). A new analytical subthreshold model of SRG MOSFET with analogue performance investigation. *International Journal of Electronics* 99 (2): 267–283.

33 Sarkar, A., De, S., Dey, A., and Sarkar, C.K. (2012). Analog and RF performance investigation of cylindrical surrounding-gate MOSFET with an analytical pseudo-2D model. *Journal of Computational Electronics* 11: 182–195.

34 Deyasi, A., Bhattacharjee, A.K., Mukherjee, S., and Sarkar, A. (2021). Multi-layer perceptron based comparative analysis between CNTFET and quantum wire FET for optimum design performance. *Solid State Electronics Letters* 3: 42–52.

35 Cui, Y., Zhong, Z., Wang, D. et al. (2003). High performance silicon nanowire field effect transistors. *Nano Letters* 3 (2): 149–152.

36 Mukesh, S. and Zhang, J. (2022). A review of the gate-all-around nanosheet FET process opportunities. *Electronics* 11: 3589. https://doi.org/10.3390/electronics11213589.

37 Yao, J., Han, X., Zhang, X.-P. et al. (2022). Investigation on the negative capacitance field effect transistor with dual ferroelectric region. *Crystals* 12 (11): 1545. https://doi.org/10.3390/cryst12111545.

38 Xia, F., Farmer, D.B., Lin, Y.-m., and Avouris, P. (2010). Graphene field-effect transistors with high on/off current ratio and large transport band gap at room temperature. *Nano Letters* 10 (2): 715–718. https://doi.org/10.1021/nl9039636.

39 Wilmart, Q., Boukhicha, M., Graef, H. et al. (2020). High-frequency limits of graphene field-effect transistors with velocity saturation. *Applied Sciences* 10: 446. https://doi.org/10.3390/app10020446.

40 Deyasi, A. and Sarkar, A. (2018). Analytical computation of electrical parameters in GAAQWT and CNTFET with identical configuration using NEGF method. *International Journal of Electronics* 105 (12): 2144–2159.

41 Takagi, S. and Takenaka, M. (2015). Ge/III-V MOS device technologies for low power integrated systems. *Proceedings of the 2015 45th European Solid State Device Research Conference (ESSDERC), Graz, Austria*, 14–18 September 2015. 20–25.

42 Teo, K.H., Zhang, Y., Chowdhury, N. et al. (2021). Emerging GaN technologies for power, RF, digital, and quantum computing applications: recent advances and prospects. *Journal of Applied Physics* 130 (16): 160902. https://doi.org/10.1063/5.0061555.

43 Mimura, T., Hiyamizu, S., Fujii, T., and Nanbu, K. (1980). A new field-effect transistor with selectively doped GaAs/n-$Al_xGa_{1-x}As$ heterojunctions. *Japanese Journal of Applied Physics* 19 (5): 225–227.

44 Paul, S., Mondal, S., and Sarkar, A. (2021). Characterization and analysis of low-noise GaN-HEMT based inverter circuits. *Microsystem Technologies* 27 (11): 3957–3965.

45 Biswas, K., Ghoshhajra, R., and Sarkar, A. (2022). High electron mobility transistor: physics-based TCAD simulation and performance analysis. In: *HEMT Technology and Applications*, 155–179. Singapore: Springer Nature.

46 Sriramani, P., Mohankumar, N., Prasamsha, Y. et al. (2023). Threshold and surface potential-based sensitivity analysis of symmetrical double gate AlGaN/GaN MOS-HEMT including capacitance effects for label-free biosensing. *Physica Scripta* 98 (11): 115036.

47 Fossum, J.G. and Zhang, W. (2003). Performance projections of scaled CMOS devices and circuits with strained Si-on-SiGe channels. *IEEE Transactions on Electron Devices* 50 (4): 1042–1049.

48 Rim, K., Hoyt, J.L., and Gibbons, J.F. (2000). Fabrication and analysis of deep submicron strained-Si n-MOSFET's. *IEEE Transactions on Electron Devices* 47 (7): 1406–1415.

49 Tsutsui, G., Mochizuki, S., Loubet, N. et al. (2019). Strain engineering in functional materials. *AIP Advances* 9 (3): 030701.

2

MOSFET: Device Physics and Operation

Ruthramurthy Balachandran[1], Savitesh M. Sharma[2], and Avtar Singh[1]

[1]*Department of Electronics and Communication Engineering, SOEEC, ASTU, Adama, Ethiopia*
[2]*Chinmaya Vishwa Vidyapeeth, Ernakulam, Kerala, India*

2.1 Introduction to MOSFET

The metal–oxide–semiconductor field-effect transistor (MOSFET) is basically an electronic device whose output voltage (V_O) depends on the gate voltage. Hence, it is called a voltage-controlled device. It is a unipolar transistor, which means that only one sort of charge carrier, which may be either holes or electrons, is required for operation. It is a FET where the input voltage controls the conduction of the device. It has four terminals: body (***B***), drain (***D***), gate (***G***), and source (***S***). The source terminal is often connected to the ***B*** also called as substrate.

The following are the differences between the MOSFETs and conventional bipolar junction transistors (BJTs).

- Like diodes, MOSFETs have got N-type and P-type whereas BJT got PNP and NPN.
- The conduction of MOSFET depends on voltage whereas current for the BJT.
- The MOSFET has a high input resistance whereas BJTs have a low input resistance.
- Applications requiring high currents typically utilize MOSFETs, whereas those requiring low currents typically use BJTs.

Figure 2.1 depicts the structure of the MOSFET.

A channel is considered as a path between its source and drain. It works by changing the width along which electrons or holes move. Silicon dioxide is used to form the MOSFET's gate (SiO_2). It is also known as an insulated gate field-effect transistor (IGFET) because the gate uses an insulating layer to create electrical isolation. In the mega ohm range ($10^6 = M\Omega$), it greatly increases its input impedance. As a result, the MOSFET has no input current [1].

Advanced Nanoscale MOSFET Architectures: Current Trends and Future Perspectives,
First Edition. Edited by Kalyan Biswas and Angsuman Sarkar.

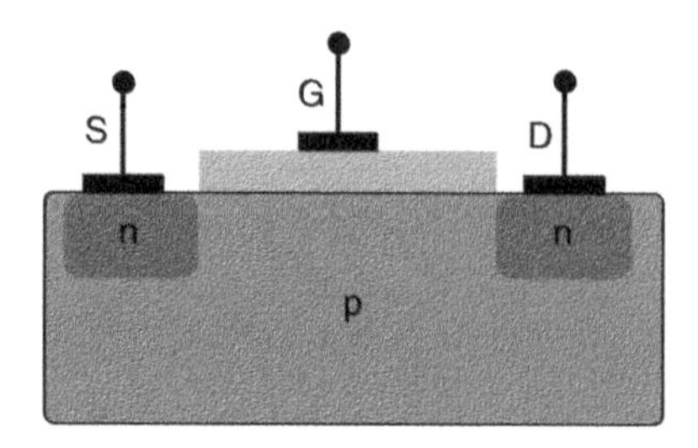

Figure 2.1 Structure of a basic MOSFET.

2.2 Advantages of MOSFET

MOSFETs began to pose a reliability challenge to bipolar transistors. They were a better choice than bipolar devices because of their fast switching, small size, and high-frequency operations. They can switch at rates of much more than a few hundred kHz. The following are the advantages of MOSFET over conventional BJT [2].

- ➢ MOSFET has no input current (I_G) because of very high input impedance.
- ➢ It is used for very high frequency (VHF) applications.
- ➢ Very little output resistance is present.
- ➢ It is really compact.
- ➢ They have two modes of operation: depletion mode and enhancement mode.
- ➢ It operates at low voltages while delivering improved efficiency.
- ➢ It is voltage-controlled, unipolar, and operates quietly. It also loses very little power.

Besides the advantages, there are a few disadvantages which are as follows.

- Because of the capacitance between the ***G*** and channel, it can be damaged by any electrostatic charge buildup.
- High voltages cannot be applied to it.
- MOSFETs cost more than BJTs.

2.3 Applications of MOSFETs

In electronic circuits, MOSFETs are primarily utilized for switching and amplification. The following are a few MOSFET applications [3].

- It is utilized in high-frequency amplifiers for quick switching and amplification of incredibly small signals.
- In DC motors, power control is accomplished by power MOSFETs.
- Due to their quick switching times, MOSFETs are the optimum choice for chopper circuits.

- Their high efficiency and low power consumption make them suitable for microcontrollers and CPUs.
- Employed in switch mode power supply (SMPS).
- Utilized in complementary metal–oxide–semiconductor (CMOS) to save space and power.
- Found usage in H-bridge circuit.
- Additionally, they are utilized in boost and buck converters.

2.4 Types of MOSFETs

The MOSFET has mainly two types

- Depletion MOSFET.
- Enhancement MOSFET.

Based on the type of channels, these types can be differentiated [4]. The classification of MOSFET based on the construction and the material used is shown in Figure 2.2.

Depletion MOSFET: Depletion-mode MOSFET could be considered as a "Normally Closed" switch. Due to the fact that they are manufactured with a built-in channel, they are also referred to as "usually ON" MOSFETs. When gate voltage is applied, the channel width is reduced, turning the MOSFET OFF. As it permits current flow at zero gate-source voltage, the depletion MOSFET symbol contains a continuous line.

Enhancement MOSFET: Enhancement mode MOSFET could be considered as a "Normally Open" switch. Although there is no channel present during

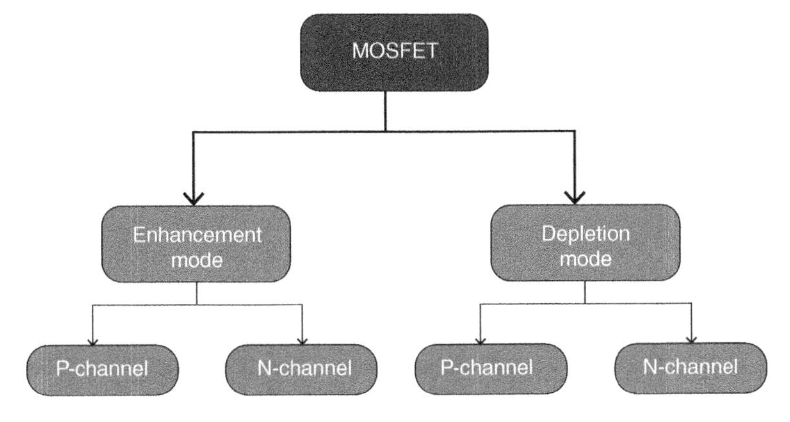

Figure 2.2 Classification of MOSFET.

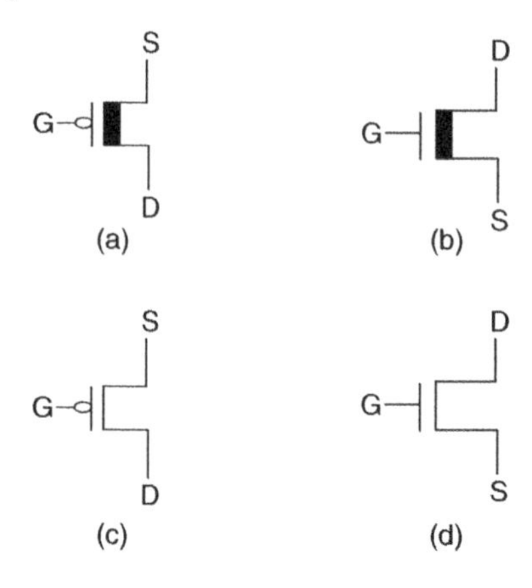

Figure 2.3 Symbol of (a) depletion mode of P-type and (b) N-type MOSFET (c) Enhancement mode of P-type and (d) N-ype.

manufacture, the enhanced MOSFET is also referred to as a "Normally OFF" MOSFET since it is induced by applying voltage.

MOSFET could be classified as per the operating principle and are listed as follows:

- P-Channel depletion MOSFET.
- P-Channel enhancement MOSFET.
- N-Channel depletion MOSFET.
- N-Channel enhancement MOSFET.

2.4.1 P-Channel and N-Channel MOSFET

In P-Channel MOSFET, the substrate is N-type extrinsic semiconductor material whereas **D** and **S** are highly doped p-regions. In P-Channel MOSFET, the current flows because of holes whereas in N-type channel MOSFET, the substrate is a P-type semiconducting material, and the **D** and **S** are highly doped n-regions. Hence, the current flows because of electrons.

A channel is created as a result of the $+V_{GS}$ making electrons flow from the highly doped N-type area and the drain part.

The symbol of depletion and enhancement mode N-type and P-type MOSFET is shown in Figure 2.3.

2.4.2 MOSFET Working Operation

The metal–oxide–semiconductor (MOS) capacitor has a metal, oxide material like SiO_2 and semiconductor-like P-type material (Figure 2.4). The oxide layer is in between the **S** and **D** terminals, and by using either + or – V_{GS}, it can be converted from one type to the other.

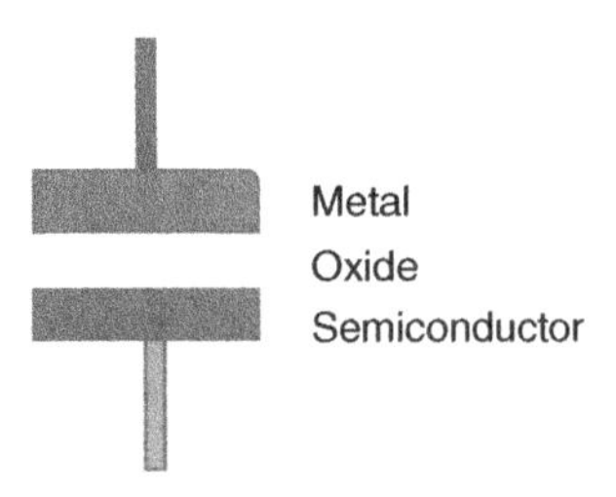

Figure 2.4 MOS capacitor structure.

The holes experience the opposite when a + V_G is supplied. It is the reason behind the holes being pushed downward across the base.

A negative charge attached to the acceptor atom is in the depletion zone. A channel is created when an electron enters the channel. A positive voltage on the channel also pulls electrons out of the highly doped "**n**" source and drain areas.

Now, the current passes and the V_G controls the channel's population of electrons. If a negative voltage is supplied a whole channel will form under the silicon dioxide layer.

2.5 Band Diagram of MOSFET

An equilibrium device condition could be made if no outside voltage is supplied to the metal oxide semiconductor capacitor. It is due to the fact that the energy level called as Fermi level of the metal and semiconductor are at equal levels. When outside voltage is supplied to the MOSFET, the V_{th} and V_{FB} act in line with the applied voltage. While threshold voltage (V_{th}) is the lowest V_{GS} necessary to construct a conducting channel. The V_{FB} is defined as the work function difference between the gate and semiconductor when there is no charge at the oxide–semiconductor interface [5].

The band diagram of the Metal–Oxide–Semiconductor system is shown in Figure 2.5.

Figure 2.6 shows the combined MOS system's energy band diagram for Metal, Oxide, and Semiconductor regions. The Fermi level of metal and semiconductor are aligned.

There are three operating regions/layers in MOSFET viz., accumulation region, deletion region, and inversion region.

2.5.1 Accumulation Layer

The applied voltage (V_g) in this instance is lower than the V_{FB}. Negative voltage is supplied to the gate (Figure 2.7).

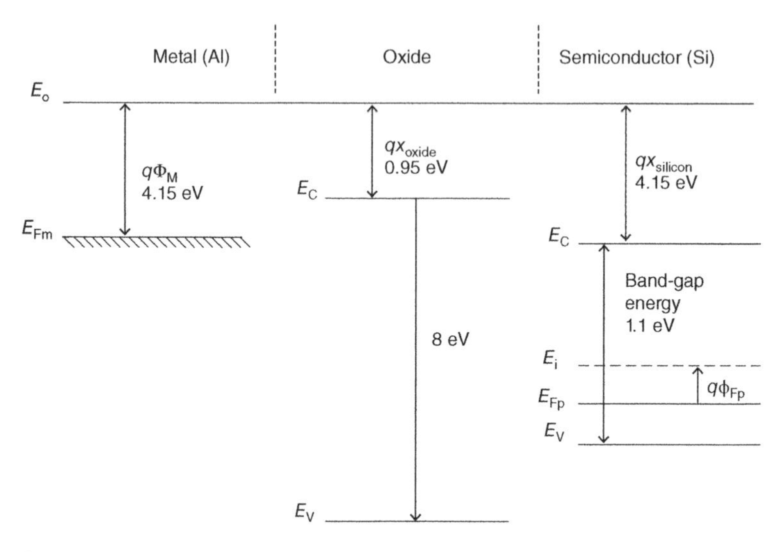

Figure 2.5 Energy band diagram of the Metal–Oxide–Semiconductor system.

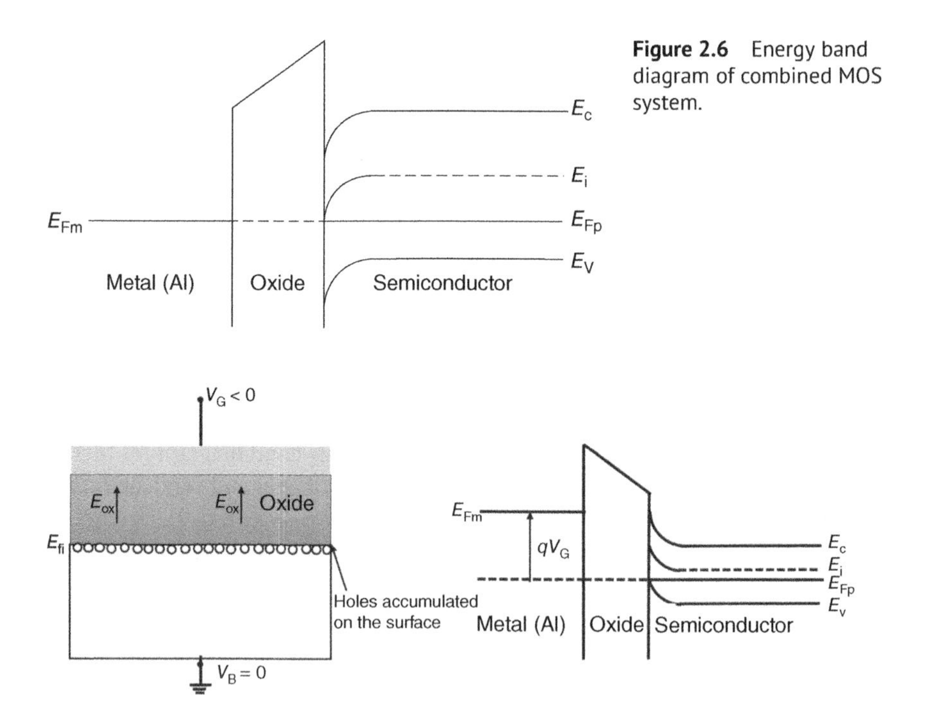

Figure 2.6 Energy band diagram of combined MOS system.

Figure 2.7 Energy band diagram in accumulation region.

where

E_c = Energy level in conduction band.
E_{fp} = Energy level at Fermi region.
E_v = Energy level in valence band.
E_i = Energy level at intrinsic semiconductor area.
Q = Electron charge.
V_G = Gate voltage.
V_s = Voltage at the surface.
V_B = Voltage at the substrate.

MOSFETs can no longer maintain equilibrium when voltage is added. By multiplying the electron charge by the voltage being applied, the metal's Fermi energy level changes. Because the applied voltage is negative, metal Fermi levels grow while semiconductor Fermi levels stay the same. Negative voltage is given to the gate; as a result, a negative charge forms at the metal-oxide junction. Meanwhile, positively charged holes move in the direction of the oxide junction and produce a positive charge there. Surface voltage is created near the junction as a result of positive charge accumulation; this energy band bending causes surface voltage, which is equal to the charge of the electron times the V_S. Energy band bending is the result of changes in the energy offset of a semiconductor's band structure close to a junction.

2.5.2 Depletion Layer

The V_G in the depletion area is higher than the flat band voltage but lower than the V_{th} (Figure 2.8).

Since a positive voltage is being provided to the gate in this instance, the Fermi energy level of the metal is falling. But, the Fermi energy level of the

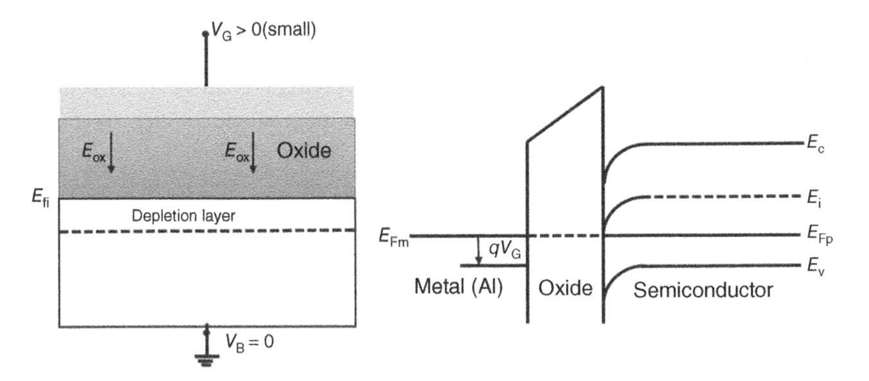

Figure 2.8 Energy band diagram in the depletion region.

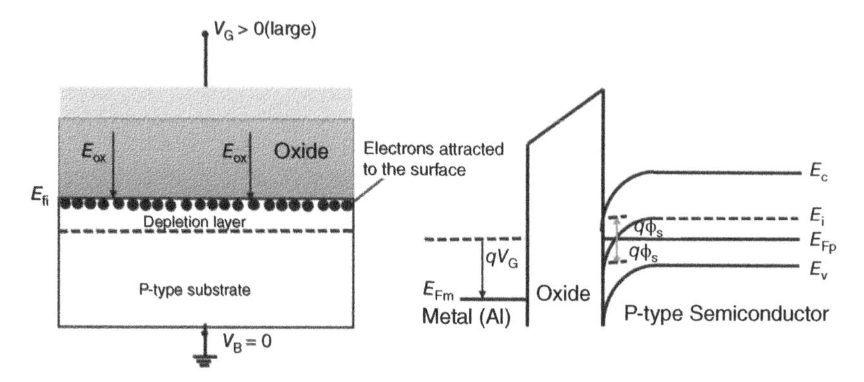

Figure 2.9 Energy band diagram in inversion region.

semiconductor is rising. As the electrons move toward the gate, a negative charge forms near the junction. Nearby holes and electrons combine to form a depletion area. In the depletion zone, a surface voltage develops, and as a result, the energy band bends there.

2.5.3 Inversion Layer

The supplied voltage in the inversion layer is higher than the V_{th}. The name inversion layer comes because the surface is inverted from P-type semiconductor to N-type semiconductor close to the junction. Due to the extremely high voltage supplied, the metal's Fermi level decreases even further. Since the V_G is positive, electrons move in its direction and gather close to the semiconductor-oxide junction, creating a surface potential. Because of this potential, energy band bending occurs. According to the band structure, the substrate close to the junction has an intrinsic energy level that is below the Fermi energy level. Hence, this region of the base behaves as an N-type semiconductor. The region of the substrate above the Fermi level, on the other hand, behaves as a P-type semiconductor. Surface inversion is an event that occurs when the electron concentration near the junction is greater than the concentration of holes. Applying a positive drain-source voltage causes current to flow through the N-type semiconductor, which serves as the channel for the current (Figure 2.9).

2.6 MOSFET Regions of Operation

Three terminal voltages can be seen in the figure above which are:

- V_{GS}: Voltage between Gate and Source.
- V_{DS}: Voltage between Drain and Source.
- V_{BS}: Voltage between Body and Source.

Based on a very small electrical signal, transistors can function as an insulator or a conductor. The MOSFET functions in three areas, much like any other transistor [6].

Cutoff Region: If the V_{GS} is less than the V_{th} and $V_{DS} \neq 0$, there will be no drain current I_D in this area and the MOSFET is OFF. Regardless of V_{DS}'s value, $I_D = 0$.

Saturation Region: If the V_{GS} is more and the output Voltage $V_{DS} \geq (V_{GS} - V_{th})$ is more, then the transistor allows a continuous current between the source and drain. The transistor functions as the switch in ON-state. Maximum drain current I_D is flowing through the MOSFET, which is completely on. The $V_{DS} > V_p$ and the $V_{GS} > -V_{th}$ in this area. The maximum drain current I_{DSS} that the MOSFET will allow depends on the V_{GS}.

Linear or Ohmic Region: If the V_{GS} is more than V_{th}, and $V_{DS} < (V_{GS} - V_{th})$, a continuous current between the source and drain is permitted by the MOSFET and it acts as if a resistor. That is why the name comes linear or Ohmic region. It is considered as ON state of a switch. The MOSFET functions as an amplifier in this region.

2.6.1 N-Channel Depletion MOSFET

The source and drain electrodes in N-Channel depletion MOSFET are mounted on thin N-type layers. The N-Channel is already present in between the source and drain region. The gate electrode is electrically insulated by a layer of insulating metal oxide. A P-type substrate serves as the foundation for the fabrication of the N-type material channel.

2.6.2 P-Channel depletion MOSFET

A P-type depletion MOSFET is built similarly to an N-Channel device; the source electrodes are located on P-type layers and a P-layer atop an N-type substrate makes up the channel. Hole-based charge carriers are employed. Over electrons, holes have one drawback. They can slow it down during operation because they are much heavier than electrons.

If there is a voltage between its **S** and **D**, it can conduct current normally. The channel width can be made wider or narrower by the V_{GS}.

The electric field will attract electrons from the N-type base, which combines with the holes to deplete the channel of charge carriers when a positive V_{GS} is applied. Both the channel's width and the current flow are decreased. The i_{GS} eventually destroys the entire channel and halts the current flow.

The P-Channel depletion MOSFET has a positive V_{th} as a result; it turns on when there is no V_{GS} and turns off when there is positive V_{GS}. More holes will be

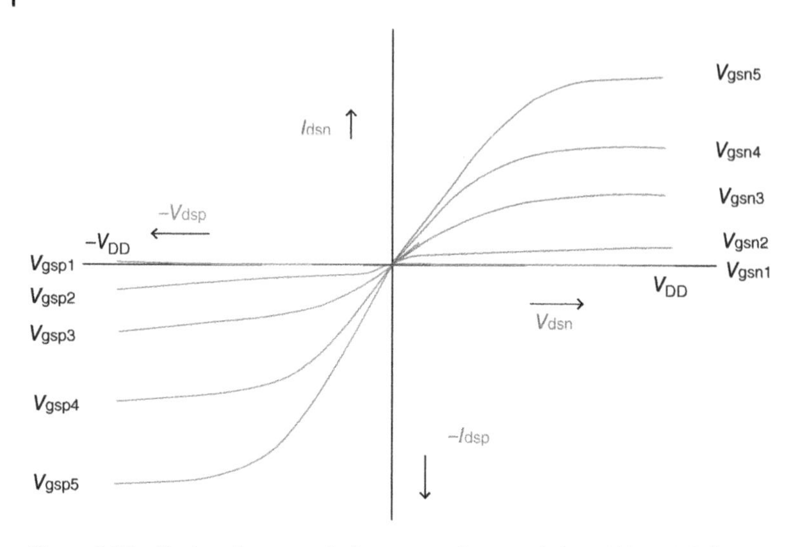

Figure 2.10 Drain voltage vs. drain current characteristics of N- and P-Channel depletion MOSFET.

induced into the channel by applying negative voltage, increasing, or enhancing the current conduction.

2.6.3 Operating Regions of P-Channel Depletion MOSFET

Cutoff Region: The gate-source voltage (V_{GS}) in this region is $+V_{th}$. Regardless of V_{DS}'s value, $I_D = 0$ does not represent a drain current. It is turned off in the MOSFET.

Saturation Region: The $V_{DS} > V_p$ and the V_{GS} $+V_{th}$ in this area. The maximum drain current I_{DSS} that the MOSFET will allow depends on the V_{GS} level.

Linear or Ohmic Region: When $V_{GS} < +V_{th}$ and $V_{DS} < V_p$ are present, the MOSFET functions as an amplifier. According to the I–V characteristics, the current I_D in this region grows with the V_{DS} while its amplification depends on the V_{GS}.

The Drain voltage vs. drain current characteristics of N- and P-channel depletion MOSFET are as shown in Figure 2.10.

2.6.4 Enhancement MOSFET

Enhancement MOSFET is manufactured without a channel. Instead, V_G is responsible to induce the channel in the substrate. The term comes from how well it conducts when the voltage is increased.

When there is no voltage applied to the enhancement MOSFET's gate, it remains switched OFF and does not conduct. Its alternate name is "Normally OFF" MOSFET because of this. Charge carriers are induced in the base by providing a V_{GS}. It creates a channel for the current between the source and the drain.

The term "enhancement MOSFET" refers to a device where the channel width is improved and the current flow is increased by applying voltage over the V_{th}.

The enhancement MOSFET has two types N-Channel and P-Channel enhancement MOSFET.

2.6.5 N-Channel Enhancement MOSFET

Like the depletion of MOSFET in structure, it is manufactured without a channel. The V_{GS} applied across the gate of the channel induces it.

N-Channel enhancement: When $V_{GS} = 0$, a MOSFET will not conduct current between its source and drain terminals because there is no path for current to travel through. An electric field is created beneath the gate layer when a $+V_{GS}$ is supplied to the gate. As a result, it pulls electrons off the P-substrate and forces holes back toward the insulating layer. Induction of a channel permits current to flow between the two electrodes.

2.6.6 P-Channel Enhancement MOSFET

This type of MOSFETs is identical in structure to P-Channel depletion MOSFETs without the channel. While it is being built, there is no channel. It is brought on by using V_{GS}. Positive charges (holes) accumulate behind the insulating layer when $-V_{GS}$ is supplied to the gate, pushing the electrons back. Between the source and drain, a channel is created by the holes accumulating together. It will now begin to conduct current if V_{DS} is applied.

Like N-Channel, it is non-conducting when V_{GS} is 0 V. When the voltage is decreased below V_{th}, the channel widens, enabling more current to pass through.

2.7 Scaling of MOSFET

Scaling theory can be used to explain how to shorten a MOSFET's channel length. The scaling parameter S ($S < 1$) is used to scale a MOSFET's size. From one generation of CMOS technology to the next, the value of S is normally in the vicinity of 0.7. A process that requires a V_{DD} of 2 V, for instance, would use a V_{DD} of 1.4 V in a next-generation process. Alternatively, $V_{DD}' = V_{DD}.S$. The scaled operation reduces the channel length to $L' = L.S$ and the width to $W' = W.S$ [7].

- For the design of VLSI chips, the MOSFET packing density must be as high as feasible for the transistor sizes to be as compact as possible. Scaling is a term used to describe the size reduction or the dimensions of MOSFETs.
- It is necessary to scale down MOSFETs in order to obtain very high packaging density.
- Scaling a MOSFET transistor involves systematically reducing the device's overall size to the extent that is permitted by the technology at hand.
- There is a new constant scaling factor called **"S"**. All the dimensions of the large-size transistor are multiplied by this scaling factor to produce the scaled device.
- It is anticipated that the operating properties of the MOS transistor will change as its size decreases.

Table 2.1 shows various parameters of MOSFET before and after resizing the dimension.

Advantages of Scaling

- Reduces chip size.
- Boosts switching speed and
- Cuts power consumption.

Disadvantages of Scaling

➢ Thickness of SiO_2 sometimes will get thick enough to lose its dielectric property.

Table 2.1 Parameters of MOSFET before and after scaling.

Specification	Before scaling	After Scaling	
		Full Scaling	**Constant Voltage Scaling**
Channel length	L	$L' = L/S$	$L' = L/s$
Channel width	W	$W' = W/S$	$W' = W/S$
Gate oxide thickness	t_{ox}	$t_{ox}' = t_{ox}/S$	$t_{ox}' = t_{ox}/S$
Junction depth	X_j	$X_j' = X_j/S$	$X_j' = X_j/S$
Power supply voltage	V_{DD}	$V_{DD}' = V_{DD}/S$	$V_{DD}' = V_{DD}$
Threshold voltage	V_{TO}	$V_{TO}' = V_{TO}/S$	$V_{TO}' = V_{TO}$
Doping densities	N_A, N_D	$N_A', N_D' = SN_A, SN_D$	$N_A', N_D' = S^2N_A, S^2N_D$
Oxide capacitance	C_{ox}	$C_{ox}' = C_{ox}/S$	$C_{ox}' = C_{ox}/S$
Drain current	I_D	$I_D' = I_D/S$	$I_D' = S.I_D$
Power dissipation	P_D	$P_D' = P_D/S^2$	$P_D' = S.P_D$
Power density	P_D/Area	P_D'/Area $= P_D$/Area$/S$	P_D'/Area $= S^2.P_D$/Area$/S$

➢ Subthreshold current – For tiny devices, the drain-source voltage also has a stronger impact on the channel's carrier flow in addition to V_{GS} voltage. By raising V_{DS} voltage, it is possible to enhance the channel's carrier and lower the potential barrier even with $V_{GS} < V_T$.
➢ Sub-threshold current is the term used to describe the current that passes through the channel at higher values of V_{DS}.
➢ Noise issue – Noise issues are a natural byproduct of scaling, which reduces the dependability of high-density chips [8].

2.7.1 Types of Scaling

Constant Field Scaling

- It is also called full scaling.
- It attempts to preserve the intensity of the MOSFET's internal electric field while shrinking the size by a factor of "*S*."
- To do this, the voltages should be scaled down correspondingly using the same **S**.
- Voltage scaling may not be very beneficial in some circumstances.
- The input and output circuits, for instance, might need a certain voltage value. As a result, numerous supply voltages and sophisticated level-shifter configurations would be needed.

Constant Voltage Scaling

➢ The external supply and terminal voltage remain unaltered, however, all MOSFET dimensions are reduced by an amount designated "*S*."
➢ The power and drain current densities grow, which may eventually result in major reliability issues for the scaled transistor. They are electro-migration, hot carrier deterioration, oxide breakdown, and electrical overstress.

2.8 Short-channel Effects

Several occurrences start to happen if the length of the transistor's channel becomes closer to the ***S*** and ***D*** junctions with the base. Due to this, the following issues are expected to occur. Polysilicon gate depletion effect, the V_{th}, roll-off, DIBL, the velocity saturation, the rise in reverse J_L, the mobility reduction, and the hot carrier effects.

A solution to the first issue is proposed that requires reinstating the metal gate structure. It is challenging to turn off the transistor fully due to the V_{th} drop. Electrostatic interaction between the source and drain makes the gate useless via the drain-induced barrier lowering (DIBL) effect. Velocity saturation reduces the current drive in the transistor. Because of the J_L, there is an increase in the power dissipation. The output current is impacted by increased surface scattering because charge carriers' mobility is reduced. The MOSFET's performances are critically reduced by Impact ionization and hot carrier effects. Hence, they behave diversely from long-channel devices. Figure 2.11 shows the Gate-metal last process.

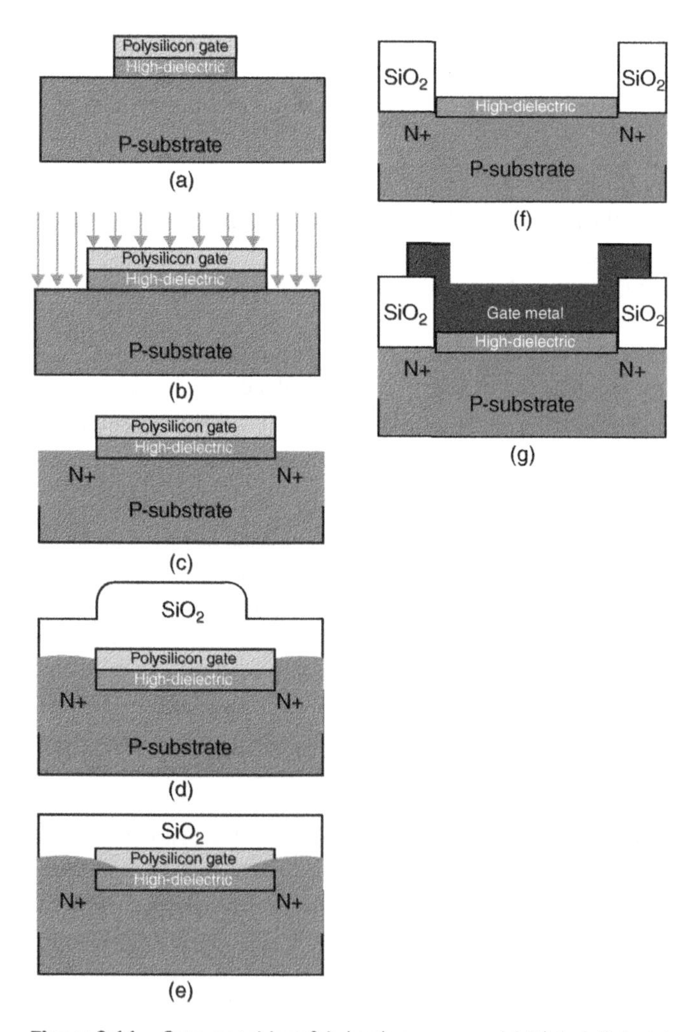

Figure 2.11 Gate metal last fabrication process. (a) High-*j* dielectric and polysilicon gate deposition. (b) Highly doped N-type **S/D** implant. (c) Sintering at very high-temperature (d) SiO_2 coverage. (e) CMP (f) Removal of dummy gate. (g) Formation of replacement gate.

Three noteworthy options include the use of high-dielectric materials, strain engineering, and a reduction in gate oxide thickness. However, the aforementioned phenomenon critically limits the accomplishment of planar CMOS transistors where the process nodes are less than 90 nm [9].

2.8.1 Drain-induced Barrier Lowering

The achievement of sub-micron MOS transistors can be significantly impacted by the effect of the V_D in the vicinity of the channel. The DIBL effect is comparable to the punch-through effect [10]. Punch-through is occasionally referred to as "subsurface DIBL" in the literature, as opposed to "surface DIBL."

A barrier could exist between the source and the channel in the weak inversion region. The height of the barrier is decided by the balance between drift and diffusion current. The barrier height may decrease when an elevated V_D is applied and hence there is a raise of I_D. As a result, both the gate voltage and the drain voltage control the drain current. For device modeling purposes, a V_{th} reduction dependent on the drain voltage can be used to account for this parasitic impact [11]. The DIBL analysis is depicted in Figure 2.12.

When examining a MOS transistor's transfer curves for the linear and saturated situations, the DIBL effect is evident (Figure 2.13). The two curves would collide in the subthreshold regime if the DIBL didn't exist. The DIBL effect, which is expressed in (mV/V) units, can be found by dividing the lateral shift of the transfer characteristics by the V_D difference of the two curves.

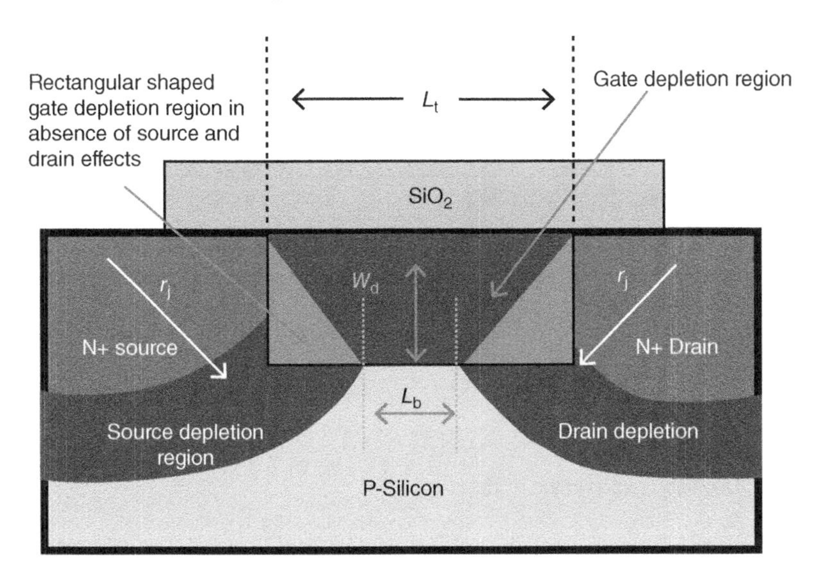

Figure 2.12 Analysis of DIBL.

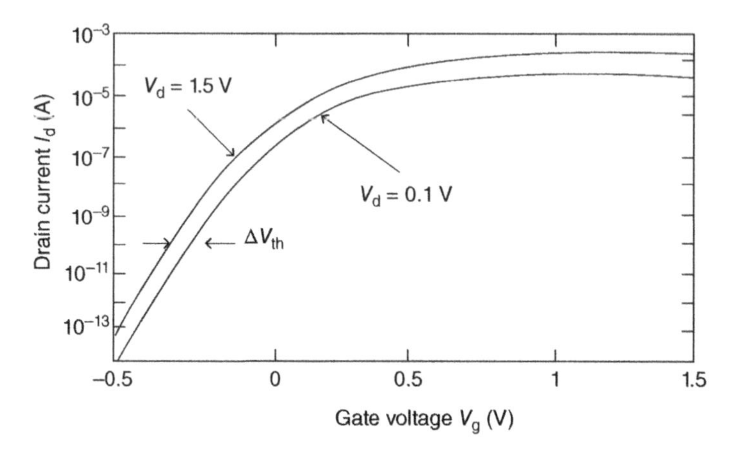

Figure 2.13 Transfer curves of device for $V_D = 0.1$ V and 1.5 V [10].

2.8.2 Gate-induced Drain Leakage

At V_D, which is much lower than the junction V_{BR}, thin SiO_2 MOSFETs provide significant gate-induced drain leakage current. It is discovered that the band-to-band tunneling that takes place in the deep-depletion layer in the gate-to-drain overlap region is the source of this current. The oxide field in the **G-D** region needs to be kept at about 2 MV/cm in order to retain the J_L at about 0.2 pA/m. This may lead to further limitations on the supply voltage/thickness of SiO_2 for scaling VLSI MOSFETs (Figure 2.14). We investigate design elements of the device that can lower gate-induced drain leakage current [12].

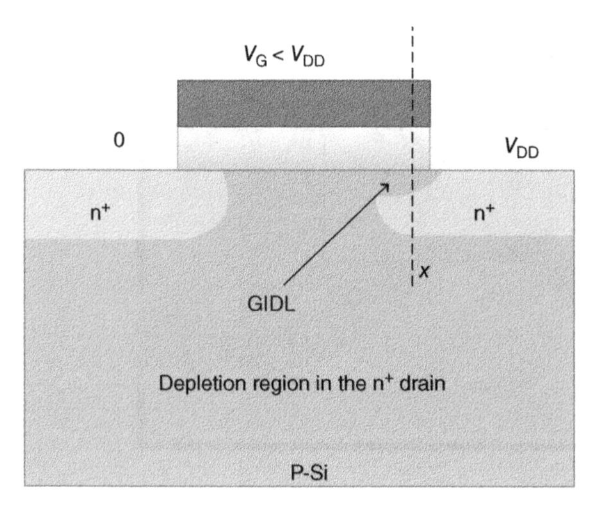

Figure 2.14 The concept of GIDL.

2.9 Body Bias Effect

In order to lower energy consumption for a longer battery life, ICs should be used in low-power mode. Every designer should be familiar with the body bias effect technique for modifying a circuit's characteristics to content both the power and production requirements [13]. The body bias effect, which can change the ON time for a P-type metal oxide semiconductor transistor, is shown in Figure 2.15.

2.9.1 Salient Feature of Body Bias

Body bias is mainly used to interactively alter the V_{th}. The voltage difference between the V_s and V_b affects the V_{th}. It could be considered as another gate that affects how a transistor conducts and is non-conducting.

When an unbiased device is in optimal condition, its V_s and V_b match. The effective V_{th} needed to switch on the device changes depending on whether the device is in a forward or reverse body bias region when the voltage of the bulk connection is changed. V_b will be greater than or less than V_s while V_s stays at its nominal voltage. This delta generally corresponds to a variation in V_{th}. This modified V_{th} provides the device a quicker than-normal t_{ON} depending on the body bias's polarity (t_{ON}) (Figure 2.15). By lowering the V_{th} needed to turn on the device ($V_{th}(f_b)$), a forward bias enables the gadget to conduct rapidly for maximum performance (t_{ONFB}), however, the device's leakage current increases as a result. Reverse body bias, on the other hand, increases the needed V_{th} for the device to switch on (V_{th}(rb)). The longer t_{ONRB} is due to the greater V_{th}, but it also reduces the leakage current, which is advantageous in terms of power efficiency.

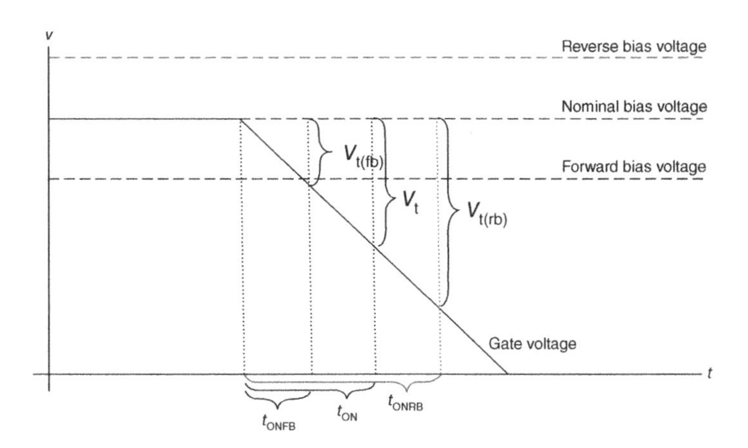

Figure 2.15 Body bias effect for a PMOS transistor.

2.9.2 Significance of Body Bias

The number of components that were wrongly connected determines whether the fault will produce reliability issues or a change in the item's capacity to meet requirements. Designers should apply exact and thorough confirmation to ensure that the voltage levels of all devices are precisely adjusted. There is a higher possibility that a device will be connected to the wrong power domain when many power domains are employed in a design. For each design size, the No. of domains, and the inclusion of specialized forward or reverse voltage areas all elevate this potential. Any forbidden connection runs the risk of damaging the equipment it is attached to. A product's overall quality quickly loses confidence without a suitable verification method, which is terrible for everyone's bottom line.

Reverse body bias raises a device's stress levels, which may eventually result in the equipment deteriorating or performing less as intended. The breakdown voltage value decreasing is one apparent manifestation of this degradation. Studies have also shown that prolonged reverse bias stress can increase a device on resistance value, which can have an impact on the operation of a circuit. Only equipment that has been specifically made to withstand this increased stress should be used in reverse body bias conditions in order to preserve a design's optimal performance and longevity.

Methodical design strategies should also be applied regarding forward-body biased conditions to prevent the greater J_L induced by applying forward-body bias to shorten a circuit's delay. Only devices meant for forward body bias use these leakage-reducing techniques, therefore any device unintentionally placed in forward bias will not have the extra circuitry. The design now uses more energy as a result. The combined effect could go over power restrictions if this condition is true for a lot of devices.

The defeat of the circuit's malfunction could be caused by the unbiased device's V_{THth} being greater than the operational point of the circuitry or by a device or sequence of devices having a longer delay than the allowable tolerances. Like this, if it is intended to be biased in reverse bias to decrease J_L, improper biasing could result in leakage values that are higher than allowed.

2.9.3 Body Bias Verification

Designers certainly require a strong bias detection process given the potential consequences of unfavorable bias circumstances. The easiest validation procedure is an apparent examination of the design, which requires the examiner to be attentive to any improper connections as well as have a full understanding of all design parts to find any devices with a body association to the erroneous area. Along with the risk of human error, exhaustively evaluating every aspect of the design through visual inspection takes time and gets more challenging to handle as the design complexity rises. It becomes much more challenging to have complete assurance that a human verification approach offers correct, complete coverage as this design complexity rises.

Therefore, it makes sense that the following step would involve using EDA technologies to locate any bias problem spots that might be present in a design. However, there might be flaws in the answer depending on the methods a tool for bias verification employs. Dynamic simulation is a method used by some programs to compute values all across a design. These procedures may have trouble offering turnaround times that support a productive design cycle. Alternate techniques depend on manually adding marker layers to the layout to find the important data. This technique not only brings back the element of human error, but it is also only applicable during the post-layout design stage, when making modifications may be challenging or expensive.

The way in which errors are conveyed is equally crucial. Body bias mistakes cannot be fixed by only detecting and reporting them. Designers must be able to identify the location of errors, understand how they fail, and have access to sufficient data to choose the right cure.

The values of voltages all over the design must first be able to be identified without the aid of SPICE simulations or drawing marker layers for an automated body bias verification solution to be effective. Once this data has been successfully recognized, it may be utilized to comprehend, validate, and compare the bias condition of a specific device to the anticipated bias state of the body (unbiased, forward-biased, or reverse-biased).

In systems with ever-increasing complexity, designers dealing with today's high-performance, low-power needs to confront a challenge in properly and swiftly validating all body bias scenarios. A design's power consumption and performance can suffer as a result of the dangers created by a device that is put into the wrong body bias condition or one that doesn't satisfy the necessary ratings for the body bias condition to which it is allocated. In order to meet strict performance and reliability criteria, accurate, quick, automated body bias verification is essential.

2.10 Advancement of MOSFET Structures

Figure 2.16 shows the Double gate (DG) MOSFET's primary structure. In this case, a DGMOSFET with a tri material is entirely depleted [14].

The structure was made with the latest technology. Three different gate materials with a range of functions are selected for n-MOSFET. The larger work-functioning component of the gate is located near the source, and the

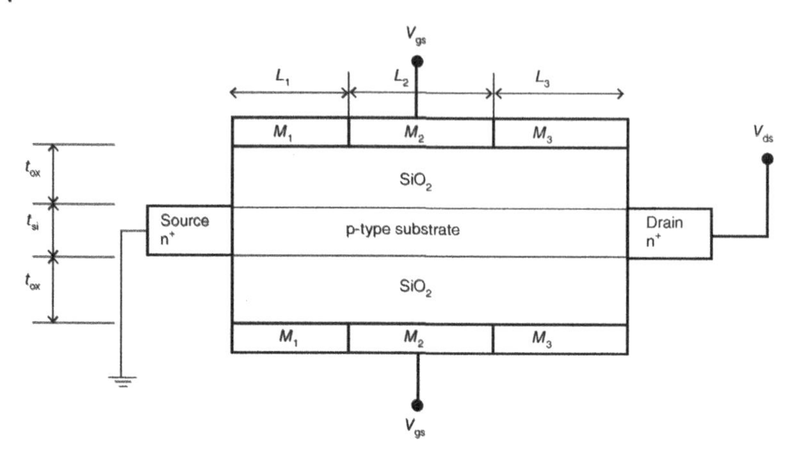

Figure 2.16 DG MOSFET's primary structure.

smaller work-functioning component is placed near the drain. Gates are used because of their impact on dopant penetration and polysilicon depletion widths. The substrate's doping concentration is assumed to be 10^{16} cm^{-3}. To enhance the performances, Figure 2.17 shows gate-all-around junction-less (GAAJ) MOSFET with S/D extension regions.

In contrast to the channel doping, the S and D extensions in this architecture are heavily doped. The level of doping is given as n++/n+/n++. The drain current is enhanced by incorporating the strongly doped extensions. When compared to a GAAJ without extensions, the MOSFET for the GAAJ has a higher current. In the strongly doped sections, the ion current's strength has increased by 70%. The reliability issue with junction less double gate (JLDG) MOSFET is covered in Figure 2.18 [15].

This gate structure provides lower current, which lowers the performance of JLDG MOSFETs. The performance of the MOSFET is impacted by the front and back gate alignment. The rear gate changing in the direction of the **S** or **D** side results in misalignment. Non-ideal results are produced by the

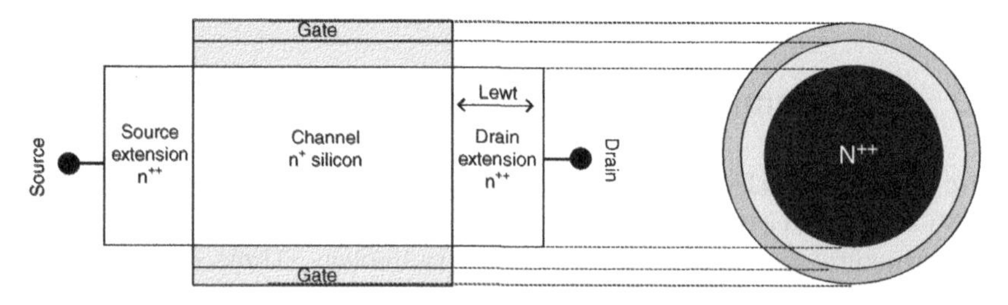

Figure 2.17 JL MOSFET with **S** and **D** extensions.

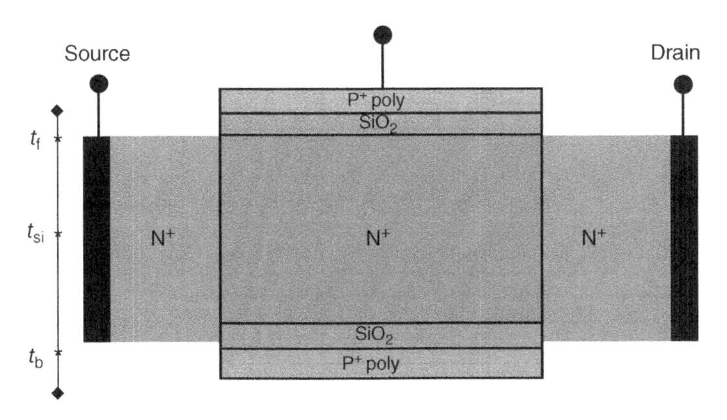

Figure 2.18 N-type junction less double gate MOSFET structure.

gate's misalignment. The device characteristics and gate work function are found to be 5.2 eV. Figure 2.19 suggests a double step buried oxide (DSBO) Silicon-On-Insulator (SOI) MOSFET.

The benefits of a MOSFET and silicon on insulator structure are combined in this arrangement. This construction is planned to prevent self-heating. It could be achieved by altering the geometry of BOX into a double-step shape and thinned SiO_2. It is simple to conduct heat exchange between the channels and substrate. Despite the consequences of self-heating, the drain current in this structure increases. Two cases are used in this article. The JLDG MOSFET end source-gate

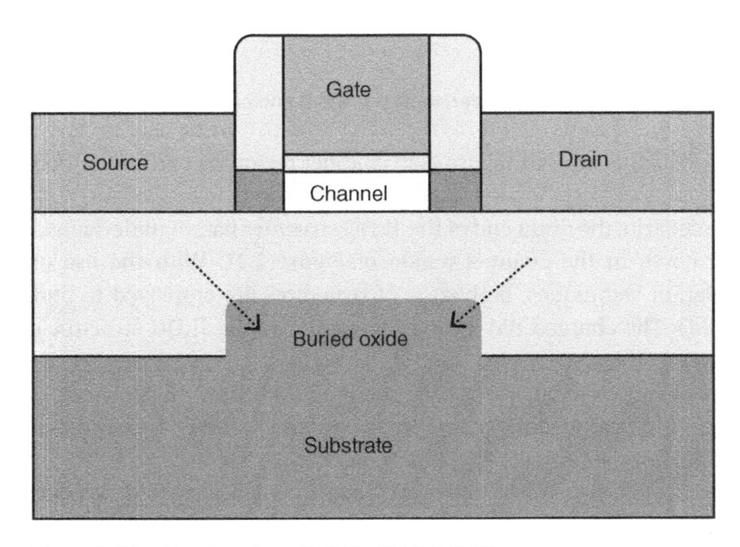

Figure 2.19 Construction of DSBO-SOI MOSFET.

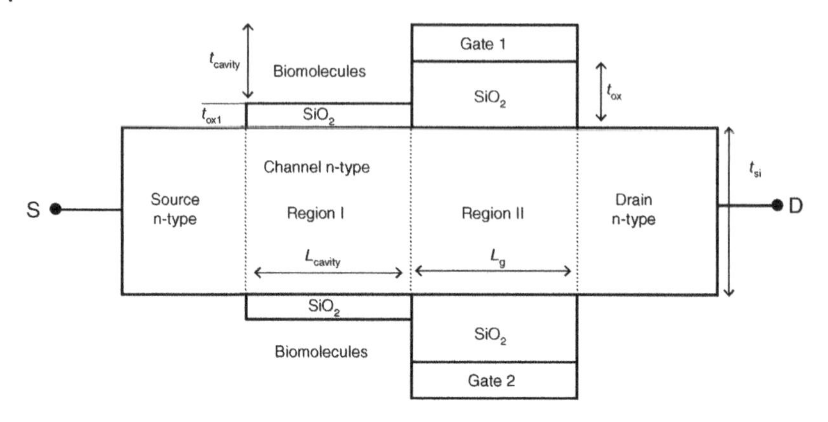

Figure 2.20 JLDG structure.

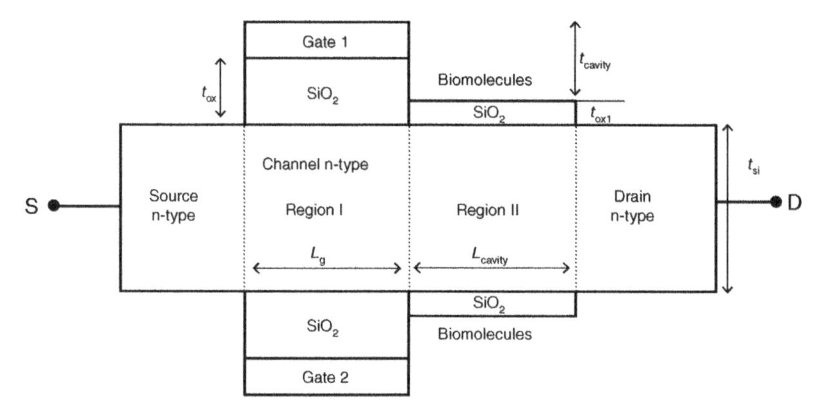

Figure 2.21 Junction Less MOSFET – Underlapping at the **D** end of the channel.

underlap region, which is seen in Figure 2.20 channel region, is part of the first scenario.

In the second scenario, the drain end of the JLDG structure has an underlapped gate region, as shown in the channel region of Figure 2.21. With the use of dielectric modulation techniques, both sorts of structures are employed to find biomolecules (BM). The charged BM have an influence on the JLDG structure's surface voltage.

When BM is positively charged, the surface potential shifts upward. However, if they are negatively charged, it shifts below. A silicon-based MOSFET to improve performance at high temperatures is shown in Figure 2.22.

For high-temperature usage, the following procedure could be adapted. A localized narrow band gap material is placed between the **S** and channel. The buried oxide (BOX) in the proposed SOI MOSFET construction stops current from leaking

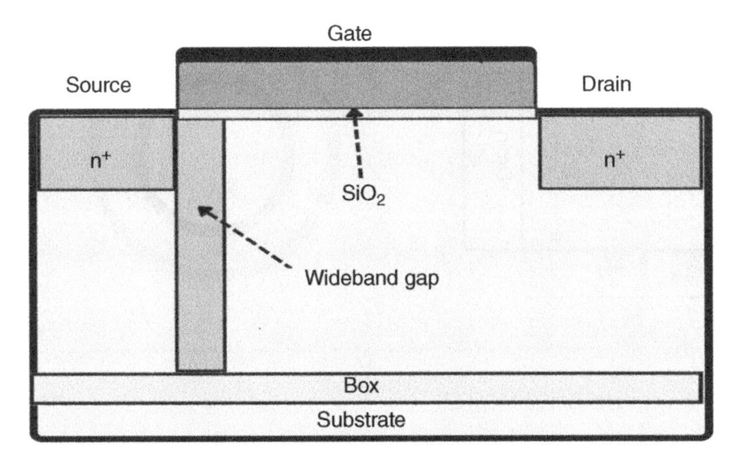

Figure 2.22 Structure of Si-supported MOSFET.

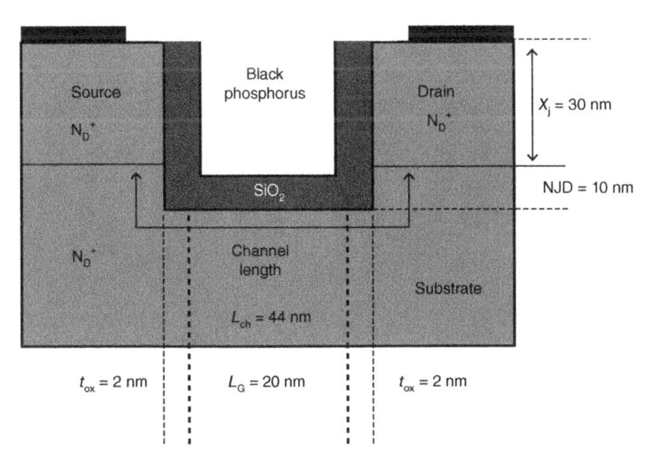

Figure 2.23 Black Phosphorous-based Junction Less Recessed Channel (BP JL-RC) MOSFET structure.

into the substrate. A sinking channel MOSFET for radio frequency applications using 45 nm technology is suggested in accordance with Figure 2.23.

Within the junction less recessed MOSFET black phosphorus is introduced where the drain current can rise by as much as 0.3 mA. The OFF current declines as the slope of the sub-threshold digital applications, two materials surrounding gate MOSFET with a 10-nm has been considered. As depicted in Figure 2.24, the benefits of 50 nm dual material surrounding gate (DMSG) MOSFETs and the multiple objective genetic algorithm (MOGA) optimization methods have been coupled.

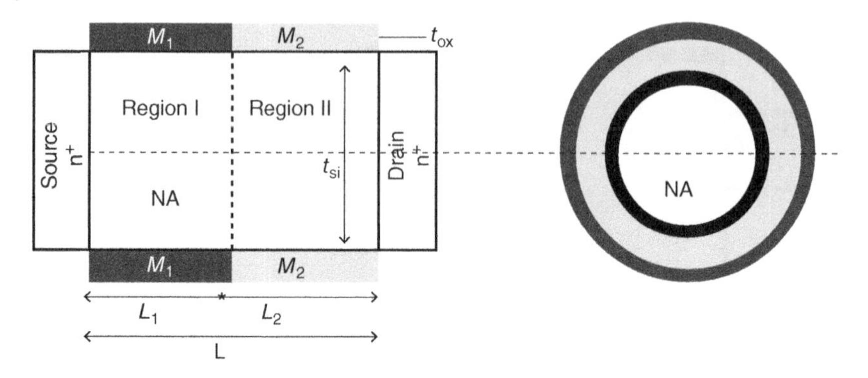

Figure 2.24 The structure of DMSG MOSFET.

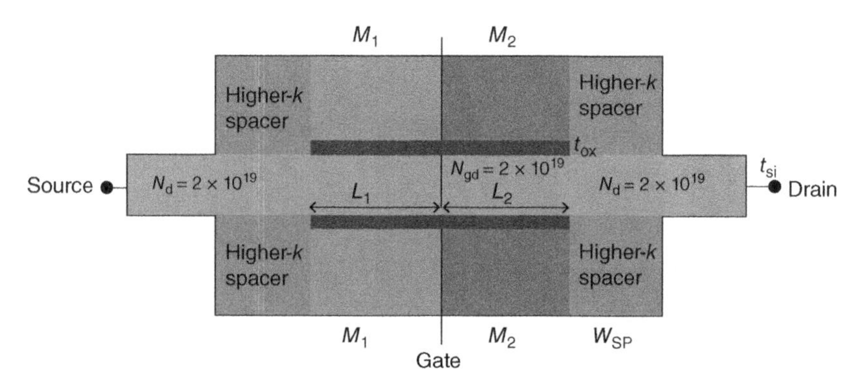

Figure 2.25 GC-DMGJLT arrangement.

The 10 nm DMSG MOSFET's electrical behavior is improved and optimized via the MOGAs method. This offers nanoscale high-speed digital applications with low power consumption. For analog applications, a graded channel dual material junction less (GC-DMGJL) MOSFET is investigated (Figure 2.25).

A GC-DMGJL MOSFET's performance is evaluated with that of a UC-DMGJL MOSFET. High transconductance and drain current are provided by the GCD-MGJL MOSFET, which also lessens short-channel effects (SCEs). The apparatus has a large doping area. Near the channel's drain, N_{gd} is doped at a higher value whereas the remaining sections are uniformly doped. Using the dielectric modulation technique, Figure 2.26 offers a JL MOSFET structure that can detect BM like DNA, enzymes and cells.

The procedure in SiO_2 etching from both the **S** and **D** sides creates a nanogap cavity. The surface potential in the channel under the nanocavity, which binds the silicon dioxide layer that exists in the cavity, is affected by the BM. Figure 2.27 discusses the 0.1 m n-MOSFET pocket construction for low-voltage applications.

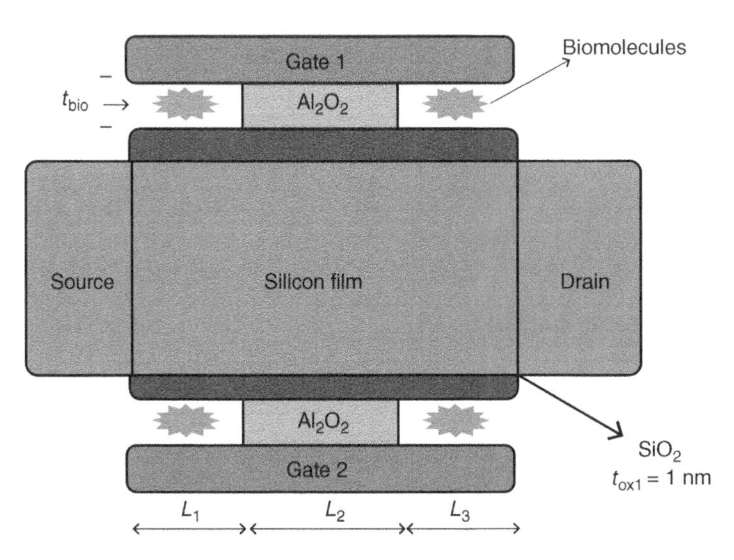

Figure 2.26 JL MOSFET for identifying biomolecules.

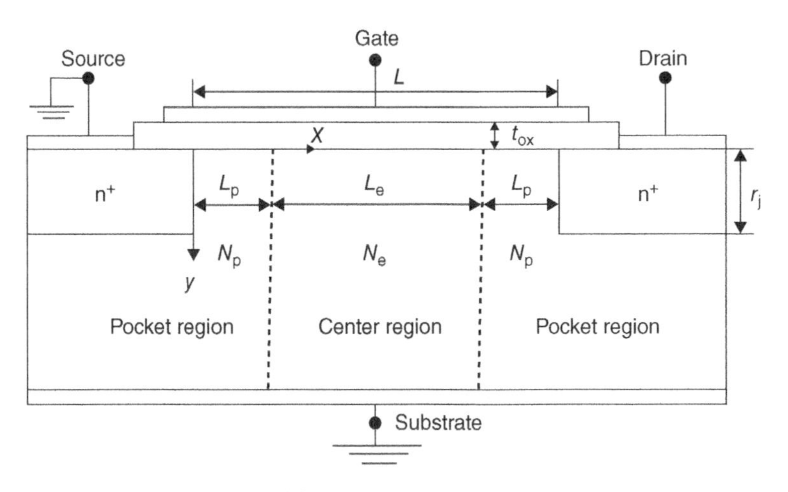

Figure 2.27 A pocket n-MOSFET arrangement.

In this structure, the central portion of this structure is generated and only weakly doped, whereas the pocket region surrounding the drain and source region is built and substantially doped. This design may meet the needs for both ON and OFF current and offers good immunity from SCEs. An electrically induced source and drain extension nanoscale SOI MOSFET architecture is shown to eliminate hot electron effects and SCEs in channels smaller than 50 nm. Despite the difficulty of producing shallow drain and source, Electrically Shallow

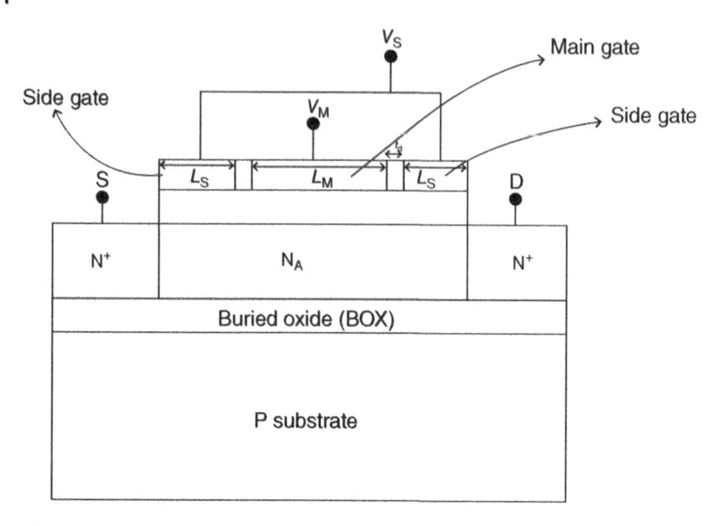

Figure 2.28 Structure of EJ-SOI MOSFET.

Junction (EJ) SOI MOSFET (Figure 2.28) may electronically form a virtual **D** and **S**.

It has three gates: one main gate, and two lateral gates. Side gates may be biased separately from the main gate. A virtual drain and source are produced as a result of these side gates' generation of inversion layers.

Figure 2.29 depicts an Asymmetric Gate Junction Less (AGJL) MOSFET with an unsymmetrical gate to enhance the device's performance.

The AGJL structure consists of two gates, with a sideways offset between them. The channel length in this design is determined by the ON and OFF states of the MOSFET. The length of the gate's overlap determines the MOSFET's channel length in the ON state. The length of the two gates added together lacking the gate's overlay determines the channel length in the OFF state. In this arrangement, the sub-threshold slope and DIBL is reduced. At the same time, the I_{ON} to I_{OFF} ratio increases.

The radio frequency performance of the indented channel with a clear gate is shown in Figure 2.30 for recessed channel (RC) MOSFET.

Calculations have been made to determine the values of g_m, the f_c, DIBL, and the maximum f_0. The construction's results are contrasted with those of the traditional indented channel metal oxide semiconductor field effect transistor. Its gate uses clear indium tin oxide. The results show a 132% increase in oscillator frequency and a 42% increase in cut-off frequency. Using the transparent gate material helps to lessen the SCEs since the gate control has improved.

With this construction, the devices are smaller and function better. The comparison of Recessed channel MOSFET and JLDG, it is found that the ON current

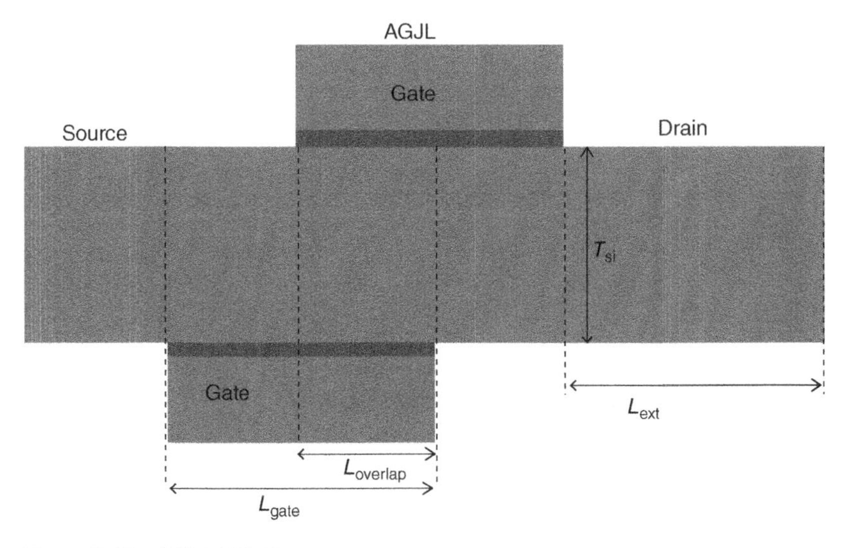

Figure 2.29 AGJL MOSFET structure.

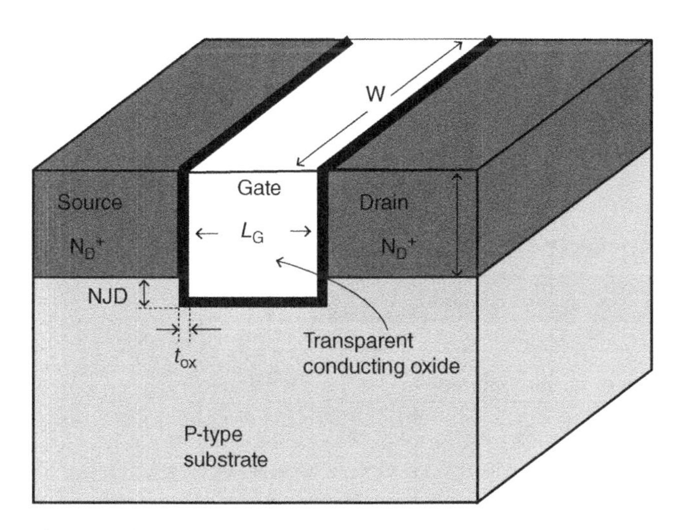

Figure 2.30 Recessed channel MOSFET.

is increasing while the OFF current is decreasing. The usual construction takes up twice the space as this one does. The suggested structure also enhances read and write operations. The I_{ON}/I_{OFF} ratio is 10^6. Figure 2.31 shows a short-channel JLDG MOSFET.

Because the channel's doping matches that of the drain and source, this configuration lacks p-n junctions. The structure's RF and analog performance has been

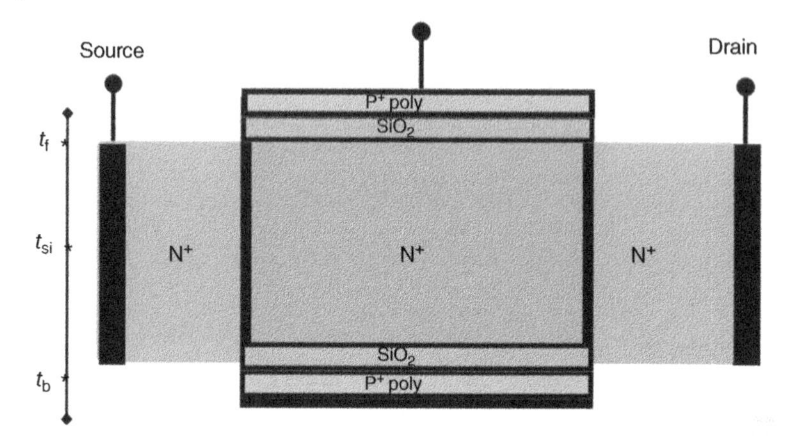

Figure 2.31 N-type JLDG MOSFET structure.

examined. The front and back gates supplied voltages are the same. The acquired ratio of I_{ON} and I_{OFF} ratio is about 4.9, and the sub-threshold slope and the DIBL value are improved parameters. The I_{ON} and I_{OFF} ratio is increased by about 17%, the sub-threshold slope is reduced by about 1.6%, and the DIBL is reduced by 4.52%, according to the results. Figure 2.32 illustrates silicon on insulator MOSFET with a (dual material gate (DMG) and short-channel that has recessed source and drain pins.

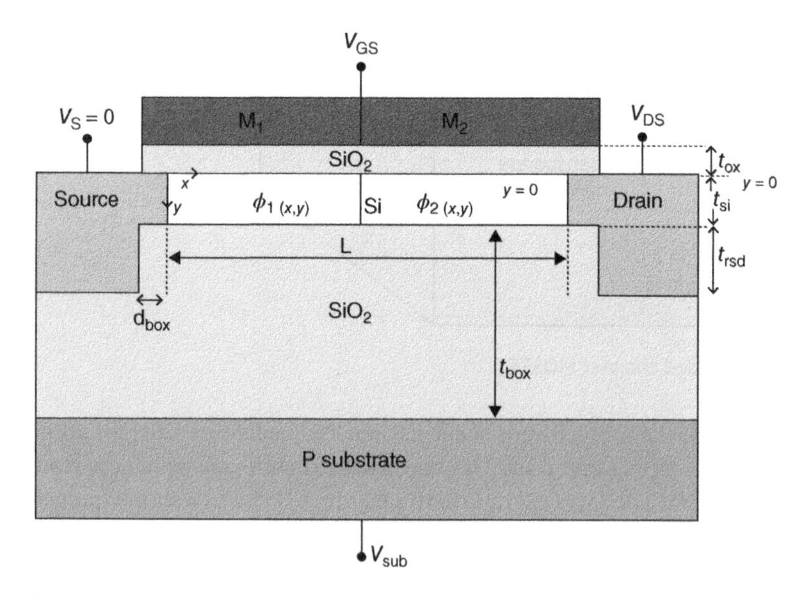

Figure 2.32 DMG with recessed **S** and **D** UTB silicon on insulator MOSFET.

This gadget offers a low DIBL value and a high ON current. The channel area is only weakly doped compared to the drain and source sections, which are both heavily doped. Polysilicon gates with an elevated work function are discovered to be the best choice for minimum OFF-state leakage. The underlying unsymmetrical gate boosts the extremity electric field, which enhances the performance of ON state transistors. In junction less transistors, the lack of a depletion area among all the electrodes enhances the transistor's capacity to drive current by increasing ON-state current. To improve sub-threshold performance characteristics and transistor switching behavior, the oxide region under the gate is proposed to be made of a high-K dielectric material. The 20 nm JLDG MOSFET, in comparison, has the highest I_{ON}/I_{OFF} ratio. It also offered the lowest sub-threshold slope. The GC-DMGJL with a 15 nm channel length produced the lowest DIBL. For low-power applications, multi-gate junction less MOSFETs with less leakage and an improved I_{ON}/I_{OFF} ratio could be selected. By adding the nano-gap cavity region, several studies have investigated the DG MOSFET's capacity for biosensing. Regarding the biomedical uses of these devices, the permittivity of the sections is dependent on changes in biological types.

Different MOSFET structures have been examined along with their fundamental nuances, measurements, and applications. Most MOSFET structural changes have been made in an effort to mitigate the SCEs, including DIBL values. Increasing the ratio of I_{ON} to I_{OFF} is the underlying idea behind these architectures. To obtain the optimal range of g_m, G, etc., the MOSFET architectures are further checked for reasonable simple/RF performance constraints. Based on correlation analyses, the JLDG MOSFET offers the highest I_{ON}/I_{OFF} proportion and the most decreased sub-threshold inclination for channel length 20 nm. The shortest double material entrance junction and a 15 nm channel length produced the lowest DIBL. The advanced MOSFET structures have been discussed in this section to be useful for improving sub-threshold performance. This shows how CMOS innovation can potentially be useful in various applications including IoT or botanical applications.

References

1 Chung, I.-Y., Park, Y.J., and Min, H.S. (2001). SOI MOSFET structure with a junctio N-type body contact for suppression of pass gate leakage. *IEEE Transactions on Electron Devices* 48 (7): 1360–1365. https://doi.org/10.1109/16.930652.

2 Choi, W., Son, D., and Kim, D. (2014). Advantages of low parasitic inductance packages of power MOSFET for server power applications. *2014 International*

Power Electronics Conference (IPEC-Hiroshima 2014 – ECCE ASIA), Hiroshima. 2914–2919. https://doi.org/10.1109/IPEC.2014.6870096.

3 Galluzzo, A., Melito, M., Musumeci, S. et al. (2000). A new high voltage power MOSFET for power conversion applications. *Conference Record of the 2000 IEEE Industry Applications Conference. Thirty-Fifth IAS Annual Meeting and World Conference on Industrial Applications of Electrical Energy (Cat. No.00CH37129)*, Rome, Italy. 2966–2973, vol. 5. https://doi.org/10.1109/IAS.2000.882588.

4 De Graaff, H.C. and Klaassen, F.M. (1990). Models for the depletio N-type MOSFET. In: *Compact Transistor Modelling for Circuit Design*, Computational Microelectronics. Vienna: Springer https://doi.org/10.1007/978-3-7091-9043-2_8.

5 Liou, J.J., Ortiz-Conde, A., and Garcia-Sanchez, F. (1998). MOSFET physics and modeling. In: *Analysis and Design of Mosfets*. Boston, MA: Springer https://doi.org/10.1007/978-1-4615-5415-8_1.

6 Uyemura, J.P. (1992). MOSFET characteristics. In: *Circuit Design for CMOS VLSI*. Boston, MA: Springer https://doi.org/10.1007/978-1-4615-3620-8_2.

7 Cham, K.M., Oh, S.Y., Moll, J.L. et al. (1988). MOSFET scaling by CADDET. In: *Computer-Aided Design and VLSI Device Development*, The Kluwer International Series in Engineering and Computer Science, vol. 53. Boston, MA: Springer https://doi.org/10.1007/978-1-4613-1695-4_15.

8 Kilchytska, V., Vancaillie, L., de Meyer, K., and Flandre, D. (2005). MOSFETs scaling down: advantages and disadvantages for high temperature applications. In: *Science and Technology of Semiconductor-On-Insulator Structures and Devices Operating in a Harsh Environment*, NATO Science Series II: Mathematics, Physics and Chemistry, vol. 185 (ed. D. Flandre, A.N. Nazarov, and P.L. Hemment). Dordrecht: Springer https://doi.org/10.1007/1-4020-3013-4_19.

9 Khanna, V.K. (2016). Short-channel effects in MOSFETs. In: *Integrated Nanoelectronics*, NanoScience and Technology. New Delhi: Springer https://doi.org/10.1007/978-81-322-3625-2_5.

10 Wolf, S. (1995). *The Submicron MOSFET*, Silicon Processing for the VLSI Era, vol. 3. Sunset Beach, CA: Lattice Press.

11 Cheng, Y., Jeng, M.-C., Liu, Z. et al. (1997). A physical and scalable I-V model in BSIM3v3 for analog/digital circuit simulation. *IEEE Transactions on Electron Devices* 44 (2): 277–287.

12 Chan, T. Y., Chen, J., Ko, P. K., and Hu, C. (1987). The impact of gate-induced drain leakage current on MOSFET scaling. *1987 International Electron Devices Meeting*, Washington, DC, USA. 718-721. https://doi.org/10.1109/IEDM.1987.191531.

13 Wittenhagen, E., Runge, M., Lotfi, N. et al. (2021). Advanced mixed signal concepts exploiting the strong body-bias effect in CMOS 22FDX®. *IEEE*

Transactions on Circuits and Systems I: Regular Papers 68 (1): 57–66. https://doi.org/10.1109/TCSI.2020.3023077.

14 Wann, C. H., Noda, K., Tanaka, T. et al. (1996). A comparative study of advanced MOSFET structures. *1996 Symposium on VLSI Technology. Digest of Technical Papers*, Honolulu, HI, USA. 32–33. https://doi.org/10.1109/VLSIT.1996.507782.

15 Aditya, M., Rao, K.S., Balaji, B. et al. (2022). Comparison of drain current characteristics of advanced MOSFET structures – a review. *Silicon* 14: 8269–8276. https://doi.org/10.1007/s12633-021-01638-8.

3

High-κ Dielectrics in Next Generation VLSI/Mixed Signal Circuits

Asutosh Srivastava

School of Computer & Systems Sciences, Jawaharlal Nehru University, New Delhi, India

3.1 Introduction to Gate Dielectrics

In the 1960s, the first successful demonstration of Silicon–Silicon dioxide-based metal–oxide–semiconductor field-effect transistor (MOSFET) was carried out [1–3]. Silicon as a substrate material has been a preferred semiconductor over the years due to its availability in abundance, being cheap, and its ability to form a very high-quality interface layer between silicon and silicon dioxide [1]. Silicon-based MOSFET has become the preferred active device as a building block for the integrated circuits (IC) after the invention of complementary MOSFET (CMOS), which has an inherent property of consuming low power and has the ability to be scaled down [4].

Semiconductor industry since its very inception has been guided by famous "MOORE's Law" which postulated that in IC, the number of transistors will double per unit area in every 24 months [5]. The ability of silicon-based MOSFET in IC's to scale down without having any negative impact on the performance of transistors has enabled digital circuits to show improvements in critical parameters such as circuit speed, power consumption/dissipation and transistor density. The continuous scaling of MOSFET has helped different types of digital blocks like processors and memories to achieve higher circuit speed, lower power consumption/dissipation, increase in overall transistor density, and lowering of overall chip cost.

The constant doubling of number of transistors per unit area is achieved by the scaling of transistors that are used as building blocks for any digital and analog/mixed signal circuits. The scaling of transistors is performed by implementing any of the three different scaling approaches, namely constant field scaling, constant voltage scaling, and generalized scaling [6–8]. In the initial years, the constant electric field approach was chosen to scale down transistors, where both the

Advanced Nanoscale MOSFET Architectures: Current Trends and Future Perspectives, First Edition. Edited by Kalyan Biswas and Angsuman Sarkar.

dimensions along with electrostatic parameters like drive voltage and threshold voltage were scaled down in the transistor. This type of scaling method gradually started to decline when the electrostatic parameters could not keep the same pace of scaling as those of dimensions. This forced the researchers to move away from the traditional type of scaling approach to a generalized type of scaling approach. This transformation started to become prominent when the feature size of transistor was reduced to submicron and below sub 90 nm technology node. The performance issues due to continuous miniaturization of transistors started to dominate in the form of short channel effects (SCEs), where the performance parameters of transistors started to worsen and led to failure of circuits. This further fuelled the transformation of scaling type in transistors to generalized scaling in which electric field parameters were not scaled in the same proportion as those of other parameters. In this generalized scaling approach, the scaling of electric field was granted some concession, and the supply voltage was not scaled down with the same factor as the channel length or feature size of the transistor [6–10].

Traditionally, the gate in the MOSFETs is made of heavily doped poly-silicon electrode (poly-Si), gate dielectric is made up of silicon dioxide (SiO_2) while Silicon bulk forms the semiconductor substrate. SiO_2 has been constantly scaled down and has been a natural choice as gate dielectric as it exhibits excellent properties, such as acting as an excellent gate insulator by blocking current flow through the gate dielectric into the channel of the MOSFET. The role of gate dielectric is primarily to achieve an inverted channel with a large charge density to obtain large current for the given supply of voltage and to restrict the emergence of SCE's.

In the pursuit to sustain Moore's Law, SiO_2 was scaled aggressively by lowering the thickness of gate dielectric along with other parameters by applying constant field or generalized scaling approach by a constant factor with every generation. The decrease in thickness of SiO_2 in planar MOSFET led to increase in gate capacitance (per unit area), which ultimately resulted in increase in drive current (per unit width of the transistor). The constant decrease in gate dielectric thickness, ultimately led the MOSFET to reach a stage where direct tunneling of charge carriers started to appear from poly-Si gate into the inversion region. This phenomenon, also known as direct tunneling, started to become prominent [10] when thickness of SiO_2 got reduced to around 1.2 nm at around 90 nm transistor technology node. The tunneling current or leakage across the SiO_2 started to increase exponentially with the further reduction in the thickness of SiO_2, thus jeopardizing the inherent advantage of the MOSFET/CMOS as a low-power device.

The tunneling leakage is a quantum mechanical phenomenon where SiO_2 acts as a barrier layer for electrons and holes in silicon substrate. The tunneling leakage current depends on many parameters [11], as shown in Eq. (3.1),

$$\mathrm{Jg} = \frac{A}{T_{\mathrm{ox}}^2} e^{-2T\sqrt{\frac{2qm^{\{*\}}}{\hbar^2}}} \left[\varphi_B - \frac{V_{ox}}{2}\right] \tag{3.1}$$

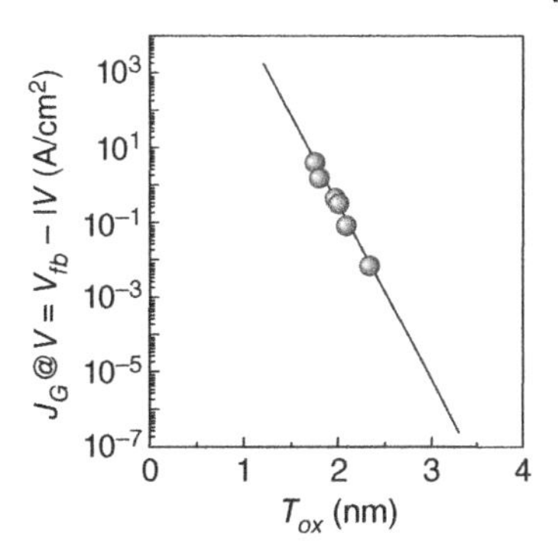

Figure 3.1 Gate leakage current density vs. SiO_2 physical thickness. Source: Adapted from Hokazono et al. [12].

where A is an experimental constant, T_{ox} is the physical thickness of SiO_2, Φ_B is the potential barrier height between the metal and the SiO_2, V_{ox} is the voltage drop across the dielectric, and m^* is the effective electron mass within the dielectric. In an ideal condition, SiO_2 is free from any defect, the potential barrier Φ_B is determined by the energy band alignment between SiO_2 and the Si substrate. But in the real world, which is non-idealistic, SiO_2 dielectric has many more defects. The electrons get trapped by these defects, forming energy levels in the SiO_2 bandgap. The electron transport is then governed by a trap-assisted mechanism such as the Frenkel–Poole emission or hopping conduction [11]. The dependence of the tunneling leakage current on the physical thickness of SiO_2 as a gate dielectric is shown in Figure 3.1 [12].

3.2 High-κ Dielectrics in Metal–Oxide–Semiconductor Capacitors

Metal–Oxide–Semiconductor (MOS) capacitors are the building blocks for MOSFET. Traditionally, MOS capacitors constituted of highly doped poly-Si acting as metal and a thin layer of oxide acting as a gate dielectric, which is on the top of doped silicon acting as a bulk semiconductor substrate. MOS capacitor is studied primarily to investigate and determine the quality of interface regions along with critical parameters like the threshold voltage and flat-band voltage in a metal–insulator–semiconductor configuration. In n-substrate MOS capacitor, the silicon forms the bulk substrate, which is doped by n-type impurities. When one applies an external bias across the MOS structure and ramp voltage from positive to negative, three regions of operation are observed within a MOS capacitor, namely accumulation, depletion, and inversion. The capacitance voltage studies provide detailed information about the workings of MOS capacitors, namely in terms of insulator thickness, maximum depletion width, flat band voltage, threshold voltage, and maximum capacitance, along with interface profile in terms of interface states and mobile ion charges between metal, oxide, and the semiconductor.

As the feature size of MOS transistor is scaled down to nano-scale, the thickness of SiO_2 gets scaled down to ultra-thin dimension of around 1 nm, which results in direct tunneling. The exponential rise in leakage current starts to dominate even after improvements in design structure of the transistor and results in failure of transistors and circuits in the IC's. The dielectric constant (kappa or κ) of silicon dioxide is 3.9. The need to scale down the thickness of SiO_2 further and the failure of circuits due to direct tunneling due to further scaling have promoted the replacement of SiO_2 with high-κ dielectrics material, where κ is dielectric constant and needs to have value higher than that of SiO_2. This replacement of dielectric material requires the researchers to study and investigate the interface quality of high-κ dielectric in a metal–insulator–semiconductor configuration so that the study can facilitate further scaling of high-κ gate dielectric thickness without any performance degradation. The need to further decrease the thickness of high-κ gate dielectric becomes necessary as it further increases the gate capacitance, which results in improvement of the drive current (per device width) of the MOSFET. The usage of high-κ dielectric further helps in reducing SCE's and reliability issues in nano-scale MOS transistors [13–15].

Another way to achieve better performance for MOS transistors, is by replacing silicon as a semiconductor substrate with higher intrinsic mobility semiconductor substrate like germanium. Researchers were motivated to study and investigate Ge-based MOS configuration, in which they observed major issues that emerged due very poor quality interface layer formation between oxide and germanium. This challenge led the researchers to investigate better interface quality for germanium-based MOS device formation by replacing existing ones with new high-κ dielectrics, using different metal gates, applying novel deposition techniques, etc. [14]. The surface passivation techniques, along with high-κ, result in high-quality MOS structure with Ge as a substrate [14].

3.3 High-κ Dielectrics in Metal Insulator Metal (MIM) Capacitors

Metal insulator metal (MIM) configurations are predominantly used for analog and mixed-signal circuits. It also has applications both in volatile and non-volatile memories. The volatile memory includes dynamic random access memory (DRAM), while non-volatile memory includes next-generation memory devices like resistive random access memory (RRAM). The usage of insulators in MIM capacitors requires constant improvement, which has motivated researchers to study, explore, and improve their properties by substituting the lower κ value dielectric materials with higher-κ value dielectric materials for improved performance and applications in analog/mixed signal circuits and memory devices [16].

3.3.1 High-κ Dielectrics for Mixed Signal Circuits

MIM configuration has also been used as a capacitor for analog/mixed signal circuits, along with DRAM applications. The technology roadmap for semiconductors demands the researchers to pursue research to explore MIM capacitors with improved performance parameters like higher capacitance density, lower leakage current density, and better capacitance–voltage linearity for analog and mixed signal circuits [17, 18]. In order to achieve this high capacitance density and other critical parameters, several high-κ dielectric materials have been studied and explored. MIM device structures with various high-κ/high-κ stacks were successfully developed for mixed signal and (embedded) DRAM applications, with promising results in terms of enhanced capacitance density, lower leakage current density, and better capacitance–voltage linearity [16–18].

3.3.2 High-κ Dielectrics as Stacks for Resistive Random Access Memories

RRAM is an emerging non-volatile memory device owing to its attributes like simple structure, low operating voltage, faster operations, and higher scalability. The device is made up of a metal oxide sandwiched between two electrodes, i.e. top and bottom electrodes. The dielectric is generally made of metal oxide with higher-κ value as compared to silicon dioxide, while the electrodes generally have a higher electrical conductivity. This MIM configuration can be integrated into a crossbar array, with a size as small as $4F2$ (where F is the minimum feature size). This can be divided into n sub-parts. A $(4F2/n)$ with vertically stacked three-dimensional (3D) architectures can be arranged and realized (where n stands for stacking layer number of the crossbar array) [19–21]. The high-κ dielectric in RRAM memory is either made up of binary oxide materials or various combinations of complex oxide materials. There is an urgent need to study and explore MIM/MOM structures as RRAM NVM to achieve the demand of reproducing excellent electrical properties. The application of various types of high-κ dielectrics/dielectric stacks in RRAM shows promising electrical results in terms of improved switching behavior, lower energy consumption, and longer retention time [19–21].

3.4 MOSFETs Scaling and the Need of High-κ

In a traditional MOSFET, due to constant scaling of SiO_2 as a gate dielectric to sustain Moore's law, there is an exponential rise in tunneling leakage current [12]. The tunneling phenomenon is majorly dependent on the two major factors, namely the thickness of the dielectric and barrier height of the dielectric. In order to sustain Moore's law, it demands constant scaling down of the thickness of gate dielectric by same factor as that of feature size in transistor. Researchers have investigated various dielectric materials with higher dielectric constant and higher barrier height to overcome the challenges of tunneling current and SCE's. The replacement of SiO_2 with high-κ gate dielectric has become a natural choice as it helps in scaling without compromising the performance of MOSFET [22–27]. The decrease in thickness of gate dielectric in form of SiO_2 in planar MOSFET becomes necessary as it leads to increase in gate capacitance (per unit area) which ultimately results in higher drive current (per unit width of the transistor). So, in order to further reduce the thickness of gate dielectric, i.e. SiO_2, without changing the electrical gate capacitance, one needs to replace SiO_2 with high-κ dielectric at a point when thickness of SiO_2 scales down to thickness around 1 nm.

The relationship between physical thickness of SiO_2 and any high-κ dielectric material with same gate capacitance is given by the following expression:

$$C = \frac{\varepsilon_0 \varepsilon_{\mathrm{SiO_2}}}{T_{\mathrm{SiO_2}}} = \frac{\varepsilon_0 \varepsilon_{\mathrm{high}-\kappa}}{T_{\mathrm{high}-\kappa}} \tag{3.2}$$

where εo is the vacuum permittivity, ε_{SiO2} is the dielectric constant of SiO_2 (=3.9), where as $\varepsilon_{\mathrm{high}\text{-}\kappa}$ is the dielectric constant of high-κ dielectric materials. The physical thickness of high-κ value of dielectric material is given by $T_{\mathrm{high}\text{-}}\kappa$. The equivalent oxide thickness (EOT) represents the theoretical thickness of SiO_2 in order to achieve the same capacitance density as that of high-κ dielectric. EOT is given by the following expression:

$$T_{\mathrm{EOT}} = \frac{\varepsilon_{\mathrm{SiO_2}}}{\varepsilon_{\mathrm{high}-\kappa}} * T_{\mathrm{high}-\kappa} \tag{3.3}$$

In traditional MOSFET configuration, poly-Si is used as a metal, while SiO_2 is used as a gate dielectric. In reality, the actual performance of gate dielectrics (as stacks) does not scale by same proportion as dielectric k value, possibly due to ignoring the effects of depletion and quantum mechanics [26]. When the thickness of SiO_2 reaches around 1 nm and the feature size of MOSFET reaches around 90 nm, not only one replaces SiO_2 with high-κ dielectric with lower thickness, thus maintaining the same electrical capacitance, but also the poly-Si gate is replaced by metal gate. The total gate capacitance is the series combination of capacitance offered by poly-Si, gate dielectric, and inversion layer formation.

Therefore, replacement of poly-Si gate to metal gate becomes necessary as it helps in getting rid of capacitance due to depletion layer formation in the poly-Si. This helps in further increasing the total capacitance of the MOS configuration in the nano-scale MOSFET, resulting in better transistor performance and reducing the SCE's [25].

The choice of high-κ dielectric in the planar MOSFET configuration depends on various critical parameters. High-κ as an alternative to SiO_2 as gate dielectric needs to possess higher dielectric constant, larger band gap, high band offset with respect to silicon, thermodynamic stability, interface quality, process compatibility, and reliability. The other important parameter for choosing high-κ dielectric depends on its impact on channel mobility [25–30]. In the process of analyzing different high-κ dielectric materials, one observes that the band gap of the dielectric decreases with the increase in the dielectric constant [25–27]. The decrease in bandgap induces smaller conduction band offset, (barrier height), resulting in an increase in tunneling leakage. This increase in tunneling leakage shall negate any reduction in leakage current that was introduced by increasing the physical thickness of the high-κ gate dielectrics [25–31]. Thus, one needs to explore, investigate, and choose a gate dielectric material that not only has higher-κ value but also has large enough conduction band offset [25].

3.5 High-κ Dielectrics in Next Generation Transistors

Planar type of transistor is the traditional transistor design that is used in the semiconductor industry to manufacture digital and mixed signals circuits and systems. The constant scaling of planar transistors beyond a threshold in order to sustain Moore's law has led to drastic degradation in performance of planar transistors and circuits. This degradation was primarily due to emergence of various SCE's, which were fatal and resulted in worsening the performance of transistors in digital and mixed signal circuits. This motivated the researchers to study, investigate, and explore further to find probable solutions to challenges thrown by scaling of planar transistors. New transistor design and architecture were studied and explored to overcome the problems of scaling and SCE's that had arisen due to constant scaling in the planar MOSFET configuration. The investigation and exploration of various new types of non-planar transistor design and architecture were performed to extend the limits of scaling without degrading the performance of transistors. The researchers were successfully able to introduce novel non-planar transistor structures like SOI and FINFETs and were successfully able to extend the scaling limits beyond the planar MOSFET's.

The constant demand to have next-generation transistors with extremely low-power digital and mixed signal chips like in processors and memories motivated researchers to study, explore, and investigate some disruptive technologies by introducing and applying new physics inside the novel transistor design. The introduction of new physics in next-generation transistors became essential as "Denard" type of scaling approach slowly came to an end. In this scaling approach, both the dimensional as well as the electric field parameters were scaled down by same factor for MOSFET [32, 33]. As the constant scaling progressed, electrostatic parameters were not scaled down in the same proportion as the dimensions of the transistors. This new challenge of further scaling electrostatic parameters like ON current post-saturation led researchers to introduce disruptive technologies into transistor design that led to successful demonstration of revolutionary new types of transistors like tunnel FET (TFET) and negative capacitance FET (NC-FET). In these new types of transistor design, transport model with new physics was introduced so that fundamental limits, namely the "Boltzmann Tyranny" could be broken. Initially, these novel types of extremely low power transistors were designed with materials with lower dielectric constant. Later, higher-κ dielectrics was introduced in these new types of transistors namely TFET and NC-FET, in order to achieve greater gate capacitance, better gate control, and effects of fringing capacitance, which resulted in higher ON current per transistor width [34].

3.5.1 Planar–Nano Scale Field Effect Transistor

Planar transistor is the classical MOSFET that has been used to design digital and mixed signals circuits in the semiconductor industry. These planar transistors have been the top choice for feature sizes above sub-nano scale as they inherit excellent properties like extremely low leakage current, high quality interface layer, and the ability to scale down constantly. In planar configuration, silicon is the chosen semiconductor, which is used as a base material, possesses an excellent chemistry with oxygen, and forms an excellent silicon dioxide gate dielectric with a high-quality interface layer. On the top of this silicon dioxide, there is a poly-Si that is deposited using cutting-edge fabrication techniques, thus forming an excellent MOS structure. The pursuit to constantly scale down planar MOSFET in terms of dimensions and electrostatic parameters by same proportion/scale has created major issues in its performance. These issues in planar transistors slowed down the pace of scaling and motivated researchers to study and explore new transistor designs and architectures beyond planar transistors.

3.5.2 Silicon on Insulator

As the feature size of MOSFET was scaled down below sub 90 nm, SCE's started to dominate. The SCEs increase due to high electric field lines that reach from drain to source region. The narrow spacing of source/drain to body-depletion regions in the channel may start to interfere with each other, leading to punch-through problem. One of the novel techniques to arrest this challenge of SCE was by modifying the planar MOSFET structure to silicon on insulator (SOI) structure. SOI essentially has a very thin silicon layer on the top of an insulator or dielectric, which is done by introducing a buried oxide (BOX) layer onto the substrate. The usage of various – high-κ dielectrics in SOI configurations helps to further reduce SCE as it helps in lowering drain electric field lines and improving the trans-conductance along with quasi-ideal subthreshold of the new type of transistor. So, the issue of punch-through effect can be greatly reduced by introducing the BOX configuration of different magnitudes, which may be referred to as fully-depleted or SOI technology.

The introduction of ultra-thin body (UTB) as a BOX structure further helps in scaling of transistors by suppressing the SCE. The replacement of SiO_2 with high-κ dielectric material in UTB-SOI and by introducing double-gate BOX structure helps in further suppressing the SCE as it helps in further lowering of drain electric field lines along with providing better electrostatic control over the channel [35–37].

3.5.3 FIN Field Effect Transistor

Researchers, while in their pursuit to scale down transistors and suppress SCE's, investigated and explored other novel types of transistor configuration, namely as a double gate MOSFET (DG-MOSFET). But, new issues due to misalignment of gates in the DG-MOSFET architecture led to capacitance overlap and emergence of source/drain resistance. Researchers were motivated to solve this new problem and proposed two different types of transistor architecture, namely tri-gate, and later FIN-FET. The next generation of non-planar transistors has shown promising results while scaling transistors below 65 nm regime. These next generation non-planar transistors with various designs and architectures have the capability to extend scaling without showing any decay in the transistor properties. Later, as the feature size of transistors reached around 10 nm, gate all around FINFET structure was proposed in order to suppress the SCE and to further extend the scaling of transistors to even smaller values for digital and mixed signal circuits. The replacement of SiO_2 with high-κ dielectrics or gate dielectric stacks showed promising results with higher gate capacitance, better gate control over the channel, lowering of SCE's, and improvement in non-planar transistor properties due to fringing effect [38–40]. The replacement of poly-Si by metal gate along with high-κ material as gate dielectric not only resulted in further scaling of transistors as well as suppression of SCE but also resulted in improvement (lowering) of bias temperature instability (BTI) [40].

3.5.4 Tunnel Field Effect Transistor

The semiconductor industry has traditionally followed the Dennard scaling approach, in which all parameters, including dimension and electrostatic (voltage), are simultaneously scaled down by same factor across the planar MOSFET structure. As the transistor feature size entered sub-nano dimensions, further scaling of transistors both in terms of dimensions and electro-static parameters by same proportion/factor came to an abrupt halt. The hindrance to constant field scaling approach began to dominate as the feature size of transistors approached below sub 90 nm. In order to maintain same level of ON current, both operating voltage and threshold voltage need to scale down by same proportion. But, scaling of threshold voltage below sub-nano scale transistors, by the same proportion as other parameters like dimensions got restricted due to exponential rise in leakage current. This exponential rise in leakage current is primarily dictated by subthreshold swing of sub-nano transistors reaching their fundamental limits of 60 mV/Dec at room temperature also known as "Boltzmann tyranny."

The inception of TFET offered a disruptive change in transport mechanism of charge carriers from the classical thermally injected mechanism to band-to-band tunneling at heavily doped interfaces [41]. Traditionally, TFET showed promising results in terms of lower off current but had a major drawback of having very low ON current. Researchers got motivated to study, investigate, and explore different techniques to overcome the challenges of low drive (ON) current and poor ambi-polar current in TFET. The replacement of dielectric in TFET with high-κ dielectric led to improvement in the drive (ON), Ambi-polar conduction, along with subthreshold swing, making it more relevant for extremely low-power digital and mixed signal applications [41–45].

3.5.5 Negative Capacitance Field Effect Transistor

The emergence of NC-FET as a disruptive transistor technology helped in overcoming the challenges arising from fundamental limits caused by Boltzmann Tyranny along with scaling challenges of transistor electrostatic parameters. The new configuration of transistor NC-FET helped in extreme lowering of power requirements of chips for digital and mixed signal circuits. Ferroelectric-based capacitors inherit and exhibit excellent properties, which include a very high value of dielectric constant and negative capacitance [46–49]. Researchers got strongly interested to study, explore, design and fabricate ferroelectric based FET with stable and static negative capacitance in a non-transient hysteresis-free region. Ferroelectric materials have been shown to be the most promising candidates to achieve negative capacitance [44–47]. Negative capacitance in a ferroelectric-based capacitor is due to the Landau double-well formation in the capacitor with energy (W), as a function of the applied charge (Q). Researchers, after studies and investigations, have proposed the application of high-κ dielectrics in combination with ferroelectric material as a NC-FET structure that was able to achieve a stable and static negative capacitance [50, 51]. NC-FET with stable and static negative capacitance could attain a higher drive current at lower leakage current due to attainment of subthreshold swing lower than 60 mV/decade, which helped in realizing extremely low-power digital and mixed signal circuits [50–51].

References

1 Atalla, M.M., Tannenbaum, E., and Scheibner, E.J. (1959). Stabilization of silicon surfaces by thermally grown oxides. *The Bell System Technical Journal* 38 (3): 749–783.

2 Kahng, D. and Atalla, M.M. (1960). Silicon-silicon dioxide field induced surface devices. In: *IRE-AIEE Solid-state Device Res. Conf*. Pittsburgh, PA: Carnegie Inst. of Technol.

3 Kahng, D. (1976). A historical perspective on the development of MOS transistors and related devices. *IEEE Transactions on Electron Devices* 23 (7): 655–657.

4 Hori, T. (1997). *Gate Dielectrics and MOS ULSIs*. New York: Springer.

5 Moore, G. E. (1965). Cramming more components onto integrated circuits. (1965-04-19). *intel.com*. Electronics Magazine.

6 Baccarani, G., Wordeman, M.R., and Dennard, R.H. (1984). Generalized scaling theory and its application to a ¼ micrometer MOSFET design. *IEEE Transactions on Electron Devices* 31 (4): 452–462.

7 Borkar, S. (1999). Design challenges of technology scaling. *IEEE Micro* 19 (4): 23–29.

8 Frank, D.J., Dennard, R.H., Nowak, E. et al. (2001). Device scaling limits of Si MOSFETs and their application dependencies. *Proceedings of the IEEE* 89 (3): 259–288.

9 Skotnicki, T., Hutchby, J.A., King, T.-J. et al. (2005). The end of CMOS scaling: toward the introduction of new materials and structural changes to improve MOSFET performance. *IEEE Circuits and Devices Magazine* 21 (1): 16–26.

10 Wong, H. and Iwai, H. (2006). On the scaling issues and high-j replacement of ultrathin gate dielectrics for nanoscale MOS transistors. *Microelectronic Engineering* 83: 1867–1904.

11 Wallace, R.M. and Anthony, J.M. (2001). High-κ gate dielectrics: current status and materials properties considerations. *Journal of Applied Physics* 89: 5243.

12 Hokazono, A., Ohuchi, K., Takayanagi, M. et al. (2002). 14 nm gate length CMOSFETs utilizing low thermal budget process with poly-SiGe and Ni salicide. *IEEE International Electron Devices Meeting*. 639–642.

13 Srivastava, A., Nahar, R.K., and Sarkar, C.K. (2010). Study of the effect of thermal annealing on high k hafnium oxide thin film based MOS and MIM capacitor. *Journal of Materials Science: Materials in Electronics* 22 (7): 882.

14 Xie, Q., Deng, S., Schaekers, M. et al. (2012). Germanium surface passivation and atomic layer deposition of high-*k* dielectrics—a tutorial review on Ge-based MOS capacitors. *Semiconductor Science and Technology* 27: 074012.

15 Xia, P., Feng, X., Ng, R. et al. (2017). Impact and origin of interface states in MOS capacitor with monolayer MoS_2 and HfO_2 high-*k* dielectric. *Scientific Reports* 7: 40669.

16 RF and Analog/Mixed-Signal Technologies for Wireless Communications (2009). *2009 International Technology Roadmap for Semiconductors (ITRS)*. Semiconductor Industry Association.

17 Padmanabhan, R., Bhat, N., and Mohan, S. (2012). High-performance metal–insulator–metal capacitors using europium oxide as dielectric. *IEEE Transactions on Electron Devices* 59 (5): 1364–1370.

18 Srivastava, A., Mangla, O., and Gupta, V. (2015). Study of La-incorporated HfO_2 MIM structure fabricated using PLD system for analog/mixed signal applications. *IEEE Transactions on Nanotechnology* 14 (4): 612–618.

19 Dou, C., Kakushima, K., Ahmet, P. et al. (2012). Resistive switching behavior of a CeO_2 based ReRAM cell incorporated with Si buffer layer. *Microelectronics Reliability* 52 (4): 688–691.

20 Ielmini, D. (2016). Resistive switching memories based on metal oxides: mechanisms, reliability and scaling. *Semiconductor Science and Technology* 31: 063002.

21 Shen, Z., Zhao, C., Qi, Y. et al. (2020). Advances of RRAM devices: resistive switching mechanisms, materials and bionic synaptic application. *Nanomaterials (Basel)* 10 (8): 1437.

22 Srivastava, A., Sarkar, P., and Sarkar, C.K. (2009). Study in variation of gate dielectric permittivity with different EOT on channel engineered deep sub-micrometer n-MOSFET device for mixed signal applications. *Microelectronics Reliability* 49 (4): 365.

23 Hiroki Fujisawa, A., Srivastava, K.K., Ahmed, P. et al. (2009). Electrical characterization of W/HfO_2 MOSFETs with La_2O_3 incorporation. *ECS Transactions* 18 (1): 39.

24 Kittl, J.A., Opsomer, K., Popovici, M. et al. (2009). High-k dielectrics for future generation memory devices (invited paper). *Microelectronic Engineering* 86 (7–9): 1789–1795.

25 Zade, D., Sato, S., Kakushima, K. et al. (2011). Effects of La_2O_3 incorporation in HfO_2 gated nMOSFETs on low-frequency noise. *Microelectronics Reliability* 51 (4): 746.

26 Kawanago, T. (2011). A study on high-k/metal gate stack MOSFETs with rare earth oxides. Doctorial Thesis. Interdisciplinary Graduate School of Science and Engineering, Tokyo Institute of Technology, Tokyo, Japan.

27 Rios, R. and Arora, N.D. (1994). Determination of ultra-thin gate oxide thicknesses for CMOS structures using quantum effects. *IEEE International Electron Devices Meeting* 613–616.

28 Robertson, J. (2000). Band offsets of wide-band-gap oxides and implications for future electronic devices. *Journal of Vacuum Science and Technology B* 18: 1785–1791.

29 Vogel, E.M., Ahmed, K.Z., Hornung, B. et al. (1998). Modeled tunnel currents for high dielectric constant dielectrics. *IEEE Transactions on Electron Devices* 45: 1350–1355.

30 Robertson, J. and Chen, C.W. (1999). Schottky barrier heights of tantalum oxide, barium strontium titanate, lead titanate, and strontium bismuth tantalite. *Applied Physics Letters* 74: 1168–1170.

31 Ma, T.P. (1998). Making silicon nitride film a viable gate dielectric. *IEEE Transactions on Electron Devices* 45: 680–690.

32 Zhao, C., Wang, X., and Wang, W. (2018). Woodhead publishing series in electronic and optical materials. In: *CMOS Past, Present and Future* (ed. H.H. Radamson, J. Luo, E. Simoen, and C. Zhao), 69–103. Woodhead Publishing.

33 Gonzalez, R., Gordon, B.M., and Horowitz, M.A. (1997). Supply and threshold voltage scaling for low power CMOS. *IEEE Journal of Solid-State Circuits* 32 (8): 1210–1216.

34 Forestier, A. and Stan, M. R. (2000). Limits to voltage scaling from the low power perspective. *Proceedings 13th Symposium on Integrated Circuits and Systems Design (Cat. No.PR00843)*, Manaus, Brazil. 365–370.

35 Schlosser, M., Bhuwalka, K.K., Sauter, M. et al. (2009). Fringing-induced drain current improvement in the tunnel field-effect transistor with high-κ gate dielectrics. *IEEE Transactions on Electron Devices* 56 (1): 100–108.

36 Bernard, E. Ernst, T., Guillaumot, B. et al. (2008). Novel integration process and performances analysis of Low STandby Power (LSTP) 3D multi-channel CMOSFET (MCFET) on SOI with metal/high-κ gate stack. *2008 Symposium on VLSI Technology*, Honolulu, HI, USA. 16–17.

37 Pilo, H., Arsovski, I., Batson, K. et al. (2012). A 64 Mb SRAM in 32 nm high-κ metal-gate SOI technology With 0.7 V operation enabled by stability, write-ability and read-ability enhancements. *IEEE Journal of Solid-State Circuits* 47 (1): 97–106.

38 Basak, A. and Sarkar, A. (2021). Quantum analytical model for lateral dual gate UTBB SOI MOSFET for analog/RF performance. *Silicon* 13: 3131–3139.

39 Jurczak, M., Collaert, N., Veloso, A. et al. (2009) Review of FINFET technology. *2009 IEEE International SOI Conference*, Foster City, CA, USA. 1–4.

40 Nirmal, D., Vijayakumar, P., Thomas, D.M. et al. (2013). Subthreshold performance of gate engineered FinFET devices and circuit with high-k dielectrics. *Microelectronics Reliability* 53 (3): 499–504.

41 Lee, K. T., Wonchang Kang; Eun-Ae Chung et al. (2013). Technology scaling on high-K & metal-gate FinFET BTI reliability. *2013 IEEE International Reliability Physics Symposium (IRPS)*, Monterey, CA, USA. 2D.1.1–2D.1.4.

42 Moselund, K.E., Ghoneim, H., Bjork, M.T. et al. (2009). Comparison of VLS grown Si NW tunnel FETs with different gate stacks. *ESSDERC 2009 – Proc. 39th Eur. Solid-State Device Res. Conf.*, Athens, Greece. 448–451.

43 Singh, A., Chaudhury, S., Kumar Pandey, C. et al. (2019). Design and analysis of high *k* silicon nanotube tunnel FET device. *IET Circuits, Devices and Systems* 13: 1305–1310.

44 Tayal, S., Vibhu, G., Meena, S. et al. (2022). Optimization of device dimensions of high-*k* gate dielectric based DG-TFET for improved analog/RF performance. *Silicon* 14: 3515–3521.

45 Karbalaei, M., Dideban, D., and Heidari, H. (2020). A simulation study of the influence of a high-*k* insulator and source stack on the performance of a double-gate tunnel FET. *Journal of Computational Electronics* 19: 1077–1084.

46 Pandey, C.K., Dash, D., and Chaudhury, S. (2019). Approach to suppress ambipolar conduction in tunnel FET using dielectric pocket. *Micro & Nano Letters* 14: 86–90.

47 Srivastava, A., Gupta, V., Katiyar, R., and Sarkar, C.K. (2006). Dielectric relaxation in pulsed laser ablated $CaCuTi_4O_{12}$ thin film. *Journal of Applied Physics* 100 (3): 34102.

48 Srivastava, A. and Sarkar, C.K. (2009). Dielectric property of $CaCu_3Ti_4O_{12}$ thin film grown on Nb-doped $SrTiO_3$(100) single crystal. *Applied Physics A: Materials Science & Processing* 97: 409–416.

49 Khan, A., Chatterjee, K., Wang, B. et al. (2015). Negative capacitance in a ferroelectric capacitor. *Nature Materials* 14: 182–186.

50 Íñiguez, J., Zubko, P., Luk'yanchuk, I. et al. (2019). Ferroelectric negative capacitance. *Nature Reviews Materials* 4: 243–256.

51 Chen, J.-J. (2021). *A Journey of Embedded and Cyber-Physical Systems*, 107–124. Springer Cham.

52 Luk'yanchuk, I., Razumnaya, A., Sené, A. et al. (2022). The ferroelectric field-effect transistor with negative capacitance. *NPJ Computational Materials* 8: 52.

4

Consequential Effects of Trap Charges on Dielectric Defects for MU-G FET

Annada S. Lenka and Prasanna K. Sahu

Department of Electrical Engineering, Nano-Electronics Lab, NIT, Rourkela, India

4.1 Introduction

The increasing need for electronic devices in practically every aspect of life drives researchers and the semiconductor industry to develop new techniques, and designs, or improve upon existing ones to increase transistor performance. The reduction of short channel effects (SCEs), lower overall power consumption, and most crucially, the overall tiny size of the device causes us to shift our focus to the innovations. In other words, we can state that the goal of our research is always to control the electrical properties in every aspect of the operation of a transistor or metal oxide semiconductor (MOS) system.

Field and voltage scaling have received the majority of attention in the previous few decades when it comes to device downscaling. The targeted interest of both the scaling techniques is dimension of the device and operating bias voltage of the devices for constant field and voltage scaling respectively. As constant field scaling influences the physical dimensions of the devices, such as the channel length, channel thickness, and oxide thickness, our focus of discussion will be on field scaling. Because the thickness of the dielectric is so important to field effect transistor performance, researchers continue to focus on the scaling of oxide thickness. The traditional gate oxide used in industry, silicon dioxide (SiO_2), must now be a porous sheet in order to meet present demand, causing undesirable gate leakage current.

To tackle this problem, researchers are turning to High-K dielectric materials. Again, researchers and industry personnel are turning to multi-gate devices to increase gate control over the channel and so eliminate the majority of SCEs. In 1984, Sekigawa and Hayasi [1] proposed a design for a double-gate MOSFET, ushering in a new era in the semiconductor industry by improving

Advanced Nanoscale MOSFET Architectures: Current Trends and Future Perspectives, First Edition. Edited by Kalyan Biswas and Angsuman Sarkar.

transistor performance by introducing additional gate control over the channel or majority charge carrier. The introduction of Multigate/Multiple Gate devices by the transistor industry is made easier by this idea (MU-G). Which design for advanced non-classical CMOS devices is currently the most popular? (AnCD). The aforementioned group includes well-known devices like the double gate, triple gate, enclosed gate, gate all around, FINFET, and multi bridge channel FET (MBCFET). Contrary to single-gate devices, multi-gate structures require more oxide layers to be designed in order to accommodate each gate electrode. It leads to an increase in the number of interface layers or, put another way, the area where the oxide and semiconductor regions meet.

The golden rule of the transistor industry, which states that the number of transistors per chip will quadruple every three years [2], was first published in a well-known article by Gordon Moore in 1965. Many researchers feel that the technology outlined above allows us to meet this rule. However, these dielectric materials occasionally have increased relaxation current in memory-based designs because of flaws introduced during production or operation. Choosing the right material for a particular application can be crucial if we take into account that the production of these layers is more difficult than that of standard SiO_2.

One of the primary causes of device failure in integrated circuits, particularly for devices operating under high stress, is the electrical failure of oxides and oxide/semiconductor interfaces. The atmosphere in space and radiation-prone areas are interest of discussion because these are highly stressful. Ionizing radiation can affect any electronic device. Any high-energy particle penetrating an insulating layer will leave behind an unwanted charge, changing how the device behaves. Since there is no atmosphere in space to induce energetic particles to collide and lose energy through collision, designing electronics for usage in space can be difficult. This is because far more intense radiation occurs in space than on Earth.

As discussed above if we were to design a multi-gate structure, we would have to add more dielectric layers for each gate metal or polysilicon, which would obviously increase the creation or presence of trap charges in these layers compared to single-gate transistors and affect channel behavior such as inversion layer buildup, threshold voltage roll-off, and gate leakage current.

It is currently unclear how exactly these changes occur, despite the fact that oxide charge brought on by ionizing radiation causes well-known changes in device properties or total ionizing dose (TID) effects. Numerous theories are based on extensive experimental data as well as, more recently, molecular quantum-mechanical first-principles computations. This study tests these theories by simulating oxide degradation in accurate device geometries and comparing the resulting decline in device attributes to experimental observations.

The enhanced low-dose-rate sensitivity effect has been quantified using the charge trapping and defect-modulated transport systems. According to measurements, equipment exposed to ionizing radiation at high dose rates shows less degradation than equipment exposed to the radiation at low dose rates. Additionally, the observed trend varies according to the quantity of hydrogen present prior to, during, and following irradiation. In order to design faster testing processes for devices that have been exposed to varied quantities of hydrogen during manufacturing and packing and that must be used in the low-dose-rate space environment, it is crucial to comprehend and take into consideration the impacts of dose rate and hydrogen.

4.2 TID Effects Overview

Materials, particularly the isolation layers used in semiconductor electrical devices, are charged by ionizing radiation. Even the highest quality oxides utilized in the IC sector include flaws, including alien chemical species and abnormalities in their molecular structure. Even in SiO_2-on-Si systems, where SiO_2 is a native oxide and well-matched to Si, material interfaces are areas with particularly high defect concentrations. Some charge trapping locations, such as bulk oxide and interface defects, can hold charge for lengthy periods, on the order of years. TID effects are the study of this charge accumulation in device oxide layers and the alterations, frequently degradation, in device and circuit properties that follow.

Depending on the specific environment of the oxide, several subsequent physical mechanisms may take place once charge is deposited in the oxide. For instance, the produced charge is affected by temperature, electric field, ambient hydrogen percentage, deposition or growth temperature and procedure, and packaging method, which in turn affects the final state of individual transistors and entire circuits. In order to comprehend specific mechanisms in particular devices, circuits, and situations, empirical or non-physical models can be utilized. However, these models are often only useful in the domain to which they have been tailored. Figure 4.1 summarizes the concept of TID in an SiO_2-on-Si structure.

However, there are many different IC deployment scenarios and mission requirements. For instance, TID radiation will have an impact on the ICs in on-Earth server farms, Earth-orbiting commercial communications satellites, military communications satellites, and Mars rovers. However, each of these missions will be operating in a very different environment and have very different operational time constraints, performance requirements, and budgets. Understanding the internal workings of the onboard electronic devices can have an impact on IC testing, prediction, and hardness assurance since there

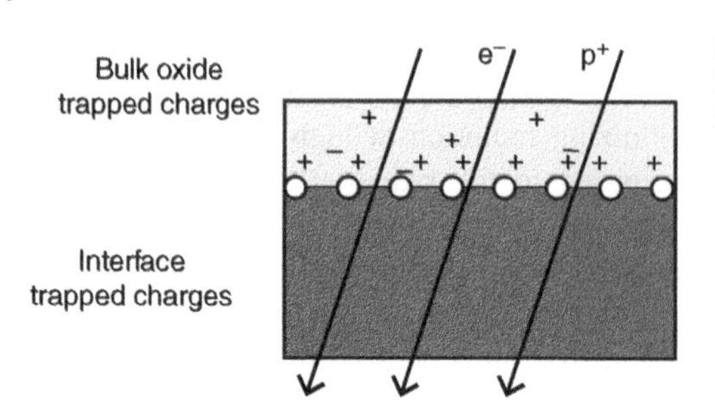

Figure 4.1 Ionizing dosage in SiO_2-on-Si structures is depicted graphically [5]/NASA/Public Domain.

are numerous physical factors that regulate device deterioration during these missions.

SiO_2 becomes ionized when energetic particles, such as electrons, protons, or other particles, interact with it, producing charge that can become trapped for a very long time. TID effects are taken into account apart from single event effects, which on the other hand, only typically occur in Si and characterize transient charge creation and dissipation on significantly shorter time scales.

Increasing knowledge of the fundamental charge transport and trapping mechanisms in device SiO_2 layers utilized in space applications is the aim of this effort. The intense electrons that encircle the Earth and other planets and must pass through them for practical orbits are our main focus. The next section gives an overview of the space environment with a focus on the electrons that are caught in planetary radiation belts. A summary of the fundamental TID effects in MOS and bipolar devices is given in the section, together with a description of the combined MOS and bipolar structure are helpful for further discussion.

4.3 Application Area of Device for TID Effect Analysis

There were nearly 6542 satellites orbiting the Earth as of January 1, 2021. Out of which 3372 satellites are active, and 3170 satellites are inactive. Many people's daily lives are impacted by communication and defense satellites. Additionally, specialized satellites and other spacecraft are employed for scientific investigations, such as Earth's oceanic monitoring or the study of our solar system. Since launching such vessels is expensive in and of itself, and because spacecraft are difficult to repair if an onboard component fails, great care and thus expense is taken to ensure reliability and success of each mission.

Every environment, including the Earth, contains radiation, but some are more harmful than others. To plan for each trip, spacecraft designers need to be aware of the conditions and component exposure times. Additionally, they must understand how various radiation types interact with the material inside their spaceship as well as with more sophisticated components, such as circuits. TID radiation impacts have been shown to create actual on-board anomalies ever since the launch of even the first satellites [3]. At best, this causes ground controllers to have an unexpected increase in work. At worst, TID effects might stop a mission.

Throughout a space mission, ground controllers continuously monitor the on-board conditions and external radiation environments in order to identify issues as fast as feasible. These controllers must determine the origin or mechanism of each defect after it is discovered and come up with reconfiguration plans [3]. These duties are made possible by a scientific understanding of the impact of radiation on gadgets.

Total TID failures are uncommon, however, this is primarily due to excessive overdesign and historically, to the high imprecision of crude shielding estimates. However, these effects continue to be a significant design phase concern, and TID-related component deterioration is extensively documented [3]. Particularly on missions where the environment was extremely harsh, like the NASA Jupiter probes, or in satellites operating for far longer than their intended lives, TID-induced degradation has been documented. The next part covers how radiation is commonly classified in testing labs and in the various space conditions in which spacecraft operate to provide context for what designers of spacecraft electronics must take into account. The discussion continues with a number of actual cases of TID-caused anomalies and system breakdowns.

For particles with mass, radiation sources are arranged in the lab according to atomic number, charge, and energy (E), and for photons, according to wavelength or energy. The most pertinent high-energy particles are those that cause charge in common IC materials. In particular, three categories are frequently taken into account: γ-rays or photons with wavelengths less than 0.01 nm, or frequencies above 10^{20} Hz; protons and neutrons with $E > 1$ MeV; heavy ions with $E > 1$ MeV Per nucleon; and electrons with $E > 100$ keV [4]. This arrangement of radiation sources is helpful in theory and in the execution of controlled tests. The focus of the current work is mostly on energetic electrons and γ-rays.

However, when it comes to space, radiation is frequently classified according to the environment or the physical source. The near-earth environment contains belts of high-energy electrons and protons trapped by the Earth's magnetic field, proton flares and photons coming from our Sun are a major concern, and radiation also comes from extra-galactic sources. As an example, the on-Earth environment contains particles with energies greatly reduced by engagement with the Earth's atmosphere [4]. This classification of radiation is beneficial for mission planning. The radiation types the spacecraft will encounter during the mission must be predicted by the designers based on the uncontrollable source of the radiation. Designers can predict the overall dose that the craft and its components will receive depending on the amount of time they will spend in each space environment by learning about those environments.

4.4 Near the Earth: Trapped Radiation

The existence of radiation in the Earth's upper atmosphere was unexpectedly first detected by the first US artificial satellite, which was launched on January 31, 1958. Early in the flight, a Geiger counter inside the satellite became saturated. Further investigation indicated that toroidal belts around the Earth are home to energetic electrons and protons. J. A. Van Allen, who had first suggested putting a Geiger counter on the satellite, gave this radiation the name "Van Allen belt(s)" [3]. Although the electrons' sources are unknown, observed abundance ratios point to both terrestrial and extraterrestrial origins [4]. The charged electrons are held in place by the Lorentz force produced by the magnetic field of the Earth [4].

Using the AE8 code developed by NASA's Jet Propulsion Laboratory (JPL), an example calculation of the trapped electron distribution that emerges from this field is shown in Figures 4.2–4.5 [5, 6].

When traveling over the South Atlantic Anomaly (SAA), a concentration of these particles caused by an offset of the Earth's magnetic axis relative to its rotation axis, the trapped protons and 1 eV to 10 MeV electrons are what primarily

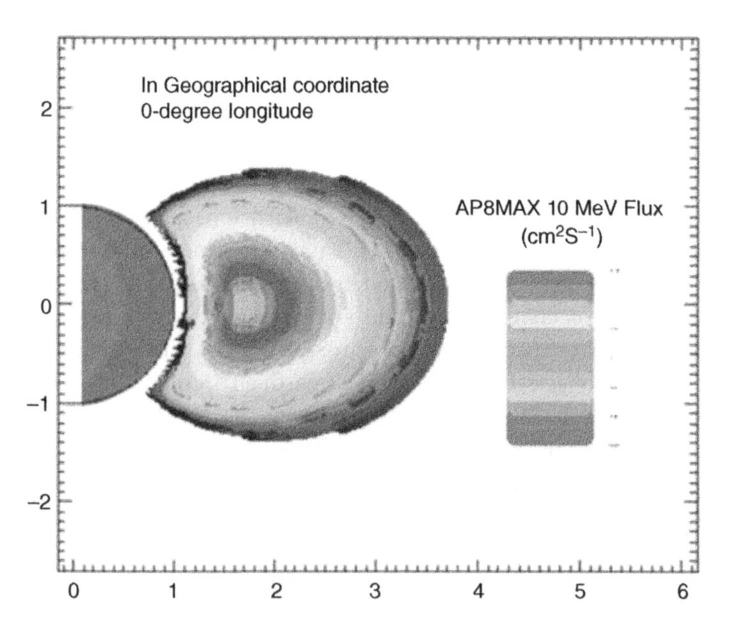

Figure 4.2 Trapped proton belt at Earth atmosphere [5]/NASA/Public Domain.

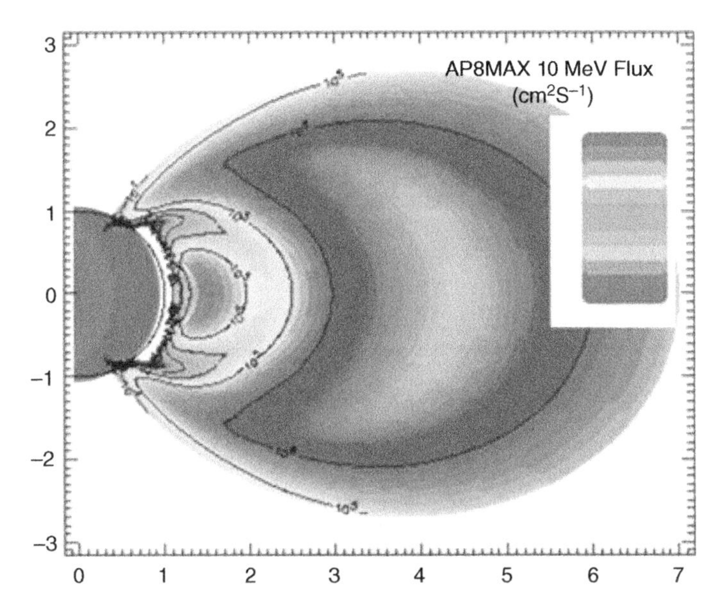

Figure 4.3 Trapped electron belt at Earth atmosphere [5]/NASA/Public Domain.

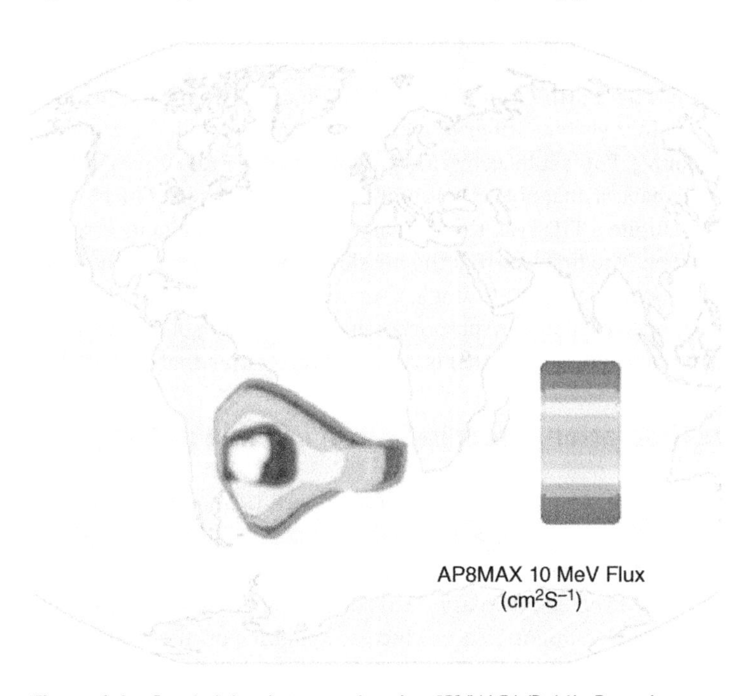

Figure 4.4 South Atlantic trapped region [5]/NASA/Public Domain.

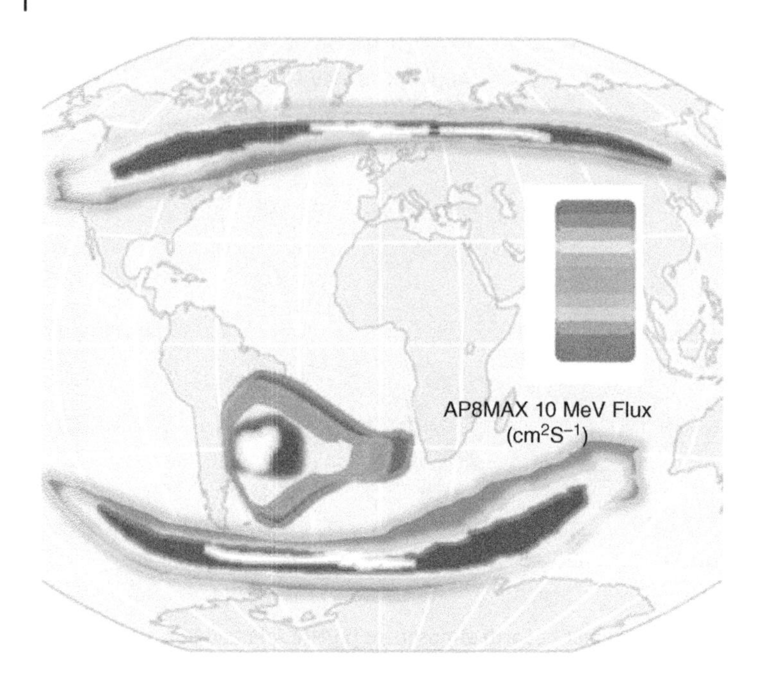

Figure 4.5 South Atlantic and electron distribution on Earth [5]/NASA/Public Domain.

contribute to TID in Low Earth Orbit (LEO) [3]. In Figures 4.4 and 4.5, the area close to Brazil is plainly visible. Other planets in the solar system have nearby belts that are similar [4]. For instance, research of Jovian UHF radio waves in 1959 revealed Jupiter to have a magnetosphere with trapped electrons. On NASA's Galileo mission to Jupiter, TID was the primary source of component error or failure for two reasons. The first was that the mission's specially created, rad-hard chips were essentially unresponsive to single-event impacts. The maximal energy and flux levels of trapped particles are proportional to the strength of the magnetic field, and Jupiter has a magnetic field that is 20 times greater than that of Earth [4].

4.5 Ionizing Radiation Effect in Silicon Dioxide (SiO_2)

Ionizing radiation is a complicated procedure that it uses to affect any material. Even considering the impact of a single incident particle is challenging since this particle may produce numerous secondary particles through a cascade process that may differ greatly from the primary particle in terms of energy and composition. The new particles might also produce secondaries on their own. The cascade process cannot be analytically described due to its complex reality, and is instead, at most, addressed as a probabilistic process utilizing high-performance computing programmers [4] (Figure 4.6).

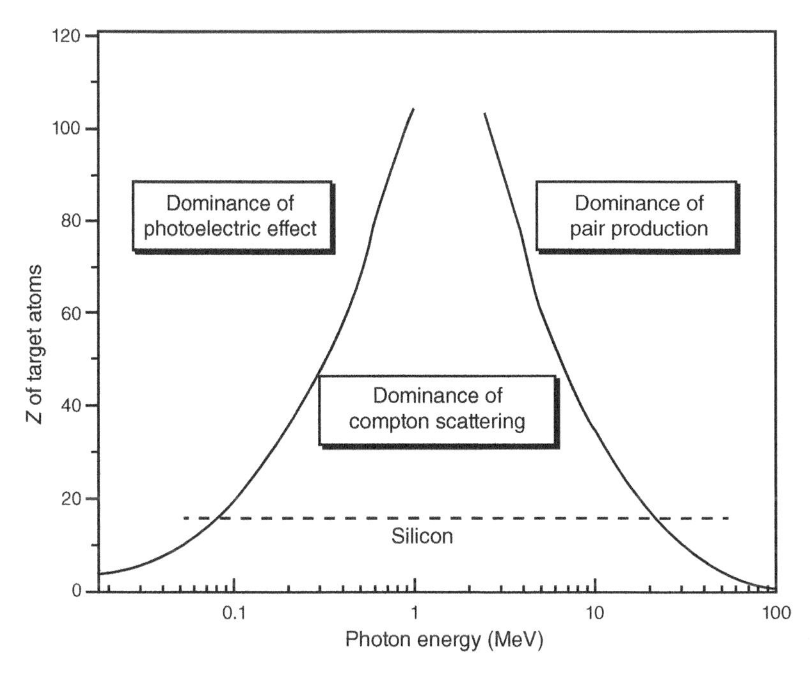

Figure 4.6 Photon-matter interaction type as a function of atomic number (*Z*) and photon energy. Source: Adapted from Garret [4].

The secondary electron, also known as the Compton electron, is typically thought of as a free electron and can range in energy from 0.5 to 1.0 MeV. These secondary electrons, which cause the SiO_2 to produce Electron Hole Pairs (EHPs), are primarily responsible for delivering the dosage to the oxide [7]. Ausman and McLean determined that, on average, one EHP is produced for every 183 eV of energy deposited in SiO_2 based on the experimental findings of Curtis et al. This value has been more precisely found to be 171 eV [8] by more recent measurements.

Within the first few picoseconds after creation, the EHPs produced by Compton electrons go through a direct first electron–hole recombination phase [7]. Due to the substantial distance between the EHPs initially formed in SiO_2 by the energetic electrons, including the secondary electrons induced by 60Co-rays, each pair's recombination can be handled independently. Geminate recombination is the name given to this approach. The high-energy secondary electrons from 60Co-rays that induce EHPs in SiO_2 are produced rather seldom, and the geminate model well matches the experimental findings [7].

The yield, which is defined as the percentage of electrons or holes that survive this initial recombination, depends on both the average separation distance between EHPs at the time of their formation and the local electric field, which acts to separate the EHPs. The incident particle type, energy, and target material all affect the average separation distance [7]. TID stands for the total energy that a particle deposits and releases as EHP in a specific material [4]. To correspond to SI values, the units of dose, in this case, the rad (SiO_2), stand for "radiation absorbed dose" in SiO_2, where 1 rad (Si) = 100 ergs/g(Si) [4].

4.6 TID Effects in CMOS

Oxide trapped charge (N_{ot}), also known as a net positive charge buildup in device oxides caused by irradiation, is often caused by the energetically advantageous capture of a hole in a neutral oxygen vacancy. As a result, the well-known oxygen vacancy defects, the E_0 centers, are formed. These hole traps can reach energy levels that are well into the SiO_2 bandgap in some cases. Both n-channel and p-channel MOSFETs experience negative threshold voltage shifts as a result of positive charge accumulation in oxides [9]. The circuit will malfunction if the V_T shift grows too great as a result of radiation's ongoing charge deposition. At some point, it will not be possible to switch either the p-channel device on or the n-channel device off.

Another SiO_2-related flaw that impairs device performance is linked to the P_b centers or "dangling bonds" that have been seen at the Si/SiO_2 interface [9]. The impact of positive interface charge on an n-channel device's IV characteristics is depicted in Figure 4.7b. The shift in threshold voltage and stretch-out ($V_{\Delta it}$) in the $I_d - V_{gs}$ response are both brought on by the interface charge, which also increases subthreshold swing.

Direct hole contact does not generate interface traps at or above room temperature, according to experiments [9, 10], and calculations using density functional theory support this conclusion [11]. Instead, the release of protons from interstitial molecular hydrogen or from defects holding hydrogen (such as a hydrogenated vacancy or Dopant-H complexes [11]) or both is required for the formation of dangling bonds at the Si/SiO_2 interface. A positively-charged dangling bond (DB_{int}^+) is left at the interface and a hydrogen molecule diffuses away as a result of protons moving by drift or diffusion interacting with hydrogen-passivated dangling bonds at the Si/SiO_2 contact [9]. The term "interface trap density" refers to the density of dangling bonds at the interface (N_{it}).

4.7 TID Effects in Bipolar Devices

When the semiconductor industry started adopting oxide isolation structures, TID effects in bipolar devices started to receive a lot of attention from the radiation

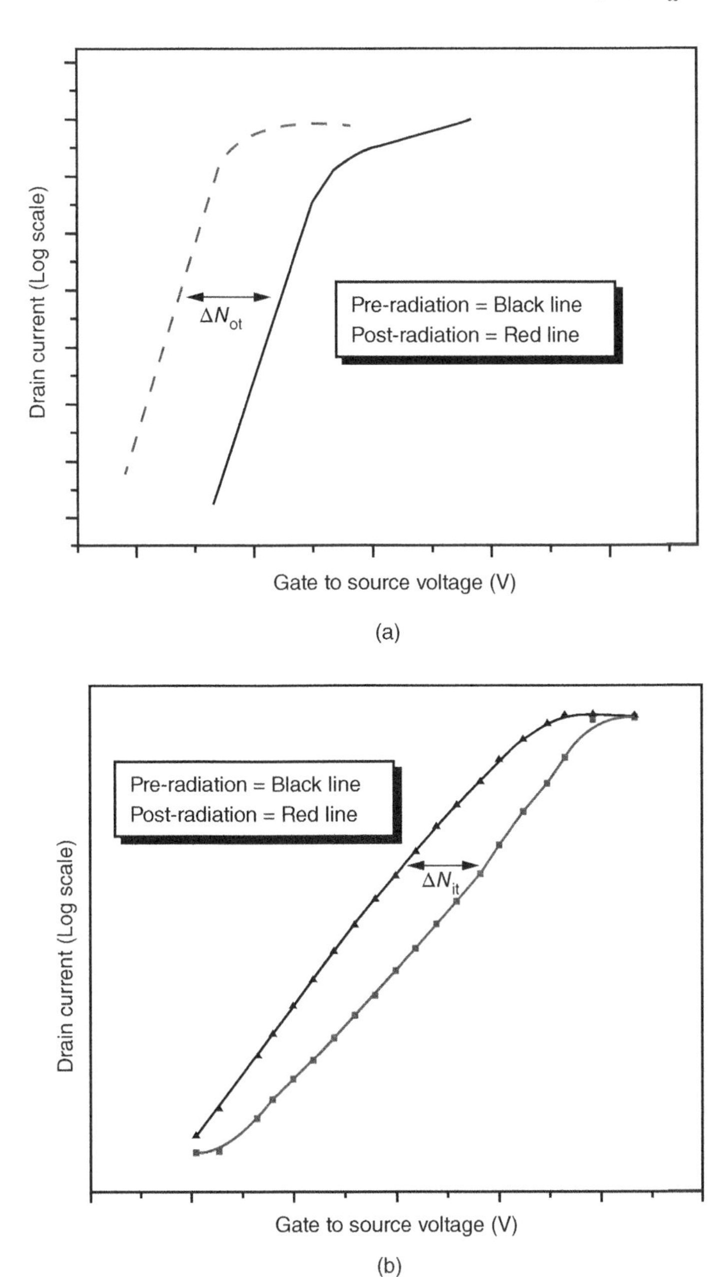

Figure 4.7 (a) Illustration showing the effect of fixed oxide trapped charge on the V_{TH} of n-channel MOSFETs. (b) Illustration showing the effect of interface trap charge on the V_{TH} of n-MOSFETs. Adapted from [9].

community [12]. In 1983, Pease et al. demonstrated that radiation-induced oxide charge had accumulated in the isolation oxides over the base portions of these structures, which explained why bipolar components that were anticipated to withstand more than 1 Mrad failed at a few krad [13]. The consequence of this charge was to open up parasitic leakage pathways. The base partially inverts when charge is trapped above it, increasing the collector current or decreasing gain.

A complex TID effect that Enlow et al. [14] found in 1991 in systems comprising bipolar isolation oxides has proven challenging to comprehend and predict. The phenomenon became known as Enhanced Low Dose Rate Sensitivity, or ELDRS, after the authors reported that bipolar device characteristics for parts exposed to low dose rates of radiation degraded more quickly than those exposed to the standard high dose rates used in hardness assurance testing [14]. Recent explanations and models for the fundamental mechanisms underlying ELDRS place the blame for the observed effect on reduced degradation at high dose rates on one or more of a number of potential causes, such as increased recombination at higher dose rates, space charge effects, or the extraordinarily slow motion of protons [12].

The initial statistics were for vertical NPN devices; however, the worst situation is for lateral PNP devices [14], where a larger portion of the base area directly interacts with SiO_2 and where the base doping is lower than in the NPN case. In surface Shockley–Read–Hall (SSRH) recombination, holes trapped on interface dangling bonds of the PNP's base isolation oxide can reunite with electrons from the n-type base region, as seen in Figure 4.8. As a result, the base contact draws in more electrons, boosting base current and lowering these devices' gain.

The results of low-dose-rate irradiation normalized to damage at a high dose rate (50 rad (SiO_2/s)) were published in 1994 by Johnston et al. The radiation world accepted this approach to describing the severity of the ELDRS impact, and it became known as the low dose rate enhancement factor (LDREF) [12] or, more colloquially, the enhancement factor (EF). According to Enlow et al. [14], the results of some early ELDRS studies appeared to be in conflict [12], but this was because several of the parts used in the studies had the same name – for instance, the National Semiconductor LM 139 – but were produced in at least eight different fabrication facilities at various times, using various manufacturing processes, and with various layouts.

A number of specialized test structures were created in conjunction with the National Semiconductor LM124 quad OP-AMP [12], a device that is sensitive to ELDRS and is frequently used in satellite design, to address these problems. One of the test structures used to derive precise N_{ot} and N_{it} values was a gated lateral PNP (GLPNP) structure. Figure 4.9 displays the GLPNP in a 2D representation. A metal gate is placed above the base isolation oxide in this arrangement. In order to mimic the space conditions where metal lines frequently cross over bipolar isolation oxides, the structure is irradiated with 0-bias on the gate. After irradiation,

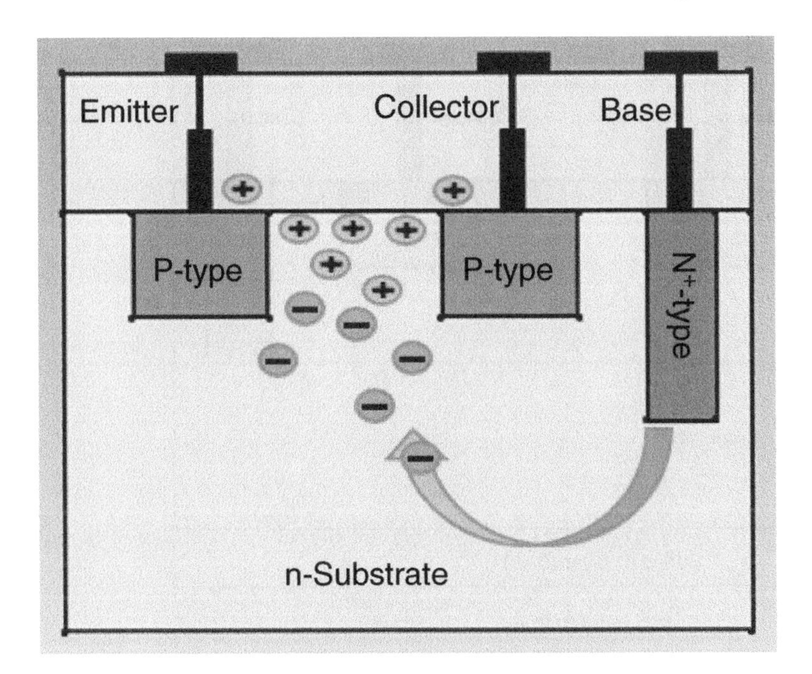

Figure 4.8 An illustration of how surface SRH (SSRH) recombination is induced by an increase in positively-charged interface traps on the Si/SiO_2 interface of the isolation oxide and n-type silicon base of PNP lateral bipolar devices, drawing more electrons from the base contact and increasing base leakage current. Adapted from [12].

researchers can run a gate sweep to transition the MOS structure's base area from accumulation to depletion and inversion.

Because maximum SSRH recombination occurs when electron and hole concentrations are equal (n = p), which happens in depletion, this additional MOS-like control of the base area is crucial. Extraction of N_{ot} and N_{it} is possible by base current analysis during the gate sweep.

Figure 4.10 displays an illustration of a gate sweep measurement made on this GLPNP by Ball et al. [15]. Electrons are removed from the surface during depletion, and finally, n = p is reached. At this voltage, a surge in the base current is observed. The height of peak current can be used to determine the density of interface traps.

$$\text{srv} = \frac{2\Delta I_{\text{B}}}{qS_{\text{Peak}} n_{\text{i}} \exp\left(\frac{qV_{\text{EB}}}{2K_{\text{B}}T}\right)} \tag{4.1}$$

$$\Delta N_{\text{it}} = \frac{\Delta \text{srv}}{\sigma V_{\text{th}}} \tag{4.2}$$

where n_i is silicon's intrinsic carrier concentration, srv is surface recombination velocity, I_B is the base current's maximum increase, q is the charge on an electron,

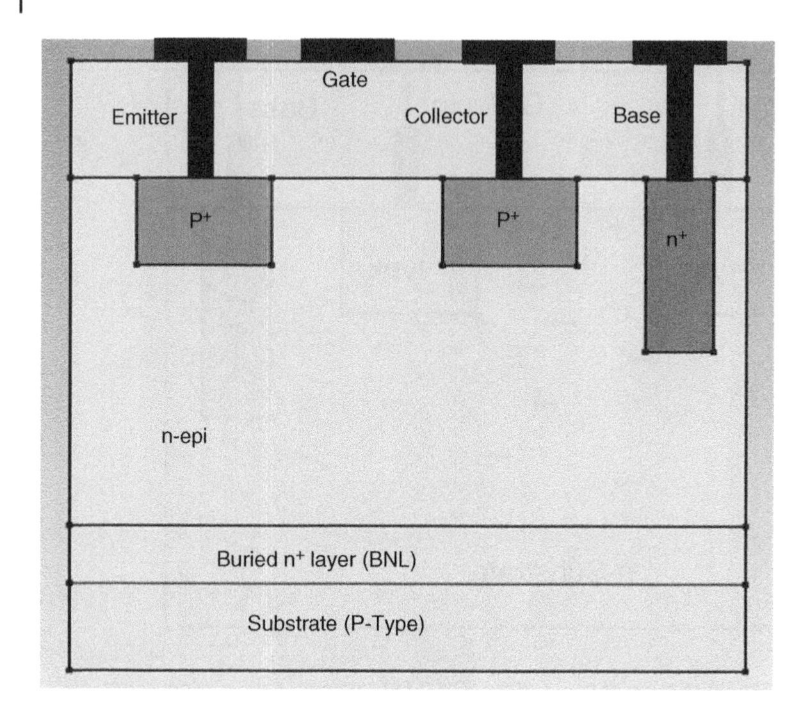

Figure 4.9 2D representation of the GLPNP test structure. Adapted from [12].

S_{peak} is the base current peak spread, V_{EB} is the emitter-base voltage, K_B is Boltzmann's constant, T is temperature, is carrier capture cross-section, and v_{th} is thermal velocity [15].

The shift in the IB–VG curve, where Cox is the oxide capacitance and V_{mg} is the mid-gap voltage shift, in this case, the shift in V_G at which the peak base current occurs, can be used to determine the oxide charge density, which is defined as the charge in the oxide that does not interact with the interface [15].

Different packaging techniques, some of which delivered trace amounts of hydrogen into the device environment, also had an impact on experimental trends. For instance, Pease et al. study compared the irradiations of two distinct packages of the AD 590 temperature transducer. For some doses, they noticed a difference in degradation of up to two orders of magnitude, with the devices placed in flat-packs containing hydrogen degrading more rapidly than those contained in metal cans, which do not contain hydrogen [16].

Although the development of N_{ot} is a related and crucial mechanism that must be taken into consideration concurrently, for a number of reasons, the interaction of interface traps (N_{it}) with the silicon base region is the mechanism that is linked to TID bipolar device degradation. This charge helps to establish the device's local electric field, which affects the drift of charged particles including

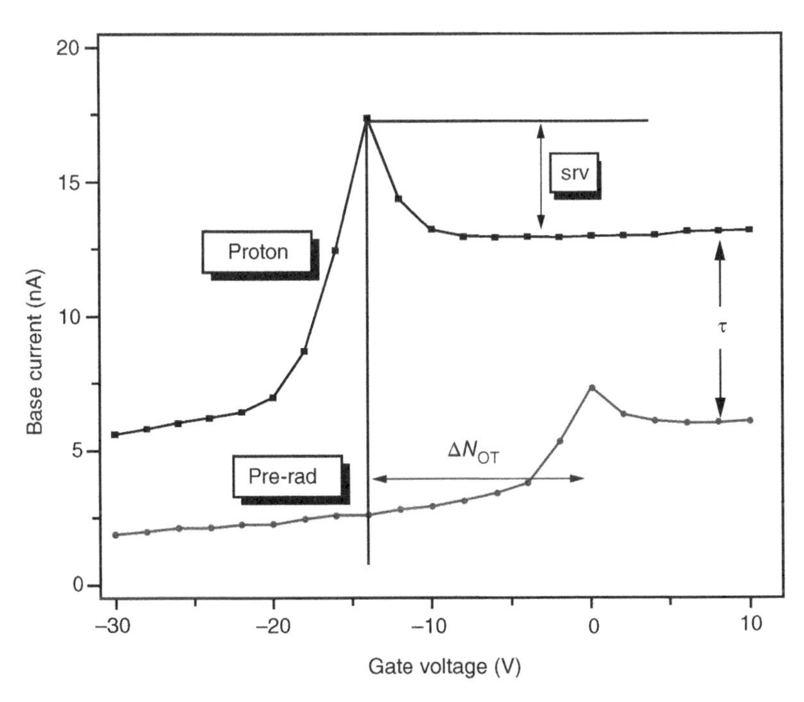

Figure 4.10 Gate sweep responses of the GLPNP base current before and after irradiation. When the gate is biased so that the channel region is in depletion, or when n = p and maximum SSRH take place, a peak is seen in the base current. This peak changes as a result of oxide charge accumulation, enabling the extraction of N_{ot}; additionally, an increase in peak height enables the extraction of interface trap charge or N_{it}. Adapted from [15].

radiation-induced electrons and holes as well as protons that, when they reach the interface, produce N_{it}. Additionally, the oxide's ability to trap holes is a crucial prerequisite for the subsequent release of protons.

Time is the last factor taken into account to study exacerbates reported in TID-induced degradation, particularly in ELDRS-sensitive components. A number of time-dependent processes, including the production and transport of EHPs on very short time scales, hole trapping on medium time scales, and proton transport and hole release on long time scales, contribute to the accumulation of oxide and interface traps. Along with trap creation, trap annealing also occurs, and it affects how much degradation is observed. If the time scales of the various mechanisms are not taken into consideration when designing tests, apparent dose rate effects rather than real dose rate effects [12]. Reaching this dose more quickly while irradiating at a high dose rate compared to a low dose rate. The center curve illustrates what occurs to the high-dose rate devices following an anneal for the same amount of time as was needed to conduct the low-dosage rate experiments.

4.8 Understanding and Modeling a-SiO_2 Physics

The E_0 centers in a-SiO_2 were examined in an excellent overview, with special emphasis on the contributions made by EPR observations and DFT computations. As a result, researchers have come to the conclusion that there are two different types of oxygen vacancy centers in amorphous SiO_2: the shallow trap with the dimer structure, which corresponds to the EPR E'_0 signal, and the deep trap, which corresponds to the EPR E^{γ}_0 signal [17]. however, there is still some disagreement regarding the nature of E_0 centers and encourage more theoretical investigation to clarify the problems.

In 1956, Robert A. Weeks presented the first report of EPR measurements of radiation-induced flaws on crystalline and amorphous SiO_2 [18]. The series was given the name E_{0n}, where n is a serial number and the prime number denotes the number of electrons, when these resonances were later connected with optical spectra data [17]. There was some disagreement about whether the signals were caused by Si "dangling bonds" or oxygen vacancies [17, 19, 20]. The E_0 center's dimer status was established in 1987 by semi-empirical computations performed on bigger clusters [21].

The 1980s saw a shift in emphasis from crystalline quartz's E_0 centers to thermal oxides E'_0 centers in electronic components. Now known as E_0, two distinct centers E'_{γ}, a shallow hole trap, and, a deep hole trap E'_{δ}, were found [22]. Figure 4.11 depicts these trap levels inside the SiO_2 band diagram and shows how holes behave in their immediate neighborhood. E_0 correlates to the E'_{γ} in crystalline SiO_2 and E'_1 in amorphous SiO_2.

The most prevalent defect in SiO_2 is oxygen vacancies (V_0), which also gives rise to both E_0 centers [23–26]. An illustration of the neutral oxygen vacancy precursor

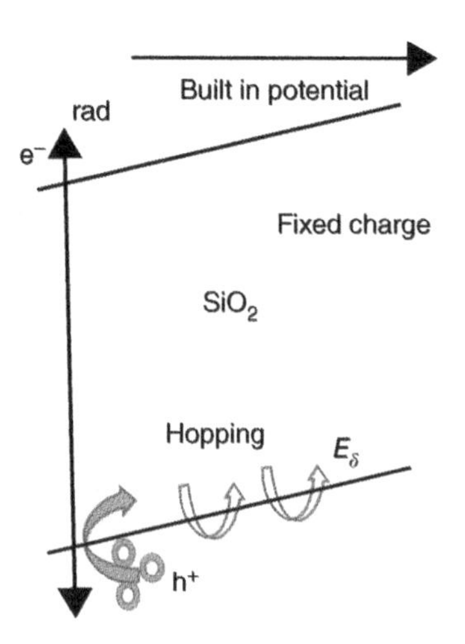

Figure 4.11 Band diagram showing the interactions between shallow and deep traps and holes caused by radiation (*E*). Due to the MOS structure's inherent electric field, holes can easily migrate toward the SiO_2 interface by hopping in and out of shallow traps. But holes, also known as fixed oxide charge, can persist for a very long time in deep traps close to the contact. Adapted from [22].

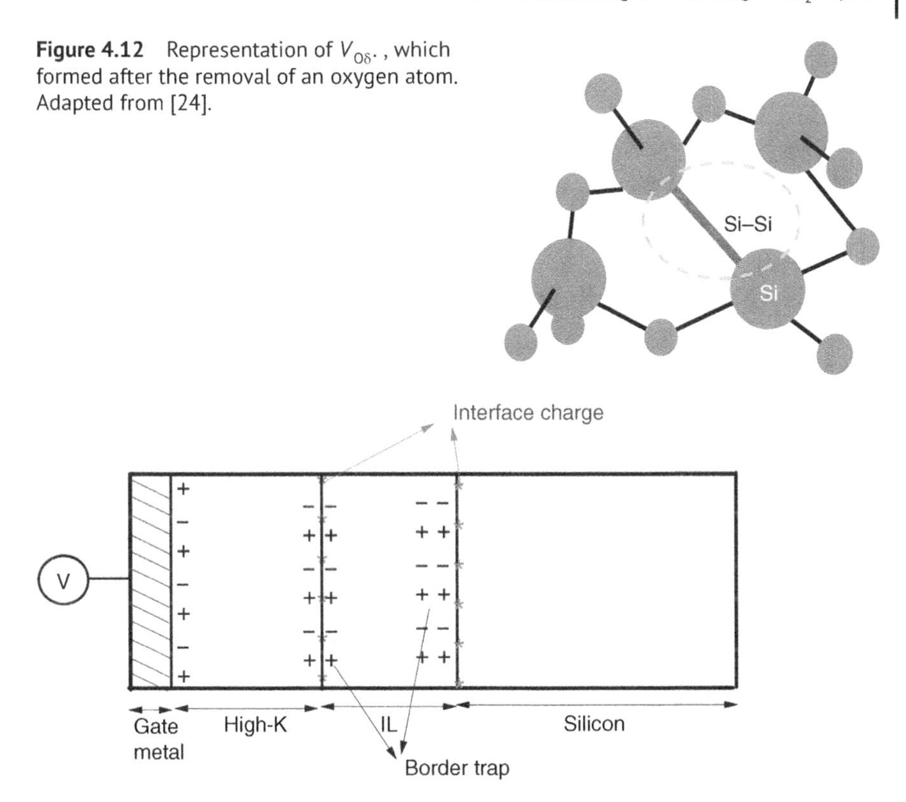

Figure 4.12 Representation of $V_{O\delta}$., which formed after the removal of an oxygen atom. Adapted from [24].

Figure 4.13 Trap charges at different interfaces in stack gate oxide structure.

structures is depicted in Figures 4.12 and 4.13, where one Si—Si bond is used in place of two Si—O bonds. Positively charged oxygen vacancies typically have the same bond structure as other vacancies, with a modest increase in the Si—Si bond length. Using electron paramagnetic resonance (EPR), this defect can be seen in the positive charge state and is designated as E'_{δ}.

The Si—Si bond can break and expand dramatically in positively charged vacancies, with one Si atom shifting to the anti-bonding position and bringing the positive charge with it to establish a bond with an oxygen atom in the network. EPR may also detect this so-called puckered state, which is designated E_0. Only in instances of positive charge and only for a small percentage of the vacancies does the puckered condition arise. Theory, however, contends that the neutral predecessors to E_0 are different from those of $E_{o\gamma}$. $V_{o\gamma}$ is a much deeper pit trap, as well. $V_{o\gamma}$ and $V_{O\delta}$ are separate species with separate trapping energies. Additionally, defects may undergo hydrogenation or double hydrogenation.

The local geometries that result in the E_0 configurations were examined by researchers who also deduced principles for atom position. Then, a computer algorithm based on these rules was used for a million-atom SiO_2 model, and the results showed that over 90% of oxygen vacancies favored E_0 (the dimer structure), while just 5% favored each of the other $E_{0\gamma}$ configurations. We can see that DFT only uses bulk computations. Conclusions concerning the concentration of E_0 species at the interface are complicated by the presence of the oxygen-poor transition area at the SiO_2/Si interface.

4.9 Hydrogen (H_2) Reaction with Trapped Charges at Insulator

This section of discussion explains how interactions between radiation-induced electrons and holes and H_2 and oxide defects lead to the release of protons. A succinct summary is provided below: Radiation causes the formation of EHPs. The test device's MOS structure, which consists of the lateral PNP base of the test structure, an isolation oxide above it, and a metal gate, has an integrated electric field that directs electrons (e) moving toward the gate and holes (h+) escaping initial recombination toward the SiO_2/Si interface. Neutral oxygen vacancy complexes entrap holes. With imprisoned holes, electrons may join once more. Intentionally or accidentally present molecular hydrogen (H_2) can fill open interstitial spaces and interact with positively charged defects to form protons and hydrogenated defects [27].

Protons may be released directly by positively charged hydrogenated faults. These reactions, as well as the computed forward and reverse energy barriers incorporated into the current model, are described in the equations below. The units for interface amounts are cm^2. The remaining concentrations are all cm^3.

$$H^+ + Si - H <=> N_{it} + H_2 \tag{4.3}$$

where N_{it} is the density of interface traps and Si—H are hydrogen-passivated dangling bonds on the interface, explaining how protons at the SiO_2/Si interface interact with Si—H complexes to create interface traps [28] and release H_2. unbiased forerunners to the Deep, semi-permanent hole traps called defects are primarily found close to the SiO_2/Si contact. This flaw can, when left neutral, catch a hole. This flaw, when positively charged, can act as a recombination center or crack molecular hydrogen to liberate a proton.

4.10 Pre-Existing Trap Density and their Respective Location

In 2009 a condensed model [29], was published which used the electron, hole, and proton continuity equations along with their proposed few main processes for the radiation response to perform electrostatic steady-state calculations on a 2D rectangle of SiO_2. Instead of considering recombination on defect sites, authors treat direct electron–hole recombination as the primary mechanism underlying the effect. The term they employed in the electron and hole continuity equations to account for this direct recombination is shown in Eq. (4.4).

$$\mathrm{RG}_{\mathrm{n,p}} = \sigma_{\mathrm{recom}}\mathrm{np} \tag{4.4}$$

where σ_{recom} is the electron–hole recombination reaction coefficient. Their hole and proton continuity equations, which are used to represent the hole-hydrogen defect reaction,

$$\mathrm{RG}_{\mathrm{H-Source}} = R_{\mathrm{DHP^+}}N_{\mathrm{DH}} \tag{4.5}$$

$$\sigma_{\mathrm{recom}} \approx \frac{q(\mu_n + \mu_p)}{\epsilon_{\mathrm{ox}}}\eta(\mathrm{H_2}) \tag{4.6}$$

where R_{DHP} is the hole-hydrogen defect reaction rate constant and N_{DH} is the concentration of hydrogen-containing defects, also includes the forward reaction of one of our Eqs. (4.4)–(4.6) above as a mechanism for producing protons (V_{OH}). Finally, the model of molecular hydrogen as a decrease in recombination efficiency, where the function (H_2) is not explicitly stated. This method, which is solely based on generic ionic recombination theory, proved successful in qualitatively capturing the TID effect.

4.11 Use of High-K Dielectric in MU-G FET

The oxide layer in the MOS capacitor was formerly made of SiO_2, which was created through dry thermal oxidation. The first discovery of the charged particle's tunneling property through the dielectric layer was made in 1965 [30]. Numerous researchers have noted that the thin SiO_2 layer experiences a tunnel current density of more than 10^{-9} A/cm^2. The superposition principle makes it simple to compute the leakage currents brought on by trap charges, which helps us get a better understanding of how an impurity in the material might produce trap charges (Table 4.1).

As was mentioned at the beginning of the section, SiO_2 has many benefits besides the leakage current that results from its low dielectric constant. Our goal is to compare several High-K materials with traditional SiO_2 because we are concentrating on the charge-trapping property of the dielectric layer here. When gate stacking is used (as shown in Figure 4.13), the High-K dielectric can typically transport bulk and interface charges up to 10^{13}/cm^2, while the Si–SiO_2 structure typically caps charge densities at 10^{10}/cm^2. When utilizing High-K,

Table 4.1 Trap charge property near the interface region.

Element	Trap energy > Si valence band energy (eV)	Interface area (cm^2)
Mg	0.54	1.1×10^{-18}
Cr	0.20	5.2×10^{-17}
Cu	0.52	6.6×10^{-10}
Au	0.97	2.2×10^{-16}

Source: Adapted from Kar and Dahlke [15].

crystal mismatch makes use of an interfacial layer, known as gate stacking, required. Additionally, this interfacial layer creates a new area where the HK and IL can share traps that are more active than the bulk trap in the carrier exchange we previously mentioned.

4.12 Properties of Trap in the High-K with Interfacial Layer

We will examine the types of potential trapped charges in each of the available regions in this phase of our discussion and attempt to do an electrical analysis of them. The local authorized state, charge state, and swapping cross-section can all be used to describe these charges. All trap characteristics can only change due to the applied gate bias; hence, if the charge state changes due to the applied electric field or external bias, the MOS capacitance may also change.

At the interface of silicon and gate insulator, transition from a covalent semiconductor to an insulator takes place. We have reason to think that the amphoteric substance, also known as amphoteric trap, is responsible for the majority of traps present in this location based on the numerous experimental verifications on the Si–SiO_2 interface. These traps can contain neutral or positive charge when they have one electron, two electrons, or no electrons, respectively. Additionally, a chemically created trap may be present in this area.

The most efficient interfacial layer for the silicon channel is either SiO_2 or SiON. Therefore, the trap in the region is caused by oxygen or nitrogen vacancies with an energy band below the Si conduction band. These can be produced by dry thermal oxidation with or without nitrogen impurities. Additionally, additional stress-induced traps may be produced as a result of the stress factor at the interface between the IL and HK.

Due to the ionic and lattice mismatch, we can locate some of the major and significant trap densities in the entire oxide layer in this area. This crucial region can also withstand several of our most potent trap extraction techniques, including charge pumping and MOS conductance (described in the following section). However, the trapping and untrapping of charges can also be influenced by the layer's width. With increasing distance from the pure silicon, the charge density rises.

The charge trapping in the body of the HK layer is mostly caused by oxygen vacancies. There are five possible O_2 vacancy states, ranging from −2 to +2. The process variable also affects the diffusion of O_2 vacancies from the HK layer to the interfacial layer that is deficient. In contrast to the HK interface, which is primarily impacted by the metal wave function and produces a high density of trap charges, the typical SiO_2 and metal (Gate) interface is stable with essentially no trap density. The main issue is that there is no such method for extracting a trap from a metal zone [31].

4.13 Trap Extraction Techniques

4.13.1 Capacitance Inversion Technique (CIT)

The main issue with the gate oxide stack technique is the leakage from the gate side caused by the IL trap density, which prevents the inversion layer from forming at the Si channel by pulling some of the majority charge carriers and so causes disruption in the monitoring of CI curve. Therefore, a few techniques are used to improve the generation of minority charge carriers, which are dominant in the number of charge participants in the generation and recombination of the trap at the IL or bulk dielectric layer, the observation of inverse CV curve characteristics using only photo illumination, and temperature excitation on the channel charge carrier using carrier excitation and injection from the source [32, 33].

4.13.2 Charge Pumping Technique (CPT)

Charge pumping technique involves non-steady state procedure, in contrast to charge inversion. The source and drain with substrate are activated by reverse bias while the gate and channel are driven by periodic pulse signals that alternate between accumulation and inversion. Measured from the substrate throughout a pulse, the charge pumping current (I_{cp}) conveys information on the trapping and untrapping of charge carriers. The cross-sectional area of exchange, trap density (D_{it}), and exchange location or point of charge exchange must all be calculated separately for the practical computation of I_{cp} [34, 35].

4.14 Conclusion

Based on extensive experimental and theoretical research into the atomic structure of SiO_2, this chapter observes the defect-carrier interaction in this material. Then, based on streamlined continuity equations, we examined earlier attempts at drift-diffusion modeling that were qualitatively successful in characterizing the trap effect. Again, simple High-K or HK gate stacks produce higher trap densities than the typical SiO_2 dielectric layer. The interfacial layer with an Hk has the advantage of crystallization matching between the Si channel and the Hk material, which also maintains the requisite EOT for the specific application but also provides a greater area for charge trapping and manufactured crystal flaws. These traps typically have different electrical characteristics than SiO_2-based traps. Thus, for the device to function and perform well, trap charge analysis when employing Hk with a multi-gate transistor is essential. Additionally, the storing capacity and efficiency with retention time for memory-based applications heavily depend on these fees. The presence of flaws or traps can produce irrational results in circuits that are significantly more exposed to intense radiation fields, such as those used in space and medicinal applications.

References

1. Sekigawa, T. and Hayasi, Y. (1984). Calculated threshold voltage characteristics of an XMOS transistor having an additional bottom gate. *Solid State Electronics* 27 (8): 827–828.
2. Moore, G. (1965). Cramming more component on to integrated circuit. *Electronics* 38: 114.
3. Ecoffet, R. (2011). On-orbit anomalies: Investigations and root cause determination. *IEEE Nuclear and Space Radiation Effects Conference Short Course*, Las Vegas, NV.
4. Garret, H.B. (2011). Spacecraft environment interactions. *IEEE Nuclear and Space Radiation Effects Conference Short Course*, Las Vegas, NV.
5. NASA (2011). The Van Allen belts. NASA Explores Earth's Magnetic "Dent". NASA's Goddard Space Flight Center. Oct. 2011. [Online]. https://svs.gsfc.nasa.gov/vis/a010000/a012300/a012379/Van_Allen_Belts.png.
6. Fennel, J. (2011). Is it true that organic matter cannot pass through the van Allen radiation belts? Aerospace Corp., [Online]. www.astronomycafe.net/qadir/ask/a11767.html (accessed 3 May 2023).
7. Ma, T.P. and Dressendorfer, P.V. (ed.) (1989). *Ionizing Radiation Effects in MOS Devices and Circuits*. Wiley.
8. Saks, N.S., Dozier, C.M., and Brown, D.B. (1988). Time dependence of interface trap formation in MOSFETs following pulsed irradiation. *IEEE Transactions on Nuclear Science* 35 (6): 1168–1177.
9. Felix, J.A., Fleetwood, D.M., Schrimpf, R.D. et al. (2002). Total-dose radiation response of hafnium-silicate capacitors. *IEEE Transactions on Nuclear Science* 49 (6): 3191–3196.

10 Felix, J.A., Schwank, J.R., Fleetwood, D.M. et al. (2004). Effects of radiation and charge trapping on the reliability of high-κ gate dielectrics. *Microelectronics Reliability* 44 (4): 563–575.

11 Fleetwood, D.M., Warren, W.L., Schwank, J.R. et al. (1995). Effects of interface traps and border traps on MOS post irradiation annealing response. *IEEE Transactions on Nuclear Science* 42 (6): 1698–1707.

12 Oldham, T.R. (2011). Space radiation environments and their effects on devices and systems: back to basics. *IEEE Nuclear and Space Radiation Effects Conference Short Course*, Las Vegas, NV.

13 Pease, R.L., Turfler, R.M., Platteter, D. et al. (1983). Total dose effects in recessed oxide digital bipolar microcircuits. *IEEE Transactions on Nuclear Science* 30 (6): 4216–4223.

14 Enlow, E.W., Pease, R.L., Combs, W. et al. (1991). Response of advanced bipolar processes to ionizing radiation. *IEEE Transactions on Nuclear Science* 38 (6): 1342–1351.

15 Kar, S. and Dahlke, W.E. (1972). Interface states in MOS structures with 20–40 Å thick SiO_2 films on nondegenerate Si. *Solid-State Electronics* 15 (2): 221–237.

16 Pease, R.L., Dunham, G.W., Seiler, J.E. et al. (2007). Total dose and dose rate response of an AD590 temperature transducer. *IEEE Transactions on Nuclear Science* 54 (4): 1049–1054.

17 Pantelides, S.T., Lu, Z., Nicklaw, C. et al. (2008). The E′ center and oxygen vacancies in SiO_2. *Journal of Non-Crystalline Solids* 354 (2-9): 217–223.

18 Weeks, R.A. (1956). Paramagnetic resonance of lattice defects in irradiated quartz. *Journal of Applied Physics* 27 (11): 1376–1381.

19 Silsbee, R.H. (1961). Electron spin resonance in neutron-irradiated quartz. *Journal of Applied Physics* 32 (8): 1459–1462.

20 Weeks, R.A. (1963). Paramagnetic spectra of E'_2 centers in crystalline quartz. *Physical Review* 130 (2): 570–576.

21 Rudra, J.K. and Fowler, W.B. (1987). Oxygen vacancy and the E'_1 center in crystalline SiO_2. *Physical Review B* 35 (15): 8223–8230.

22 Griscom, D.L. and Friebele, E.J. (1986). Fundamental radiation-induced defect centers in synthetic fused silicas: atomic chlorine, delocalized E′ centers, and a triplet state. *Physical Review B* 34 (11): 7524–7533.

23 Blöchl, P.E. (2000). First-principles calculations of defects in oxygen-deficient silica exposed to hydrogen. *Physical Review B* 62 (10): 6158–6179.

24 Lu, Z.-Y., Nicklaw, C.J., Fleetwood, D.M. et al. (2002). Structure, properties, and dynamics of oxygen vacancies in amorphous SiO_2. *Physical Review Letters* 89 (28): 285505.

25 Tuttle, B.R. and Pantelides, S.T. (2009). Vacancy-related defects and the E_0 center in amorphous silicon dioxide: density functional calculations. *Physical Review B* 79 (11): 115206.

26 Lenahan, P.M. and Dressendorfer, P.V. (1984). Hole traps and trivalent silicon centers in metal/oxide/silicon devices. *Journal of Applied Physics* 55 (10): 3495–3499.

27 Stahlbush, R.E., Mrstik, B.J., and Lawrence, R.K. (1990). Post-irradiation behavior of the interface state density and the trapped positive charge. *IEEE Transactions on Nuclear Science* 37 (6): 1641–1649.

28 Rashkeev, S.N., Fleetwood, D.M., Schrimpf, R.D., and Pantelides, S.T. (2001). Defect generation by hydrogen at the Si-SiO_2 interface. *Physical Review Letters* 87 (16): 165506.

29 Chen, X.J., Barnaby, H.J., Adell, P. et al. (2009). Modeling the dose rate response and the effects of hydrogen in bipolar technologies. *IEEE Transactions on Nuclear Science* 56 (6): 3196–3202.

30 Card, H.C. and Rhoderick, E.H. (1971). Studies of tunnel MOS diodes I. Interface effects in silicon Schottky diodes. *Journal of Physics D: Applied Physics* 4 (10): 1589.

31 Tersoff, J. (1984). Schottky barrier heights and the continuum of gap states. *Physical Review Letters* 52 (6): 465.

32 Poon, T.C. and Card, H.C. (1980). Energy and electric field dependence of Si-SiO_2 interface state parameters by optically activated admittance experiments. *Journal of Applied Physics* 51 (12): 6273–6278.

33 Kar, S. and Varma, S. (1985). Determination of silicon-silicon dioxide interface state properties from admittance measurements under illumination. *Journal of Applied Physics* 58 (11): 4256–4266.

34 Brugler, J.S. and Jespers, P.G. (1969). Charge pumping in MOS devices. *IEEE Transactions on Electron Devices* 16 (3): 297–302.

35 Groeseneken, G., Maes, H.E., Beltran, N., and deKeersmaecker, R.F. (1984). A reliable approach to charge-pumping measurements in MOS transistors. *IEEE Transactions on Electron Devices* 31 (1): 42–53.

5

Strain Engineering for Highly Scaled MOSFETs

Chinmay K. Maiti[1], *Taraprasanna Dash*[2], *Jhansirani Jena*[2], *and Eleena Mohapatra*[3]

[1] *SouraNiloy, Kolkata, India*
[2] *Department of ECE, Siksha 'O' Anusandhan (Deemed to be University), Bhubaneswar, India*
[3] *Department of ECE, RV College of Engineering, Visvesvaraya Technological University, Bengaluru, India*

5.1 Introduction

The planar silicon metal–oxide–semiconductor field-effect transistor (MOSFET) is the most significant invention in the field of microelectronics. J. E. Lilienfeld introduced the idea of field-effect conductivity modulation and the MOSFET in 1928. Moore's prediction has held over the last six decades and will continue as long as the cost of a transistor continues to drop in price [1]. Since over 20 years ago, the strain has been employed to extend Si scaling, and it is a crucial component of contemporary semiconductor technology. To effectively improve the drive current by improving the mobility of carriers, strain techniques have been introduced with the 90 nm technology node [2]. The essential idea behind strained-Si is that externally applied stress or strain will change the equilibrium lattice constant of silicon. Silicon electronic band structure is modified as a result of the altered lattice constant, leading to improved electronic characteristics.

Strain engineering involves uniaxial and biaxial strain induction in silicon wafers. It is anticipated that engineering the elastic strain/stress in the transistor channel will enhance device properties by enhancing charge carrier mobility. Currently, process-induced strain and substrate-induced strain are the main avenues for strain engineering [3]. The semiconductor modeling approach is one of the approaches that can shorten the manufacturing period and reduce the cost of the process. Technology computer-aided design (TCAD) is now utilized for manufacturing, yield optimization, and process and device design. TCAD modeling can also be used to evaluate cutting-edge device concepts and

Advanced Nanoscale MOSFET Architectures: Current Trends and Future Perspectives,
First Edition. Edited by Kalyan Biswas and Angsuman Sarkar.

innovative device architectures for MOSFET devices that are based on strain engineering technology [4]. Transistor structures have changed from planar to three-dimensional (3D) over the last 20 years. Multi-gate 3D structures are expected to replace fully depleted silicon-on-insulator (FDSOI) and other scaled planar bulk devices. Strain engineering has its limitations as the stress generated cannot be infinite and may lead to various detrimental effects such as defect nucleation, dislocation generations, or microstructural changes, which would lead to performance degradation, such as leakage, open or shorts, and finally affect device integrity. Thus, studies on stress/strain effects on the electrical performance of nanoscale semiconductor devices are essential.

A semiconductor material, known as an engineered substrate, can be created and introduced into the traditional silicon manufacturing process, producing products that are distinct and could not have been created with only silicon substrate. The physical (bandgap and mobility), electrical (carrier transport), and mechanical properties get altered in semiconductor materials when strain is introduced into them. Since the 1950s, the impacts of stress and strain on silicon and silicon technology have been an important study [5]. The process simulation is an intriguing complementary strategy given the difficulty, expense, and other experimental limitations of estimating the stress profile in the silicon channel alone using experimental techniques.

Innovation has always been a crucial component of device scaling when introducing new device architectures, but integrating new materials is facing significant challenges to meet the strict requirements of the International Technology Roadmap for Devices and Systems (IRDS) [6]. Transistor designs have advanced over the past 20 years, moving from planar, conventional single-gate field-effect transistors (FETs) to multi-gate FETs, whose behavior can only be completely explained by the appropriate carrier transport phenomena. The best solution for extreme scaling with the fewest incremental processing expenses is multi-gate FET technology. The FinFETs are the multi-gate device that, from a fabrication standpoint, has the best chance of becoming widely used.

Recent reports [7] describe a 3 nm complementary metal–oxide–semiconductor (CMOS) process from Samsung that uses gate-all-around (GAA) transistors, where the gate completely encloses the channel. The channel has a multi-bridge channel architecture and is made up of horizontal nanosheets. Scaling beyond FinFETs is expected to be possible thanks to stacked nanosheet, nanowire, and GAA transistors. Particularly for more-than-Moore (MtM) applications, such as utilization in the Internet of Things (IoT) systems, virtual reality, and data-intensive cloud processors, these ultra-low power devices are required (expanding battery life) [8].

For many years, a debate has existed as to the relative merits of silicon-based devices vs. more exotic compounds. In particular, the challenge to silicon technology has been from III-V alloys, especially for the devices used in high-frequency and optical applications. Alloy compositions such as gallium arsenide, aluminum gallium arsenide, and indium phosphide offer notable advantages such as higher charge carrier mobilities and suitable characteristics for service in the optical regime. With a primary focus on possible 10 nm and beyond device alternatives like double-gate FETs (DGFETS) and tri-gate fin FETs (TG-FinFET) designs, the results may be utilized to predict strain-enhanced performance in advanced devices with high-mobility material platforms. High electron mobility transistors (HEMTs) made of AlGaN/GaN stand out for their outstanding benefits in high-power, high-temperature, and high-frequency applications. Modern AlGaN/GaN HEMTs include built-in internal stress that could affect performance and reliability. Novel enhancement mode GaN MOS-HEMT with a high-k TiO_2 gate dielectric and a 10 nm T-gate length exhibits significant improvement in high frequency and power applications and make it suitable for next generation device [9–14].

This chapter presents a TCAD methodology for strain-engineered MOSFETs at advanced technology nodes with improved design, through parametric yield and systematic process variability simulation [15]. The merits and limitations of the uniaxial and biaxial strain designs for MOSFETs are discussed to improve channel mobility. With an emphasis on embedded $Si_{1-x}Ge_x$ or $Si_{1-y}C_y$ in S/D areas, dual-stress liners (DSL) and stress memory technique (SMT) have been used to create strain in transistor structures. Strain engineering has gained significance in material science. However, research and industry have focused only on substrate-induced strain. On the other hand, process-induced strain is an unavoidable outcome of the device fabrication process. Semiconductor devices consist of a range of materials that have mismatched thermo-elastic and lattice properties. When such devices undergo temperature changes during fabrication, such mismatched properties result in an inherent process-induced strain. This in turn degrades the performance of the device through the involuntary modification of material properties. This chapter will outline carefully thought-out methods for modifying process-induced strain to increase carrier mobility or change the band gap in nanodevices from indirect to direct.

The definitions of stress/strain and elastic parameters are provided in Section 5.2 of this chapter. A thorough analysis of strain engineering techniques currently in use in CMOS technology may be found in the reference [15]. On carrier mobility in Si n- and p-MOSFETs, the impacts of strain-induced band splitting, band warping, subsequent carrier repopulation, altered conductivity effective mass, and altered scattering rate will be discussed. Descriptions of the various mobility models used in the simulation are provided in the section simulation approach. The significance of both local and global strain approaches is discussed. The TCAD modeling of strain-engineered MOSFETs is covered in strain mapping and mechanical strain modeling in Section 5.2, along with a few case studies in Section 5.3.

5.2 Simulation Approach

Simulation of full MOSFET operation requires analysis of an inherently 3D structure. The simulation package used for this work is the 3D finite element device simulator Victory device [16] from Silvaco International. Under the proper boundary conditions, the Victory device can solve Poisson's equation and the drift-diffusion (DD) equations for electrons and holes. The CVT mobility model used in these simulations accounts for acoustic and optical phonon scattering, surface roughness scattering, and velocity saturation. Typical mesh sizes are on the order of 40 000 points, which includes approximately 2000 points in the width cross-section plane multiplied by twenty planes in the length direction. Reflective boundary conditions are imposed in the lateral direction of the width cross-section. The top boundary conditions are determined by the source, drain, and gate ohmic contacts. A large number of data post-processing routines are used to extract and manipulate charge distributions and other internal data of specific interest. The desired published channel and source/drain (S/D) doping profiles are fitted using complementary error function (erfc) and Gaussian analytical models. The FDSOI design, which has a gate length in the sub-100 nm range, is susceptible to non-local phenomena such as velocity overshoot. The energy balance (EB) transport model is preferred in this regard over the traditional DD model. For electrons, the EB model is appropriate.

5.2.1 Strain Mapping

Strain technology uses mechanical stress to alter the silicon band structure and lower the effective mass and carrier scattering rates. To continue downscaling, it can be implemented to boost carrier mobility. The strain in the channel region is now produced using the lattice mismatch between SiGe and Si in the (S/D) region. Due to the lack of nanoscale probing to quantify the mechanical stress in contemporary semiconductor processes, stress is also challenging to define [17]. It is essential for accurate strain engineering and a precise prediction of the performance of strained-induced devices to investigate and understand the stress/strain profile. High-resolution polarized Raman spectroscopy is commonly used to precisely examine the evolution of strains at the nanoscale. The effects of strain and stress relaxation in Si/SiGe multilayer structures utilized to create GAA transistors using advanced transmission electron microscopy (TEM) techniques. It is also challenging to incorporate this technique into ultra-small architectures since the dimensions of transistors are getting smaller. Thus, stress effects must be established through simulation to evaluate the impact of process variations because stress has a significant impact on transistor characteristics in modern processes. It is crucial and difficult to quantify stress and strain at the micro- and nanoscales.

Understanding local stress and strain variations is necessary for semiconductor manufacturing methods, including integrated circuits (ICs) and micro- and nanoelectromechanical systems (MEMS and NEMS), to improve system performance and design. While a variety of characterization techniques, such as electron backscatter diffraction (EBSD) and micro-Raman, are employed to map and quantify strain at nano scales ex-situ, there have been few attempts to do this with TEM. This is significant because, when the size of the device approaches the typical length scales associated with the material, the mechanical behavior of crystalline material can vary. A crystalline substance has a single dislocation-dependent value for a property that is size-independent. Yield stress, hardness, brittle-to-ductile transition, and fracture toughness are among the properties that are affected since it has been demonstrated that each of these metrics is dependent on internal and, in some cases, exterior length scales. There have been several examples of characterization techniques that can measure the amount of strain in nano-scaled transistors.

5.2.2 Mechanical Strain Modeling

By connecting stress and strain, it is possible to describe the effects of mechanical stress using an understanding of elasticity. Robert Hook established their connection for the first time. Hook's law states that for Hookean elastic solids, the stress tensor and strain tensor are linearly connected across a specific range of deformation. The general form of this linear relationship between these tensors is given by:

$$\sigma_{ij} = C_{ijkl}\varepsilon_{kl} \tag{5.1}$$

where C_{ijkl} represents the 4th-order elastic stiffness tensor with 81 ($= 3^4$) components and ε_{kl} is the strain tensor. By taking into account the symmetries required for both the strain and stress tensors under equilibrium, C_{ijkl}, can be restricted to a tensor of 36 components. The engineering shear strain is divided in half by the off-diagonal components. The strain and stress tensors can be represented as vectors after the contracted notations because the off-diagonal strain components are converted to engineering shear strains.

$$\begin{bmatrix} \varepsilon_{xx} & 2\varepsilon_{xy} & 2\varepsilon_{xz} \\ 2\varepsilon_{yx} & \varepsilon_{yy} & 2\varepsilon_{yz} \\ 2\varepsilon_{zx} & 2\varepsilon_{zy} & \varepsilon_{zz} \end{bmatrix} = \begin{bmatrix} \varepsilon_{xx} & \gamma_{xy} & \gamma_{xz} \\ \gamma_{yx} & \varepsilon_{yy} & \gamma_{yz} \\ \gamma_{zx} & \gamma_{zy} & \varepsilon_{zz} \end{bmatrix} \tag{5.2}$$

where γ is the notation used for engineering shear strain. The components can be renumbered as the following:

$$\left.\begin{aligned} \begin{bmatrix} \sigma_{xx} & \sigma_{xy} & \sigma_{xz} \\ \sigma_{yx} & \sigma_{yy} & \sigma_{yz} \\ \sigma_{zx} & \sigma_{zy} & \sigma_{zz} \end{bmatrix} &= \begin{bmatrix} \sigma_1 & \sigma_6 & \sigma_5 \\ \sigma_6 & \sigma_2 & \sigma_4 \\ \sigma_5 & \sigma_4 & \sigma_3 \end{bmatrix} \\ \begin{bmatrix} \varepsilon_{xx} & \gamma_{xy} & \gamma_{xz} \\ \gamma_{yx} & \varepsilon_{yy} & \gamma_{yz} \\ \gamma_{zx} & \gamma_{zy} & \varepsilon_{zz} \end{bmatrix} &= \begin{bmatrix} \varepsilon_1 & \varepsilon_6 & \varepsilon_5 \\ \varepsilon_6 & \varepsilon_2 & \varepsilon_4 \\ \varepsilon_5 & \varepsilon_4 & \varepsilon_3 \end{bmatrix} \end{aligned}\right\} \tag{5.3}$$

The material property matrix, also known as the stiffness matrix (S), is formed from elastic tensor constants. The inverse of S is known as the compliance matrix (C). For linear elastic isotropic materials in which the physical properties are direction-independent, C can also be written as Eq. (5.4).

$$\begin{bmatrix} \varepsilon_1 \\ \varepsilon_2 \\ \varepsilon_3 \\ \varepsilon_4 \\ \varepsilon_5 \\ \varepsilon_6 \end{bmatrix} = \begin{bmatrix} S_{11} & S_{12} & S_{13} & S_{14} & S_{15} & S_{16} \\ S_{12} & S_{22} & S_{23} & S_{24} & S_{25} & S_{26} \\ S_{13} & S_{23} & S_{33} & S_{34} & S_{35} & S_{36} \\ S_{14} & S_{24} & S_{34} & S_{44} & S_{45} & S_{46} \\ S_{15} & S_{25} & S_{35} & S_{45} & S_{55} & S_{56} \\ S_{16} & S_{26} & S_{36} & S_{46} & S_{56} & S_{66} \end{bmatrix} \begin{bmatrix} \sigma_1 \\ \sigma_2 \\ \sigma_3 \\ \sigma_4 \\ \sigma_5 \\ \sigma_6 \end{bmatrix} \tag{5.4}$$

Hence, for these materials, Hooke's law includes only two independent variables. Hooke's law for the isotropic medium in terms of stiffness form is given by

$$\begin{bmatrix} \sigma_{xx} \\ \sigma_{yy} \\ \sigma_{zz} \\ \sigma_{yz} \\ \sigma_{zx} \\ \sigma_{xy} \end{bmatrix} = \frac{E}{(1+v)+(1-2v)} \begin{bmatrix} 1-v & v & v & 0 & 0 & 0 \\ v & 1-v & v & 0 & 0 & 0 \\ v & v & 1-v & 0 & 0 & 0 \\ 0 & 0 & 0 & \frac{1}{2}-v & 0 & 0 \\ 0 & 0 & 0 & 0 & \frac{1}{2}-v & 0 \\ 0 & 0 & 0 & 0 & 0 & \frac{1}{2}-v \end{bmatrix} \begin{bmatrix} \varepsilon_{xx} \\ \varepsilon_{yy} \\ \varepsilon_{zz} \\ \varepsilon_{yz} \\ \varepsilon_{zx} \\ \varepsilon_{xy} \end{bmatrix} \tag{5.5}$$

where E is Young's modulus. v is the Poisson's ratio, which is the proportion of a loaded specimen's transverse to longitudinal strains. For a cubic crystal, the connections between the compliance–stiffness constants are written as,

$$C_{11} = \frac{S_{11} + S_{12}}{(S_{11} - S_{12})(S_{11} + 2S_{12})} \tag{5.6}$$

$$C_{12} = \frac{-S_{12}}{(S_{11} - S_{12})(S_{11} + 2S_{12})} \tag{5.7}$$

$$C_{44} = \frac{1}{S_{44}} \tag{5.8}$$

$$S_{11} = \frac{C_{11} + C_{12}}{(C_{11} - C_{12})(C_{11} + 2C_{12})} \tag{5.9}$$

$$S_{12} = \frac{-C_{12}}{(C_{11} - C_{12})(C_{11} + 2C_{12})} \tag{5.10}$$

$$S_{44} = \frac{1}{C_{44}} \tag{5.11}$$

For Si, the elasticity matrix can be simplified and expressed in terms of the three elements (C_{11}, $C_{12,}$ and C_{44}) only as a member of the cubic class (diamond structure).

5.2.3 Piezoresistivity Effect

Several techniques have been formulated to model mobility under stress [18]. Due to its ability to offer the mobility relationship under strain and no-strain situations, the piezoresistive technique has been widely used. Due to the material's piezoresistivity effect, the carrier mobility enhancement tensor extended in the stress up to the second order. The piezoresistance (π) coefficients are defined in the form of piezoresistivity and stress as:

Piezoresistance effect in germanium and silicon

$$\pi = \frac{\Delta R/R}{T} \tag{5.12}$$

where R is the "original resistance" of the sample, i.e. $R = (\rho l)/(wh)$, ΔR is the change of resistance, and T is applied stress, respectively. After combining all the relative changes in the semiconductor sample, the $\Delta R/R$ is

$$\frac{\Delta R}{R} = \frac{\Delta l}{l} - \frac{\Delta w}{w} - \frac{\Delta h}{h} + \frac{\Delta \rho}{\rho} \tag{5.13}$$

In Eq. (5.13), the first three terms of the right side represent the geometrical shape changes of the sample in stressed conditions, whereas the other term, $\Delta\rho/\rho$ is the resistivity dependence on stress. Because the change in stress-induced resistivity in semiconductor devices is greater than the change in geometrically induced resistance, the changes in stress-induced resistivity now determine the piezoresistivity. In general, resistivity ($\rho = 1/\sigma$) and stress (T) are second-rand and second-rank tensors respectively. In the linear response, under arbitrary stress stage, the change in resistivity ($\Delta\rho_{ij}$) is expressed in the piezoresistance (π_{ijkl}) tensor as

$$\frac{\Delta \rho_{ij}}{\rho} = -\frac{\Delta \sigma_{ij}}{\sigma} = \sum_{k,l} \pi_{ijkl} \tau_{kl} \tag{5.14}$$

where the summation considers all the axis points of x, y, and z. Again, rewriting $\Delta\rho_{ij}$ into a vector form in terms of compliance tensor and stiffness tensor, where $i = 1, 2, 3, \ldots 6$. The above Eq. (5.14) can be expressed as

$$\frac{\Delta \rho_i}{\rho} = \sum_{k=1}^{6} \pi_{ik} \tau_k, \tag{5.15}$$

In the cubic structures, three independent π-elements are considered, such as $\pi_{11}, \pi_{12}, \pi_{44}$,

$$\pi_{ik} = \begin{pmatrix} \pi_{11} & \pi_{12} & \pi_{12} & 0 & 0 & 0 \\ \pi_{12} & \pi_{11} & \pi_{12} & 0 & 0 & 0 \\ \pi_{12} & \pi_{12} & \pi_{11} & 0 & 0 & 0 \\ 0 & 0 & 0 & \pi_{44} & 0 & 0 \\ 0 & 0 & 0 & 0 & \pi_{44} & 0 \\ 0 & 0 & 0 & 0 & 0 & \pi_{44} \end{pmatrix} \tag{5.16}$$

The principal crystal axis along with other axis represents different piezoresistive effects such as π_{11} (with the same principal-crystal axis), π_{12} (directed along with one perpendicular crystal axis i.e. transverse effect) and whereas, π_{44} represents stress on an electric field of out-of-plane due to the change of the current induced because of in-plane shear stress.

5.2.4 Strain Induced Carrier Mobility

Effective mobility is influenced by the charge carriers' effective mass and drift velocity. Given that the DG model is obtained from the Wigner equation using the theory of moments, the DD model for short-channel devices has some compatibility with advanced transport models. The impact on hole mobility becomes more complex than that of the electron, as in the strained device, the carriers get redistributed among the sub-valleys, causing mobility enhancement. Particularly, the electron mobility enhancement can be redistributed into the sub-valleys (Δ_2 and Δ_4). Due to the lowering of sub-valleys (Δ_2 as compared to Δ_4), the maximum number of electrons could be occupied in the Δ_2 sub-valley. Hence, the corresponding effective mass of the enhanced electron mobility can be expressed as longitudinal (m_{nl}) and transverse (m_{nt}) effective mass. The transverse component of effective mass is quite lower than the longitudinal component of it; it also reduces intervalley scattering. So the electron mobility under strain can be stated as,

$$\mu_{n,ii} = \mu_{n0}\left[1 + \frac{1 - m_{nl}/m_{nt}}{1 + 2(m_{nl}/m_{nt})}\left[\frac{F_{1/2}\left(\frac{F_n - F_n - \Delta E_{C,i}}{kT}\right)}{F_{1/2}\left(\frac{F_n - F_n - \Delta E_C}{kT}\right)} - 1\right]\right] \tag{5.17}$$

where μ_{n0} represents the mobility of electrons without any strain and F_n denotes the quasi-fermi level of electrons. ΔE_C is the maximum of $\Delta E_{C,i}$. The index "i" denotes the direction (e.g. μ_{n11} are the electron mobility in the x-axis direction of the crystal system and corresponding $\Delta E_{C,1}$, represents the two-fold sub-valley along the x-axis). At higher stress, electron mobility increases with stress nonlinearly and gets saturated, whereas hole mobility is linear with increasing stress.

5.3 Case Study

5.3.1 Stress/Strain Engineering in Bulk-Si FinFETs

In the semiconductor sector, a front-end manufacturing facility houses the main production process for creating ICs. The manufacturing of semiconductor devices generally involves more than several hundred processing steps, months of processing time, and the uncertainty of the tool's performance and yield. Due to the thermal coefficient of expansion (TCE) mismatch, stress develops in semiconductor devices. This results in mechanical failure and degradation of device performance. The virtual wafer fabrication (VWF) approach to design and the advancement of new technologies now place a greater emphasis on the TCAD. The TCAD can be used to improve productivity, save design costs, and provide superior technology and device designs. Process design/simulation and device simulation make up the bulk of the TCAD. Process simulation is similar to the virtual manufacturing of semiconductor devices. It starts with a wafer substrate and simulates most of the important process steps like ion implantation, deposition, oxide growth and diffusion, and etching. A modern process simulator can simulate both front-end-of-the-line (FEOL) and back-end-of-the-line (BEOL) processes with reasonable accuracy. At the moment, in the semiconductor industry, the primary technology booster is strain engineering. Geometric scaling is a challenging procedure since the IRDS 2020 for Semiconductors has established several standards for CMOS device downscaling. To improve the electrical performance of logic devices, various stress engineering methods have been devised. Local mechanical stress control and uniaxial stress are more advantageous for band structure engineering and are technologically easier to apply [19].

There are various stress-transfer techniques developed to introduce uniaxial tension in the channel of a device. First, it has been suggested that compressive stress in p-type MOSFETs be produced by embedding SiGe stressors (eSiGe) in the (S/D) regions [2]. Filling the (S/D) with a SiGe material with a greater lattice spacing than silicon results in significant compressive stress in the channel and an increase in drive current [20]. Another choice is to add a stress material to the top of the gate stack after silicide synthesis. Successful integration of dual contact etch stop liner (CESL) has been described thanks to the engineering of compressive or tensile stress in nitride films with process conditions as uniaxial strain boosters (eSiGe and CESL), which have been widely used to improve the performance of logic devices at technology nodes ranging from 90 to 45 nm [2]. Several researches focus on optimizing these various strategies to increase the stress level caused, with the overall goal of combining the effects of various stressors. Simulated results have been compared to three different experimental configurations and several experimental characterizations in the instance of eSiGe stressors. The stress predicted by the various characterization methodologies is in relatively excellent agreement, despite certain approach constraints brought on by the plane strain assumptions for two-dimensional (2D) finite element method (FEM) simulations [19].

The addition of carbon to the silicon–germanium binary compound provides a powerful tool in the design of bandgap-engineered materials and devices utilizing such materials. Proper incorporation of carbon into the silicon matrix results in a net strain reduction when compared to the silicon–germanium alloy. This alleviation in the macroscopic strain of an epilayer allows the growth of thicker, defect-free films for a given germanium content or the potential for increased germanium content for a given film thickness. The prospect of creating both a wide and narrow band gap material through the introduction of carbon into silicon and the binary silicon–germanium compound offers virtually unlimited possibilities in the creation of bandgap-engineered materials to be used in the optimization of a multitude of device designs.

FinFET technology has become the best choice among nonplanar devices and can be extended, scaling down to a 7 nm technology node with a relatively minimal increase in processing costs. The most likely adopted candidate of the multigate devices from a fabrication perspective is the FinFET. The FinFETs are mainly in two types based on the type of substrate, i.e. bulk-Si FinFETs and SOI-FinFETs. The fabrication steps of bulk-Si FinFET are different from planar MOSFETs, as has been reported by several research groups. However, process step optimization is still an open-ended concern. In this case study, the tri-gate device process simulations are carried out using the 3D Victory process tool in 2D and 3D geometry planes [21]. The Victory stress tool has been used to simulate stress [22]. The electrical behavior of the device has been analyzed using the Victory device with default transport models [16]. Figure 5.1 shows the stress distribution (von Mises, *xx*- and *yy*-direction) in Si-fin after the completion of the device processing (Table 5.1).

5.3.1.1 Performance Analysis of Bulk-Si FinFET

In device simulation, the Victory device simulator has been used for the p-FinFET simulations by implementing various quantum-corrected models. To comprehend the electrical response of FinFET devices, many device simulations have been carried out. When a device is scaled down to the nanoscale range, the quantum effect affects the device's characteristics. The quantum effect is ignored in the classical DD transport model. The probability density function can be used to describe an electron's position. The Poisson solution with the self-consistent Schrodinger equation can also be used for calibration when a density gradient (DG) is required. Acoustic phonons, surface roughness scattering, and intervalley phonons all have an impact on electron and hole mobility. The DG model includes the quantum correction potentials for holes and electrons. However, the convergence is much better in the Bohm quantum potential (BQP) model than in the DG model. Further, the carrier temperature is incorporated into simulations using "hot carrier transport equations" (HCTE). A position-dependent quantum potential is introduced. Correct quantum effects are provided by the BQP model without the need for an explicit solution to the Schrodinger equation. The impacts of ballistic carrier transport are taken into consideration using the EB model. It results in

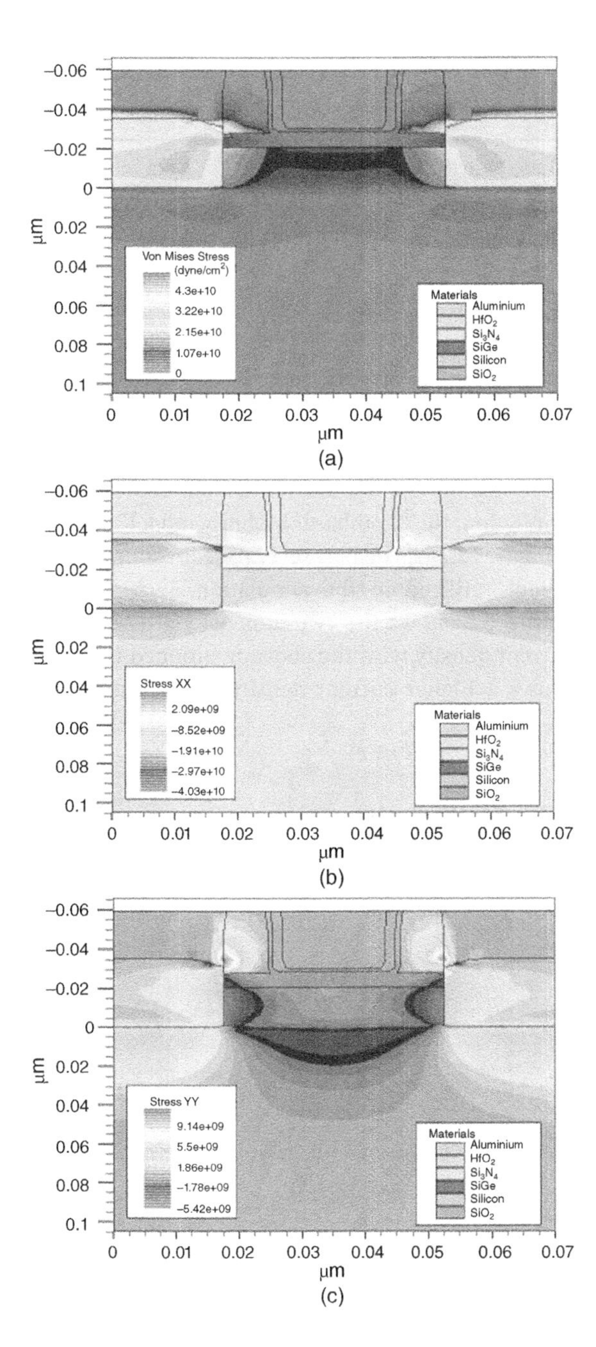

Figure 5.1 Stress distribution (von Mises, *xx*- and *yy*-direction) in Si-fin after completion of the device processing.

Table 5.1 Geometrical details of the virtually fabricated FinFET as shown in Figure 5.1.

Geometrical parameters	Value
Gate oxide thickness	1 nm
Gate length	7 nm
Fin width	30 nm
Fin height	50 nm
Substrate doping	$1 \times 10^{15}/\text{cm}^3$
S/D doping	$1 \times 10^{20}/\text{cm}^3$

lowering the output impedance of transfer characteristics and notably raising the drain current. For the transverse field-dependent mobility investigation in device simulation, the "Lombardi Mobility model" (CVT) is employed.

Figures 5.2–5.6 display several electrostatics of the FinFET for various transport models utilized in the simulation, including BQP + HCTE and BQP + CVT. Figure 5.2 shows the two-dimensional (2D) contour (XZ planes) plot of total current density with the above-mentioned two models. The BQP + HCTE model shows a higher current density. The electron concentration plot (XZ planes)

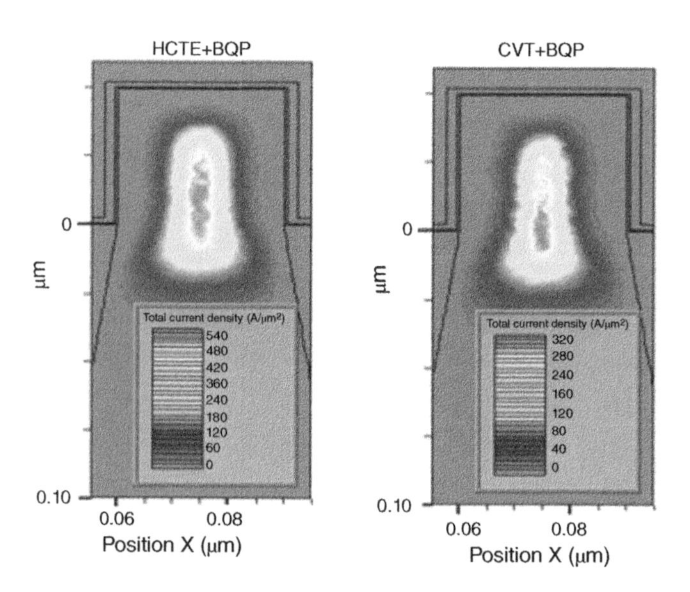

Figure 5.2 Two-dimensional contour (XZ planes) plot of total current density with respective models. Source: Jena [23].

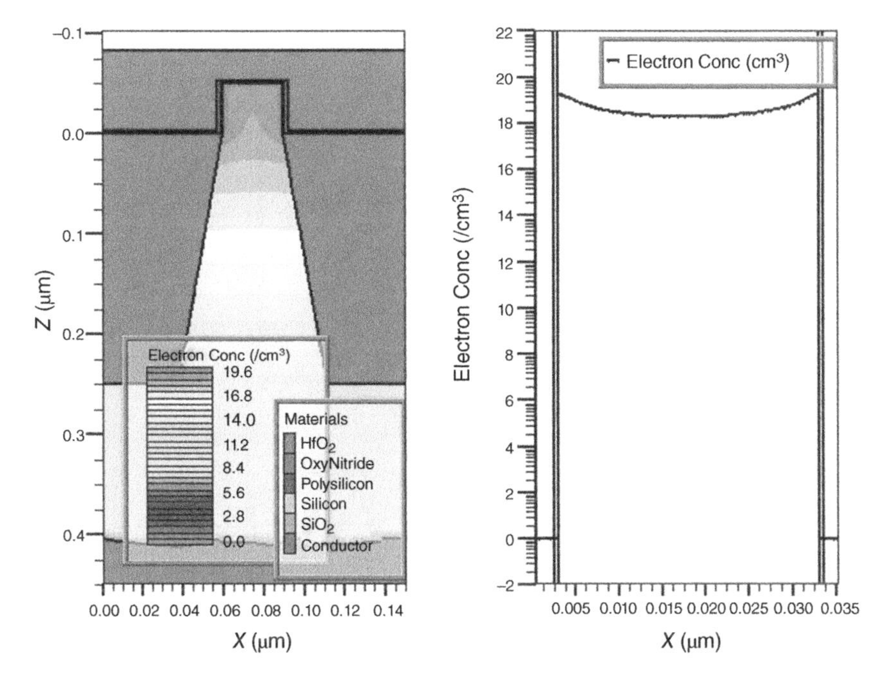

Figure 5.3 The 2D and 1D electron concentration profile of the EB–BQP model is shown at cutline $Z = -0.02\,\mu\text{m}$. Source: Jena [23].

is also shown in Figure 5.3 In the EB–BQP model, the maximum numbers of charge carriers are involved in the current conduction due to ballistic transport with quantum correction effects than in other models. Its 1D profile is shown at cutline $Z = -0.02$ microns in the XZ plane of the devices. It shows the electron concertation variation across the x-axis. Figures 5.4a,b show the 2D and 1D mobility profiles of electron and hole at the cutline of $Z = -0.02\,\mu\text{m}$ in the just top and middle part of the fin area. The electron and hole coordinate points were chosen from (0.058, −0.02) to (0.092, −0.02) in the XZ plane, respectively. The channel carrier mobility has been observed to be about 480 and $220\,\text{cm}^2/\text{V-s}$ for electrons and holes, respectively.

Figure 5.5 shows the comparison of output characteristics, taking into account four transport models. The drain current increases the most when the EB model is applied since it applies the ballistic transport equations. As the quantum confinement of the carriers occurs in the BQP model, it causes the current drop in the channel due to impeding carrier mobility. Hence, the drain current decreases by about 40% and 50% compared to energy balance (EB) and DD models, respectively. While without consideration of quantum effects, the DD and EB models show an increment in current of 35% and 44%, respectively.

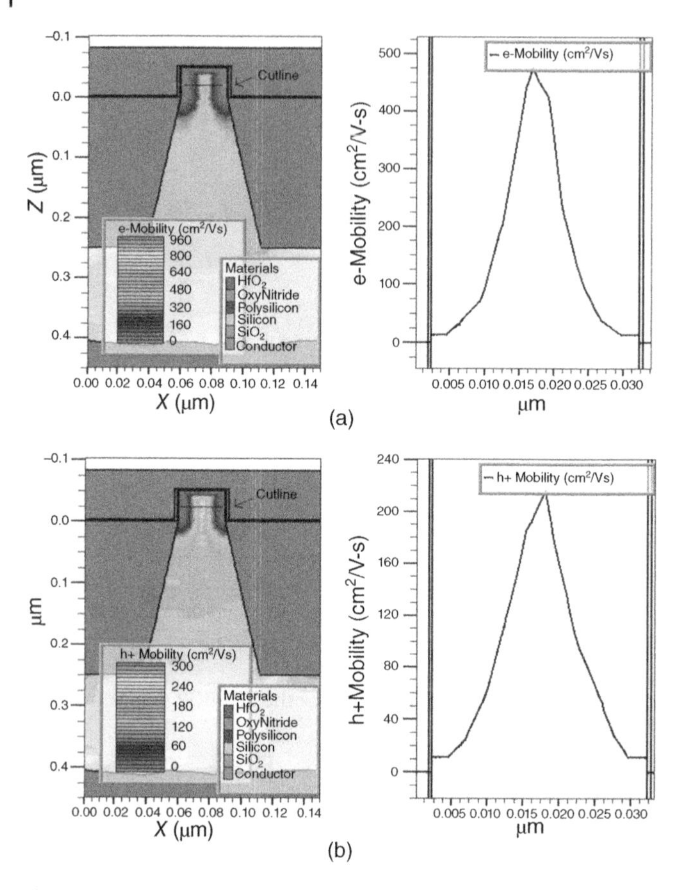

Figure 5.4 The mobility profile of (a) electron and (b) hole of the BQP model is shown in the XZ plane at cutline $Z = -0.02\,\mu$m. Source: Jena [23].

5.3.1.2 Effects of Fin Geometry Variations

It is observed that the fins have rounded tops and sloped sidewalls and are not precisely rectangular. A FinFET structure with a very similar pattern to the TEM image (considering Fin as the channel) is constructed on a wafer with a 100-surface orientation and is oriented along the 110-axis. The device in Figure 5.6 with two fin top widths is the subject of the investigation that follows (column 1). The fin bottom width is 15 nm, doping is 1×10^{17} cm^{-3} in the channel, and the contour plots are made at $V_d = 0.7$ V and $V_g = 0$ V. The electrostatic potential's fin shape dependency is depicted in Figure 5.6 (column 2). It demonstrates that fin geometry is dependent on the potential. The electron concentration distributions throughout the channel region in the on-state ($V_d = 0.5$ V and $V_g = 0.5$ V, respectively) are shown in Figure 5.6 (column 3). At extremely tiny fin,

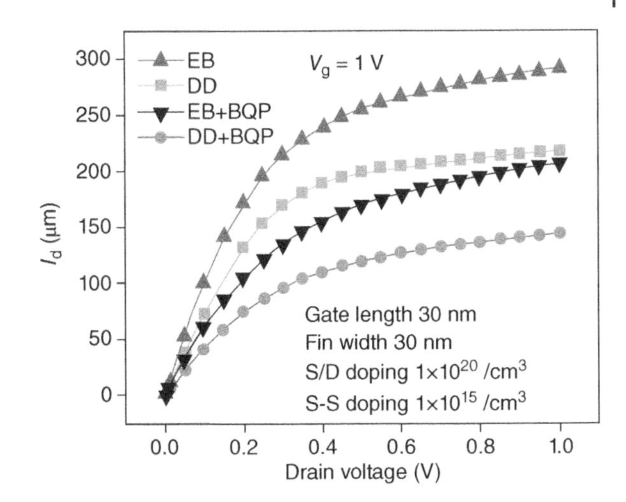

Figure 5.5 Output characteristics ($I_d - V_d$) comparison for different models used in the simulation. Source: Jena [23].

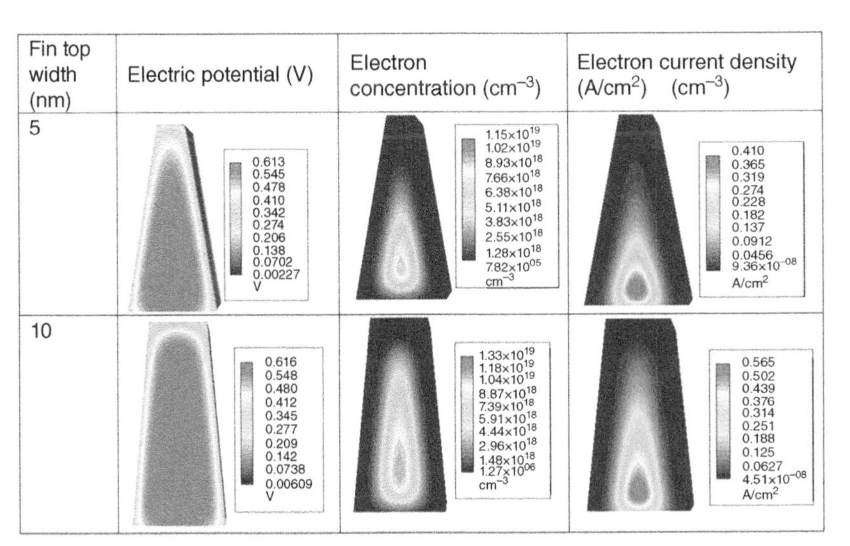

Figure 5.6 3D Contour profiles of different electrical parameters: potential, electron concentration, and current density profiles by varying the fin top width at channel doping of 1×10^{17} cm^{-3} and channel/substrate orientation <100>/<100>. Source: Jena [23].

the highest electron density is driven into the fin volume (such as the triangular) due to the quantum confinement effect. According to Figure 5.6 column 4, the bottom of the fin has the highest electron current density when simulated with various fin widths. Figure 5.6 shows that the electrostatic potential increases with the increase in fin top width. The corner effect causes the change in electrostatic potential. The structural confinement causes an increase in sharpness (i.e. a

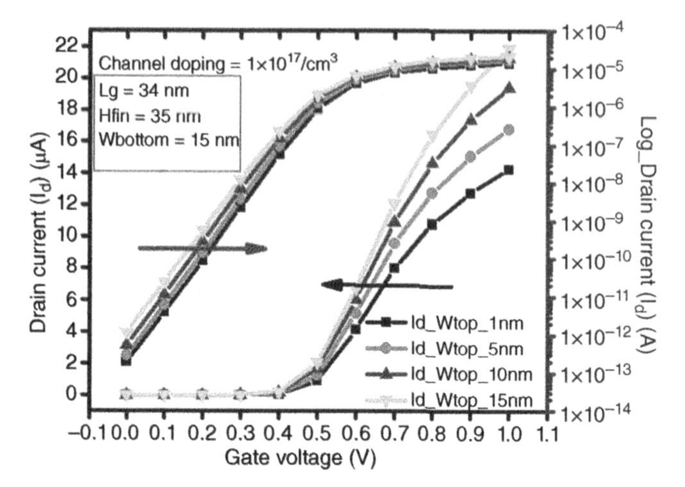

Figure 5.7 Transfer characteristic curves ($I_d - V_g$) of fin width variations at $V_d = 0.05$ V. Source: Jena [23].

decrease in fin top width), which results in a fall in electron concentration because the electron current density reduces.

Through the simulations of the 3D devices, the electrical performance of the bulk p-FinFET was ascertained. The major objective is to reduce the leakage current by varying the fin width (top) and channel doping. The optimization of the fin body is also carried out to investigate how the shape of the fin affects leakage current. Figure 5.7 shows that with a channel doping of 1×10^{17} cm^{-3}, the drain current increases with the fin top width of 1~15 nm. It achieves a value of about 22 μA of maximum current at 15 nm fin top width.

5.3.2 Nanosheet

Advanced FETs, like GAA and Nanosheet (NS) FETs, are the upcoming devices to switch from conventional FinFETs, to realize high-performance applications at low power in forthcoming nodes and to maintain all the aspects of challenges and requirements of Moore's Law. Optimization of vertically-stacked horizontal GAA bulk-Si NS transistors has been reported [24, 25]. It is time to explore novel device topologies through predictive simulation since 5 nm technology won't be available for mass production until 2025 and will likely be a FinFETs (perhaps a GAA Si NW or other similar types of devices). For GAA NS field effect transistors (GAA-NSFETs) at the 3 nm node, parasitic-aware design technology co-optimization (DTCO) has also been proposed [26–29].

For more advanced CMOS devices, the GAA silicon nanosheet field effect transistor is increasingly regarded as a worthwhile alternative to FinFET architecture [30]. Recently, it was shown that top-down nanofabrication procedures work best with ultrathin, globally strained silicon layers (nanomembranes) to produce strained Si nanosheets [31]. Intriguingly, the formation of free surfaces during the nanoscale patterning of biaxially strained nanomembranes – a critical step in the fabrication [32] of strained nanosheet – causes a local relaxation of strain as a result of the rearrangement of lattice atoms close to the freshly formed edges. The creation of nanosheets with a thickness smaller than 10 nm has been made possible by recent developments in characterization and integration. Although it is possible to create nanosheet transistors with superior electrostatic control, doing so results in quantum confinement phenomena and gives nanosheet electrical properties that are distinct from those of bulk silicon. Figure 5.8a,b show the cross-sectional view of three stacked nanosheet transistors.

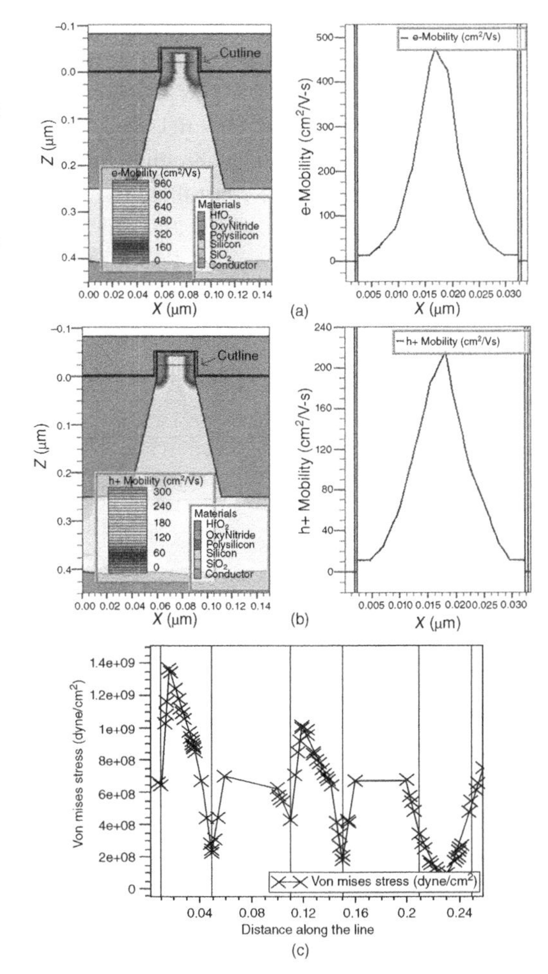

Figure 5.8 (a) Cross-sectional view of three stacked nanosheet transistors. (b) Von Mises stress distribution is found to vary from $8.5e^{+6}$ dynes/cm^2 to $1.4e^{+9}$ dynes/cm^2. (c) Location dependence of von Mises stress in three nanosheets. Maximum peak ($1.4e^{+9}$ dynes/cm^2) stress is found to develop in the top nanosheet.

Von Mises stress distribution is found to vary from $8.5e^{+6}$ dynes/cm^2 to $1.4e^{+9}$ dynes/cm^2. The location dependence of von Mises stress in three nanosheets is also shown. It is observed that the maximum peak ($1.4e^{+9}$ dynes/cm^2) stress is found to develop in the top nanosheet. Figure 5.8c shows the 1D cutline profile of the stress distribution.

The nanosheet FET comprises a silicon body within a rectangular sheet-type shape acting as the conducting channel; whilst the gate dielectric layers and gate material wrap around the semiconductor to form the MOS structure. The device geometry parameters at the 7 nm technology node are defined in Table 5.2 (following the ITRS guideline). The simulated device stacked NSFET (SNSFET) is a 3D structure consisting of an n-type nanosheet transistor with a sheet-type cross-section having a 12 nm gate length, as shown in Figure 5.9.

Future CMOS lateral GAA nanosheet transistor evaluation using TCAD is reported in the following. Using 3D numerical device simulations, a physical evaluation of the lateral GAA NSFET at the 3 nm technology node of the newly defined beyond-Moore IRDS [6] is presented (Figure 5.9). An experimental record number of seven stacked channels in GAA nanosheet transistors employing a replacement metal gate technique, inner spacer, and self-aligned contacts provides outstanding gate controllability and very high current drivability [35]. We investigate how the width and thickness of nanosheets affect electrical performance and present crucial design principles required for vertically stacked nanosheet FETs [36] (Figure 5.10).

Table 5.2 Device details of SNSFETs used in the simulation.

Device parameters	Values
Gate length (L_g)	12 nm
Nanosheet width (W_{NS})	10 nm
Nanosheet thickness (H_{NS})	5 nm
High-k (HfO_2) thickness	1.5 nm
Oxide thickness (T_{OX})	0.5 nm
Epi-length	14 nm
Epi-top-width	8 nm
Epi-middle-width	25 nm
Epi-bottom-width	14 nm
Spacer length (L_{SP})	7 nm
Channel-doping (N_{CH})	10^{16} cm^{-3}
S/D-doping	10^{20} cm^{-3}

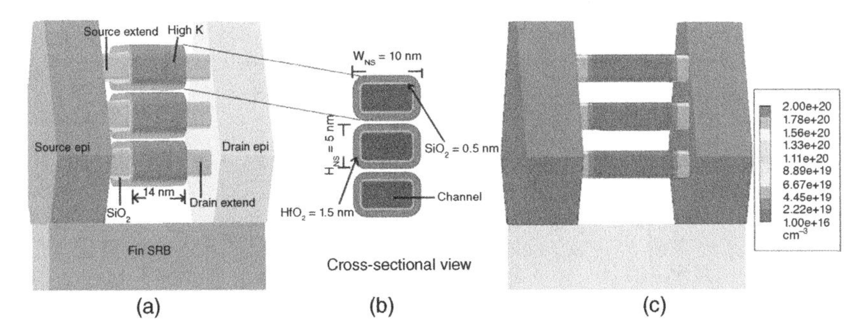

Figure 5.9 (a) A schematic representation of 12 nm long Stacked NSFETs (SNSFETs), (b) the cross-sectional view of three sheets, and (c) The net doping profile plot. Source: Mohapatra [33].

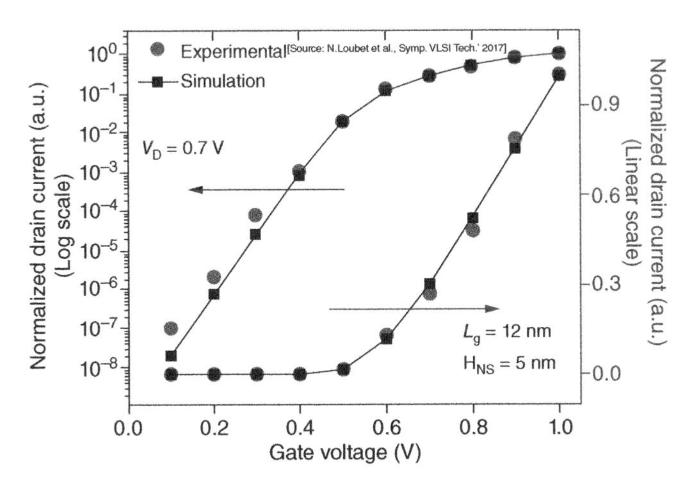

Figure 5.10 Calibration of transfer characteristics with simulated and reported experimental data [34], both in linear and log scales. Source: Mohapatra [33].

5.3.2.1 Impact of Mechanical Stress

By applying external mechanical stress, the device's performance can be improved. A four-point bending technique is used to apply external mechanical stress to FETs. By applying external stress, the effects on drain current have been investigated. VSP Solver and MINIMOS-NT tools have been used to include electronic transport and to obtain electrical characteristics [37, 38]. The increment in drain current with the variation of mechanical stress is shown in Figure 5.11

The increase in drive current concerning the increase in stress can be observed in Figure 5.11 The current increases from ~137 to ~373.73 μA, with an increase in stress from 0 GPa to +3 GPa, respectively. Figure 5.12 shows the variation of threshold voltage and subthreshold swing (in Figure 5.12a), and ON and OFF state current (in Figure 5.12b) concerning stress variation.

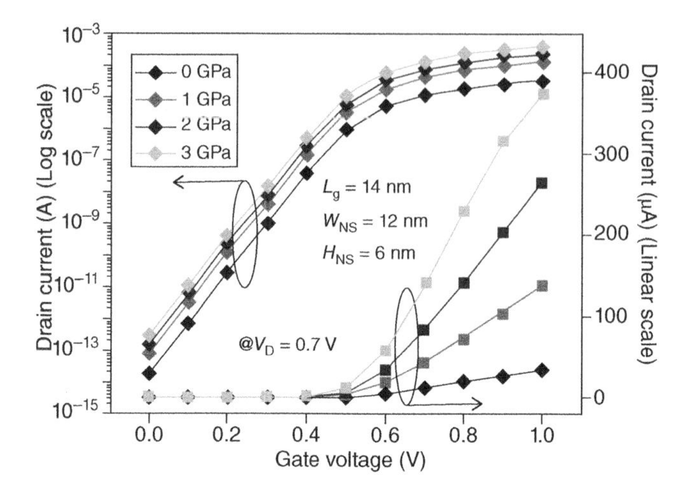

Figure 5.11 Transfer characteristics of the nanosheet under the effects of uniaxial stress along the <100> channel direction. Source: Mohapatra [33]

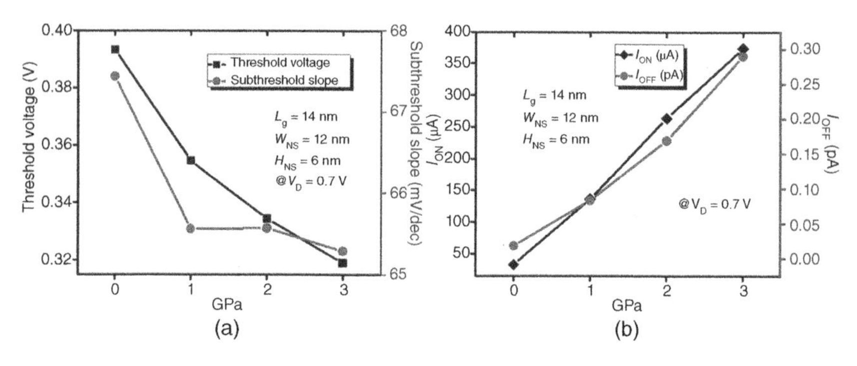

Figure 5.12 The variation of (a) threshold voltage and subthreshold slope and (b) ON and OFF current with different stress in GPa. Source: Mohapatra [33].

5.3.2.2 Strained Engineering with Embedded Source/Drain Stressor

Here, the impact of S/D $Si_{1-x}Ge_x$ material is studied. The channel is preferred as pure silicon; the S/D $Si_{1-x}Ge_x$ is varied from 0% to 50%. Compressive stress is induced in the channel, which enhances carrier mobility. The transfer characteristics curves ($I_D - V_G$) are presented in Figure 5.13 The increment in drain current can be observed for both linear ($V_D = -0.05$ V) and saturation ($V_D = -0.7$ V) drain biases. The enhancement in drain current with a higher Ge percentage indicates the rise in stress level in the SiGe S/D region, leading to an improvement in hole mobility in the channel. The percentage of enhancement in drive current is

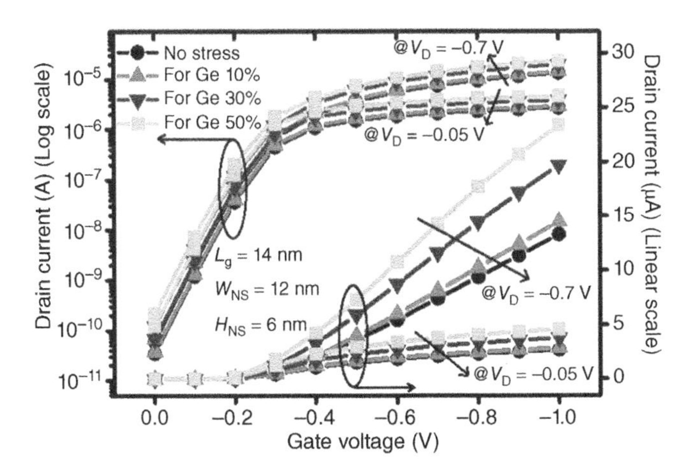

Figure 5.13 I_D-V_G characteristics of stress-enhanced NSFET devices with a change in Ge parentage. Source: Mohapatra [33]

approximately 75% in the stress-enhanced device (at Ge = 50%) compared to that of the Si channel device.

5.3.3 Extremely Thin SOI MOSFETs

Since most IoT applications need to be self-sufficient, energy management and adaptability become crucial issues. The development of mobile devices with sophisticated information processing capabilities has resulted in exponential market growth for systems on chip (SoC), which must operate at extremely high speeds while using very little power. However, because parasitic effects are amplified as MOS transistors on silicon get smaller, building such systems at the nanoscale involves several difficulties [39]. Due to its greater control of the short-channel effect (SCE) and little dopant fluctuation, ETSOI has received substantial research as a potential solution for the continuing development of CMOS technology. FDSOI technology delivers great performance at a lower production cost, particularly for RF applications. For the bulk of IoT applications, they must be self-contained; therefore, energy management and adaptability become critical concerns. As a result, one needs to use a technology that is ultra-low leakage, ultra-low power, ultra-low voltage, and cost-effective in such ultra-low power IoT applications [40–42].

In this context, the UTBB FDSOI technology is an intriguing solution that offers a balance of high performance and low leakage. We can reduce leakage or increase transistor speed by introducing a second voltage to the bottom of the transistor. Furthermore, because FDSOI is a planar technology, it is less expensive than 3D FinFET technology. The goal of this study is to show how

FDSOI technology delivers size and power reductions, increased performance and functionality, and low cost. As a result, UTBB FDSOI technology is now the most prominent technology in the IoT and wearable markets. Due to its improved short channel control, inherent low device variability due to the undoped channel, and compatibility with standard planar CMOS, FDSOI with extremely body (ETSOI) is a feasible choice for future CMOS technology with considerable device performance enhancement. High-k/metal-gate ETSOI devices are well suited for low-power applications due to their low GIDL and low threshold voltage fluctuation.

By employing faceted RSD, stress coupling between stress liners and the channel is accomplished, further improving performance. However, one issue with ETSOI performance is the inability to use embedded stressors like e-SiGe because the silicon layer is too thin. Facet-regulated in-situ doped selective Si epitaxial growth (SEG) and solid-phase diffusion-combined novel raised source/drain (RSD) structures have been reported [43]. The formation of sidewall spacers before and after selective epitaxial silicon deposition in S/D regions has also been reported in RSD MOSFET structures [44]. In this case study, two strain approaches have been combined to improve performance: (i) increased stress liner effect with faceted RSD, and (ii) SiC RSD for n-channel and SiGe RSD for p-channel MOSFETs. Drive current improvements in n-MOSFET and p-MOSFETs have been observed using the improved manufacturing flow and stress boosters. For the simulation, the gate length considered is 26 nm. Due to the device's symmetry, the stress simulation has been done for half of the device. VictoryStress used the 2D structure that was derived from the ATHENA simulation. Since the structure is 2D, the isotropic model for stress simulations is employed in the simulation. In simulation, Si, SiGe, and SiCs elastic parameters are taken into account. This simulation run generates stress distributions for different stressors. Ultra-shallow junction formation is critical for continued scaling of MOSFETs while minimizing SCEs.

The National Technology Roadmap for Semiconductors forecasts that the 100 nm CMOS generation will require junction depths of 20–40 nm. Conventional low-energy ion implantation schemes suffer from ion channeling and the difficulty of annealing damage within a low thermal budget. Based on quantum mechanical simulations calibrated with measured data, the effect of body-thickness scaling on strain-induced carrier-mobility enhancement in thin-body MOSFETs with high-k/metal gate stacks is presented to give insight into device performance enhancement trends for future technology nodes. In a UTB SOI transistor, an SOI isolation layer separates the thin active device layer and the main substrate. Due to the full depletion of the body, there is no room for unwanted current paths to form under the channel, like in a traditional bulk device. In the case of vertical RSD, the initial structure, RSD stressor, stress-liner,

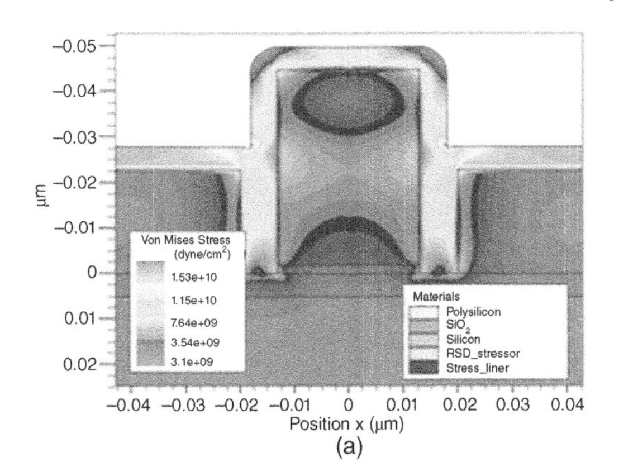

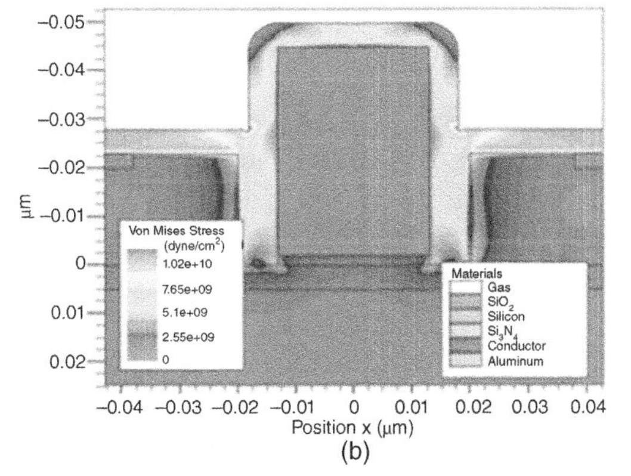

Figure 5.14 Von Mises stress distribution in vertical RSD the initial structures. (a) Von Mises stress distribution is found to be higher and varies from $3.1e^{+7}$ dynes/cm^2 to $1.53e^{+10}$ dynes/cm^2 for RSD stressors. (b) Von Mises stress distribution is found to be lower and varies from $2.55e^{+9}$ dynes/cm^2 to $1.02e^{+10}$ dynes/cm^2 for Si_3N_4 stressor.

and poly gate stressor have been simulated. Von Mises stress distribution in the basic structures is shown in Figure 5.14

In the case of faceted RSD, the initial structure, RSD stressor, stress-liner, and poly gate stressor have been simulated, and the von Mises stress distribution in these structures is shown in Figure 5.15

The acquired stress profiles show that the ultimate stress distribution is fairly varied, ranging from uniaxial to biaxial within the same structure. This non-uniform distribution of stress should be taken into account when fabricating and designing strained-engineered devices. In the simulation, we applied the strain-dependent mobility enhancement models nhance and phance in addition to the standard silicon mobility models (CVT and SRH). The second-order mobility enhancement tensor (determined by VictoryStress) is used by the mobility enhancement models directly from the low-field mobility. As can be observed in Figure 5.16, the faceted RSD structure improves stress transmission, leading to a 40% increase in hole mobility and a 15% improvement in drive current.

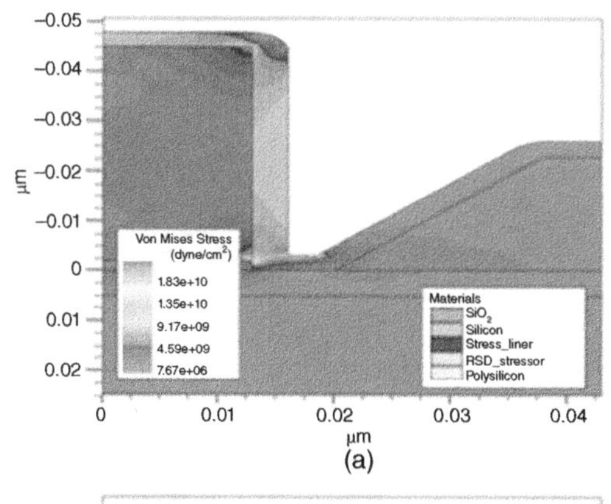

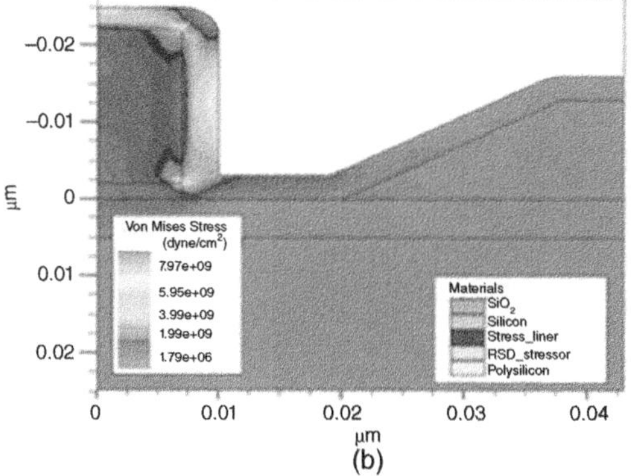

Figure 5.15 Geometry-dependent stress distributions in faceted RSD stressor. Von Mises stress distribution in faceted RSD the initial structures. (a) Von Mises stress distribution is found to be higher and varies from $7.67e^{+6}$ dynes/cm^2 to $1.83e^{+10}$ dynes/cm^2 for closely spaced faceted RSD stressors. (b) Von Mises stress distribution is found to be lower and varies from $1.7e^{+6}$ dynes/cm^2 to $7.97e^{+9}$ dynes/cm^2 for remotely spaced faceted RSD stressors.

Figure 5.16 Comparison of I_d–V_d characteristics for vertical and faceted RSD ETSOI MOSFETs at $V_g = -3$ V.

5.4 Conclusions

To provide design guidelines for performance improvement in nanoscale devices, strain engineering, as well as cutting-edge performance booster techniques/structures for advanced devices, like tri-gate FinFETs, GAA nanosheet transistors, and extremely thin silicon on insulator (ETSOI), have been considered. TCAD process/device simulations have been used to study the electrical performance of innovative facet-controlled selectively epitaxial-grown Si and solid-phase diffusion-enhanced elevated source/drain structured MOSFETs. RSD, S/D extension engineering, and high-k/metal-gate structures have been investigated. We developed a simulation framework that accounts for the stress/strain profile generation connected to the 3-dimensional nanoscale devices using Silvaco tools. Furthermore, our findings serve as the foundation for more precise calculations and modeling of the stress behavior of nanodevices and the resulting changes in their electrical characteristics. A summary of recent technology nodes utilizing strain engineering is presented, along with an outlook for future direction.

In case study 1, predictive TCAD simulations have been used for the theoretical understanding of stress tuning in several surface orientations, channel directions, and electrical characterization of advanced bulk-Si FinFETs. In case study 2, predictive TCAD simulations have been used in this work to compare the performance of "bulk-Si channel," "Si-channel with S/D stressor," and "uniaxially strained-SiGe channel" NSFETs at the 3 nm technology node. It is demonstrated that the uniaxial compressive strain resulting from lattice mismatch is more efficient in improving the device performance of strained-SiGe channel NSFETs. This work explains the stress behavior of an ultrathin silicon nanosheet that is directly based on oxide. In conclusion, we demonstrated a novel TCAD-based technique for mapping the stress in a specific region of a strain-induced device. In case study 3, we analyze various methods employed to enhance the hole and electron velocities for the FDSOI MOSFETs. The main parameters that have been tweaked and need to be optimized are the orientation, strain, and channel material. 2D and 3D device simulations were used to evaluate the electrical performance of devices with RSD regions.

References

1 Lilienfeld, J. E. U.S. Patent 1,745,175 (filed in 1926, issued in 1930), U.S. Patent. 1,877,140 (filed in 1928 issued in 1932), and U.S. Patent 1,900,018 (filed in 1928, issued in 1933).

2 Thompson, S.E., Armstrong, M., Auth, C. et al. (2004). A 90-nm logic technology featuring strained-silicon. *IEEE Transactions on Electron Devices* 51 (11): 1790–1797: https://doi.org/10.1109/TED.2004.836648.
3 Maiti, C.K. (2021). *Stress and Strain Engineering in Semiconductor Devices at Nanoscale*. USA: CRC Press (Taylor and Francis).
4 Maiti, C.K. and Armstrong, G.A. (2007). *Technology Computer Aided Design for Si, SiGe and GaAs Integrated Circuits*. The Institution of Engineering and Technology, IET.
5 Hall, H.H., Bardeen, J., and Pearson, G.L. (1951). The effects of pressure and temperature on the resistance of junctions in germanium. *Physical Review* 84 (1): 129: https://doi.org/10.1103/PhysRev.84.129.
6 IEEE (2020). International Roadmap for Devices and Systems (IRDS): Executive Summary.
7 Bae, G., Bae, D.-I., Kang, M. et al. (2018). 3nm GAA technology featuring multi-bridge-channel FET for low power and high performance applications. *2018 IEEE International Electron Devices Meeting (IEDM)*, Jan. 2018, vol. 2018-Dec. 28.7.1–28.7.4. https://doi.org/10.1109/IEDM.2018.8614629.
8 Hwang, L. and Horng, T.J. (2017). MM and MTM for mobility. In: *3D IC and RF SiPs: Advanced Stacking and Planar Solutions for 5G Mobility*, 1–65. IEEE: https://doi.org/10.1002/9781119289654.ch1.
9 Zineddine, T., Zahra, H., and Zitouni, M. (2019). Design and analysis of 10 nm T-gate enhancement-mode MOS-HEMT for high power microwave applications. *Journal of Science: Advanced Materials and Devices* 4 (1): 180–187: https://doi.org/10.1016/J.JSAMD.2019.01.001.
10 Roy, A., Mitra, R., Mondal, A., and Kundu, A. (2022). Analog/RF and power performance analysis of an underlap DG AlGaN/GaN based high-K dielectric MOS-HEMT. *Silicon* 14 (5): https://doi.org/10.1007/s12633-021-01020-8.
11 Cheng, W. C., Lei, S., Li, W. et al. (2019). Improving the drive current of AlGaN/GaN HEMT using external strain engineering. *2019 Electron Devices Technology and Manufacturing Conference, EDTM 2019*, Mar. 2019. 374–376. https://doi.org/10.1109/EDTM.2019.8731108.
12 Shervin, S., Kim, S.H., Asadirad, M. et al. (2015). Strain-effect transistors: theoretical study on the effects of external strain on III-nitride high-electron-mobility transistors on flexible substrates. *Applied Physics Letters* 107 (19): https://doi.org/10.1063/1.4935537.
13 Das, S., Dash, T.P., Jena, D. et al. (2021). Strain-engineering in AlGaN/GaN HEMTs: impact of silicon nitride passivation layer on electrical performance. *Physica Scripta* 96 (12): https://doi.org/10.1088/1402-4896/AC3EF9.
14 Das, S., Mohapatra, E., Choudhury, S. et al. (2021). Stress-engineered AlGaN/GaN high electron mobility transistors design. In: *2021 Devices for Integrated Circuit (DevIC)*. IEEE: https://doi.org/10.1109/DevIC50843.2021.9455852.

15 Maiti, C.K. and Maiti, T.K. (2012). *Strain-Engineered MOSFETs*. CRC Press.

16 Silvaco International (2018). *VictoryDevice User Manual*.

17 Maeder, X., Mook, W.M., Niederberger, C., and Michler, J. (2011). Quantitative stress/strain mapping during micropillar compression. *Philosophical Magazine* 91 (7–9): https://doi.org/10.1080/14786435.2010.505178.

18 Smith, C.S. (1954). Piezoresistance effect in germanium and silicon. *Physical Review* 94 (1): 42–49: https://doi.org/10.1103/PhysRev.94.42.

19 Krzeminski, C.D. (2012). Stress mapping in strain-engineered silicon p-type MOSFET device: a comparison between process simulation and experiments. *Journal of Vacuum Science & Technology B* 30 (2): 022203: https://doi.org/10.1116/1.3683079.

20 Chidambaram, P.R., Bowen, C., Chakravarthi, S. et al. (2006). Fundamentals of silicon material properties for successful exploitation of strain engineering in modern CMOS manufacturing. *IEEE Transactions on Electron Devices* 53 (5): 944–964: https://doi.org/10.1109/TED.2006.872912.

21 Silvaco International (2018). *VictoryProcess User Manual*.

22 Silvaco International (2018). *VictoryStress User Manual*.

23 Jena, J. (2021). Design and simulation of strain-engineered trigate FinFETs at 7 nm technology nodes. PhD Thesis, SOA University, Bhubaneswar.

24 Gundu, A.K. and Kursun, V. (2021). Optimization of 3D stacked nanosheets in 5 nm gate-all-around transistor technology. *International System on Chip Conference*, 2021, vol. 2021-September. 10.1109/SOCC52499.2021.9739517.

25 Huang, Y.C., Chiang, M.H., Wang, S.J. et al. (2021). TCAD-based assessment of the lateral GAA nanosheet transistor for future CMOS. *IEEE Transactions on Electron Devices* 68 (12): https://doi.org/10.1109/TED.2021.3124472.

26 Sun, Y., Wang, M., Li, X. et al. (2022). Improved MEOL and BEOL parasitic-aware design technology co-optimization for 3 nm gate-all-around nanosheet transistor. *IEEE Transactions on Electron Devices* 69 (2): https://doi.org/10.1109/TED.2021.3135247.

27 Zhang, Q., Gu, J., Xu, R. et al. (2021). Optimization of structure and electrical characteristics for four-layer vertically-stacked horizontal gate-all-around Si nanosheets devices. *Nanomaterials* 11 (3): 646. https://doi.org/10.3390/NANO11030646.

28 Tayal, S., Ajayan, J., LeoJoseph, L.M.J. et al. (2021). A comprehensive investigation of vertically stacked silicon nanosheet field effect transistors: an analog/RF perspective. *Silicon*: https://doi.org/10.1007/s12633-021-01128-x.

29 Li, C., Liu, F., Han, R. et al. (2021). A vertically stacked nanosheet gate-all-around FET for biosensing application. *IEEE Access* 9: https://doi.org/10.1109/ACCESS.2021.3074906.

30 Gundu, A.K. and Kursun, V. (2022). 5-nm gate-all-around transistor technology with 3-D stacked nanosheets. *IEEE Transactions on Electron Devices* 69 (3): https://doi.org/10.1109/TED.2022.3143774.

31 Minamisawa, R.A., Süess, M.J., Spolenak, R. et al. (2012). Top-down fabricated silicon nanowires under tensile elastic strain up to 4.5%. 3: 1096: https://doi.org/10.1038/ncomms2102.

32 Hashemi, P., Gomez, L., and Hoyt, J.L. (2009). Gate-all-around n-MOSFETs with uniaxial tensile strain-induced performance enhancement scalable to sub-10-nm nanowire diameter. *IEEE Electron Device Letters* 30 (4): 401–403: https://doi.org/10.1109/LED.2009.2013877.

33 Mohapatra, E. (2021). Gate-all-around stacked-nanowire/nanosheet transistors for sub-7 nm technology nodes. PhD Thesis, SOA University, Bhubaneswar.

34 Barraud, S., Lapras, V., Previtali, B. et al. (2017). Performance and design considerations for gate-all-around stacked-nanowires FETs. *2017 IEEE International Electron Devices Meeting (IEDM)*, Jan. 29.2.1–29.2.4. https://doi.org/10.1109/IEDM.2017.8268473.

35 Barraud, S., Previtali, B., Vizioz, C. et al. (2020). 7-Levels-stacked nanosheet GAA transistors for high performance computing. *Digest of Technical Papers - Symposium on VLSI Technology*. Jun. 2020. https://doi.org/10.1109/VLSITechnology18217.2020.9265025.

36 Mohapatra, E., Dash, T.P., Jena, J. et al. (2021). Design study of gate-all-around vertically stacked nanosheet FETs for sub-7 nm nodes. *SN Applied Sciences* 3 (5): https://doi.org/10.1007/s42452-021-04539-y.

37 GTS Framework (2020). *MINIMOS-NT User Manual.*

38 GTS Framework (2020). *VSP User Manual.*

39 Maiti, C.K. (2017). *Introducing Technology Computer-Aided Design (TCAD) Fundamentals, Simulations, and Applications.* Pan Stanford: CRC Press (Taylor and Francis), USA.

40 Berthier, F., Beigné, E., Heitzmann, F. et al. (2016). UTBB FDSOI suitability for IoT applications: investigations at device, design and architectural levels. *Solid State Electronics* 125: 14–24: https://doi.org/10.1016/j.sse.2016.09.003.

41 Skotnicki, T. (2011). *Competitive SOC with UTBB SOI*. 1–61.

42 Nier, O., Rideau, D., Niquet, Y.M. et al. (2013). Multi-scale strategy for high-k/metal-gate UTBB-FDSOI devices modeling with emphasis on back bias impact on mobility. *Journal of Computational Electronics* 12 (4): 675–684: https://doi.org/10.1007/s10825-013-0532-1.

43 Nakahara, Y., Takeuchi, K., Tatsumi, T. et al. Ultra-shallow in-situ-doped raised source/drain structure for sub-tenth micron CMOS. *1996 Symposium on VLSI Technology. Digest of Technical Papers*. Jun. 1996. 174–175. https://doi.org/10.1109/VLSIT.1996.507841.

44 Rodder, M. and Yeakley, D. (1991). Raised source/drain MOSFET with dual sidewall spacers. *IEEE Electron Device Letters* 12 (3): 89–91: https://doi.org/10.1109/55.75721.

6

TCAD Analysis of Linearity Performance on Modified Ferroelectric Layer in FET Device with Spacer

Yash Pathak[1], Kajal Verma[1], Bansi Dhar Malhotra[2], and Rishu Chaujar[1]

[1] *Department of Applied Physics, Delhi Technological University, New Delhi, India*
[2] *Department of Biotechnology, Delhi Technological University, New Delhi, India*

6.1 Introduction

Benefits of silicon, such as small mass, low cost, good carrier ability, and maximum wafer diameter, have been demonstrated since the creation of semi-conductor generation. The use of a silicon on insulator (SOI) created transistor is recommended for decreasing short channel effects (SCEs) and encouraging complementary metal–oxide–semiconductor (CMOS) technology to function at the nanoscale [1–5]. After reviewing the qualities of materials that can be employed to improve the capabilities of device architectures. When FE material is combined with the gate oxide of a typical MOSFET, it exhibits the negative capacitance experience, which implies that as applied voltage rises, charge decreases, and contrary to the literature [6, 7], it is referred to as negative capacitance field effect transistors (NCFET).

Another method for lowering device leakage current is to use high-k dielectrics such as gate oxide instead of silicon oxide. Which gives immunity to the device. SOI structures have been studied using novel design architectures and technology during the past two to three decades. Double-gate, multi-gate, FINFET, NCFET, tunnel FETs, and other types of FETs are a few examples [8]. We tested a novel structure using these two techniques, in which we employed both ferroelectric and high-k dielectric materials to improve the device structure's performance. The subthreshold swing (SS) value of ferroelectric materials can be lowered beyond the limit value of 60 mV/dec thanks to the negative capacitance phenomenon. The SS value, in simple terms, is the voltage required to alter the current by the first order of scale. As a result, the amount of energy used is reduced. The work of a high-k dielectric as the gate oxide also helps to reduce leakage current in the tuned range. Other SCEs demonstrate that this factor improves performance [9, 10].

Advanced Nanoscale MOSFET Architectures: Current Trends and Future Perspectives,
First Edition. Edited by Kalyan Biswas and Angsuman Sarkar.

This article suggests a circuit innovation using a NCFET to address nonvolatile processing. A potential beyond-CMOS transistor is the NCFET [11, 12]. The NCFET has garnered interest for both logic and memory applications because of its steep switching, configurable hysteresis, and strong scalability [13]. The potential for an NCFET device to be innovative is the source of the invention in this study. A negative capacitance material layer is positioned at the gate, also referred to as a ferroelectric FET (FeFET) or an NCFET [11]. A few NCFET architectures, such as a MOSFET with an externally attached negative capacitor [14] and a MOSFET with an integrated negative capacitor at the gate have been described. Numerous ferroelectric substances, including PbTiO, BaTiO, Pb(ZrTi)O, and HfZrO, have demonstrated negative capacitance. Negative capacitance technically refers to negative differential capacitance, whose charge falls as the applied voltage rises within a particular voltage range.

NCFET (spacer) provides significant advantages over NCFETs, such as enhanced sub-threshold swing, decreased spilling current, lowered drain-induced barrier lowering (DIBL), higher exchanging %, and so on, as evidenced by the reconstructed data. Due to their adaptability in both exchanging and memory applications, Fe-FETs (ferroelectric FETs) have received a lot of interest recently [15]. These parameters are very important for the device stability and should be managed to get the device properly working. One architecture that comes into the existence appreciably eliminated SCEs is NCFET. It gives the minimum value of SS with the specialty of negative capacitance phenomenon. Due to the NC phenomenon, SS value is less than the Boltzmann's tyranny limit i.e. 60 mV/dec, and SS $\geq$60 mV/dec for the conventional NCFET, but for NCFET, this value comes below this limit, and also the DIBL value is reduced in the NCFETs.

6.2 Simulation and Structure of Device

Figure 6.1a depicts the evolution of the suggested (NCFET [spacer]) device, while (b) discusses the regular device (NCFET). With the exception of the multiple levels, size, space, and development are comparable throughout the interaction in both NCFETs and NCFET (spacer). The materials used in both layouts appear to be silicon with a height of 46 nm for the body, SiO_2 for the gate oxide at a height of 0.1 nm, aluminum with a height of 2.0 nm for the substrate, and aluminum with a height of 3 nm for the source and drain. The gate is made of metal, and a modern ferroelectric sheet made of HfO_2FE with an 8 nm height is used. In two designs, the materials are indistinguishable. The difference between an NCFET (spacer) and an NCFET, is two levels, with Spacer1/Spacer2 having a height of 3.0 nm for NCFET (spacer) alone. NCFET (spacers) are made up of semiconductors, oxide layers, insulators, metals, ferroelectrics, and metals. The Visual TCAD test system

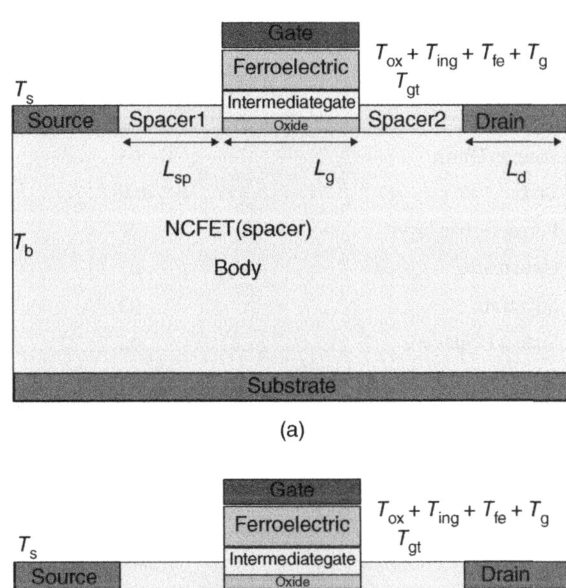

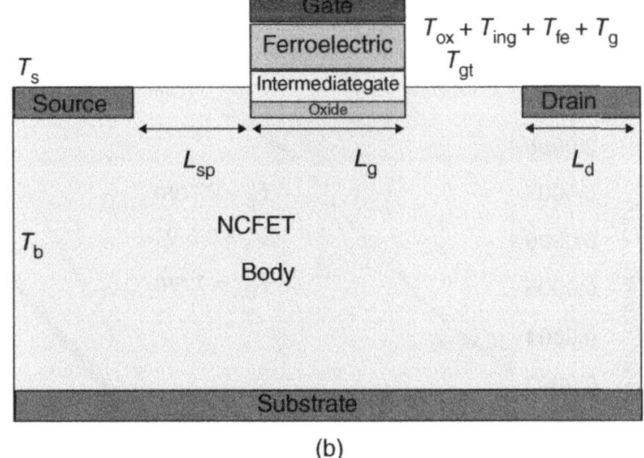

Figure 6.1 Sketch of (a) NCFETs with spacer (Proposed structure) (b) NCFETs without spacer (conventional structure).

is used to perform each of the outputs. In NCFET, Table 6.1 imitates mesh size, component, and value (spacer). NCFET (spacer) and NCFETs fixation and profile doping are revealed. Body doping in NCFET and NCFET (spacer) is consistent with 1e+16 cm^{-3} doping fixing. Table 6.1 shows the results.

6.3 Results and Analysis

The plots of gate voltage V_{gs} (V) and I_d (A) for NCFET and NCFET (spacer) (NCFETs with spacer) at drain voltage = 0.50 V are shown in Figure 6.2. At drain voltage = 0.50 V, the same figure appears in the log range. In comparison to the NCFET, the NCFET (spacer) has a lower leakage current I_{off} and a higher I_{on}.

Table 6.1 The Parameter of the proposed device (NCFET (spacer)).

Parameter's	Thickness/Lengths (nm)	Mesh's size (μm)
Source/Drain	03	0.01
Body	046	0.05
Ferroelectric layer	8	0.0008
Gate oxide	01	0.0003
Substrate	02	0.01
Spacer1/Spacer2	03	0.001
Intermediate gate	02	0.001
Gate	03	0.01

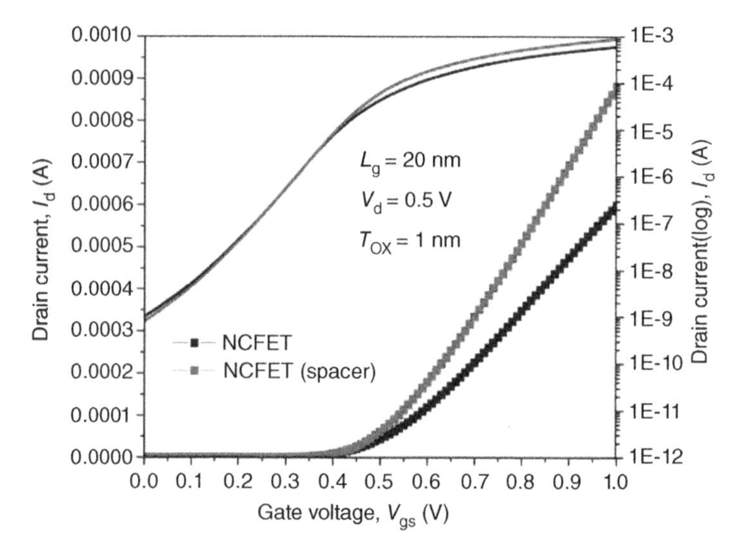

Figure 6.2 The curve of drain current and Log scale vs gate voltage for NCFET (spacer) and NCFETs.

It is a comparison of traditional (conventional) devices (NCFET) and proposed devices (NCFET (spacer)), as reflected in Figure 6.1.

The adjustment of the sideways electric field (E) travel from the gate edge to the drain edge causes a decrease in output conductance; this occurs as a result of the increase in decorating E (Figure 6.3) [16]. The diagram of transconductance and gate voltage in NCFET (spacer) and NCFET at channel voltage = 0.5 V. The layout of NCFET (spacer) is enhanced than that of NCFETs, revealing that the

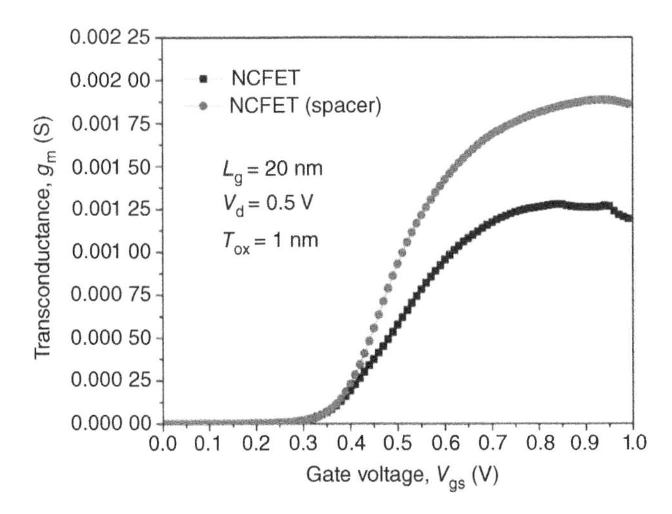

Figure 6.3 Curve of gate voltage and Transconductance for NCFET (spacer) and NCFETs.

current flow is greater than that of NCFETs. The NCFET (spacer) transconductance increase shows advancements in upgrading, electronic versatility, electronic speed, and current thickness [16–21].

$$g_m = \frac{\partial I_d}{\partial V_{gs}} \tag{6.1}$$

$$g_d = \frac{\partial I_d}{\partial V_d} \tag{6.2}$$

Figure 6.4 indicates the plot of second and third orders of transconductance (g_{m2} and g_{m3}) vs. gate voltage for NCFETs and NCFET (spacer) at $V_d = 0.50$ V. In NCFET (spacer), g_{m2} and g_{m3} are lower than in NCFETs at high voltage, revealing a shorter I_{off} (current of leakage) and improved distortion [4, 22–24].

The linearity characteristics of a device must be analyzed since they play a significant influence in all microwave and radio frequency (RF) parameters. Applications for linearity include third-order input intercept point (IIP3), third-order voltage intercept (VIP3), second-order voltage intercept (VIP2), 1-dB compression point, and third-order intermodulation distortion (IMD3) as described in [25]. The device's better input conductance, better gate biasing that reduces peak, improved distortion and noise, and improved channel region gate controllability.

$$\text{VIP2} = 4 \times \frac{g_{m1}}{g_{m2}} \tag{6.3}$$

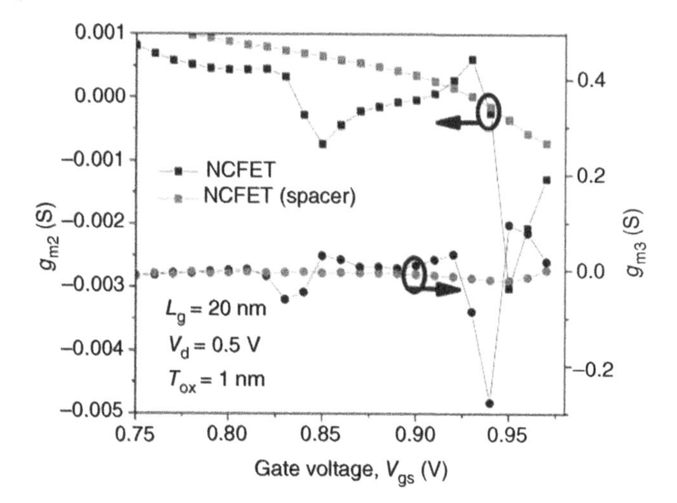

Figure 6.4 Curve of second order and third order of transconductance vs. gate voltage for NCFET (spacer) and NCFETs.

$$\text{VIP2} = \sqrt{24 \times \frac{g_{m1}}{g_{m3}}} \tag{6.4}$$

$$\text{IIP3} = \frac{2}{3} \times \frac{g_{m1}}{g_{m3} \times R_s} \tag{6.5}$$

$$\text{IMD3} = \left[\frac{9}{2} \times (\text{VIP3})^3 \times g_{m3}\right]^2 \times R_s \tag{6.6}$$

$$\text{1-db Compression Point} = 0.22\sqrt{\frac{g_m}{g_{m3}}} \tag{6.7}$$

Figure 6.5 indicates the VIP3 and VIP2 linearity parameters of NCFET (spacer) and NCFETs. In the meantime, VIP3 and VIP2 are inversely proportional with the third order of transconductance (g_{m3}) and second order of transconductance (g_{m2}), thereby decreasing the value of g_{m2} and g_{m3} and optimizing the use of linearity. Additionally, the highest VIP3 and VIP2 levels have been demonstrated. All these values are attained by using a greater NCFET (spacer), which supports the device's better linearity [26, 27].

Figure 6.6 IIP3 and 1-dB compression vs. V_{gs} for NCFET (spacer) and NCFETs. Figure 6.6 exhibits the graph of 1-dB compression and IIP3 in relation to gate bias. The third order of intercept points, known as IIP3, represent the device's input conductance and provide more gate control over the channel area of NCFET (spacer). 1-dB compression point reflects the operation of the microwave. Figure 6.7 IMD3 reveals improved gate biasing reduces IMD3 peak for NCFET (spacer) and reveals less disturbance. The improved I_{on}/I_{off} ratio, which is reflected in the lowered SS

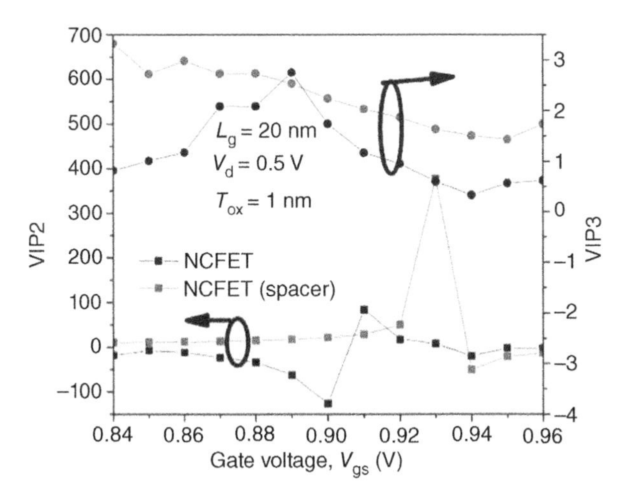

Figure 6.5 The curve between second order and third order voltage intercept point vs. gate voltage at drain voltage 0.5.

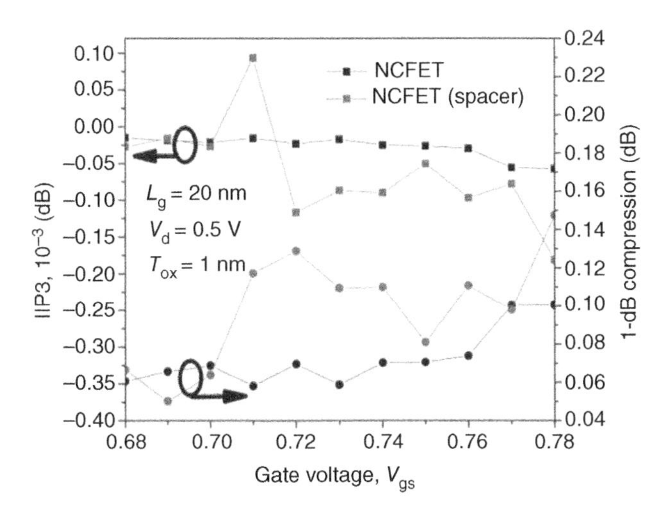

Figure 6.6 The IIP3 and 1-dB compression point for NCFETs (spacer) and NCFETs at V_d = 0.50 V are shown against gate voltage.

(swing of sub-threshold), suggests a better current of leakage (I_{off}) and a decline in DIBL in Table 6.2. NCFETs (spacer) have a lower I_{off} (current of leakage) than NCFET. All of the information was gathered using the Cogenda Visual TCAD program.

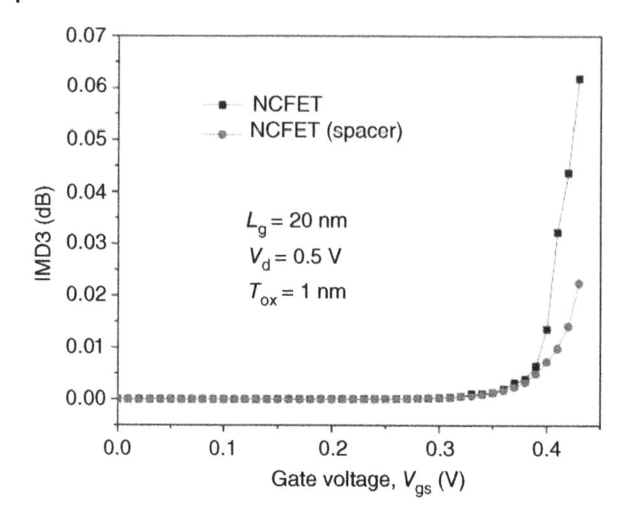

Figure 6.7 The IMD3 for NCFETs (spacer) and NCFETs at V_d = 0.50 V are shown against gate voltage.

Table 6.2 Constraint of analog and digital implementation for NCFETs and NCFET (spacer).

S. No.	Device name	V_{th}	Ioff (A/μm)	Ion (A/μm)	I_{on}/I_{off}	Subthreshold swing (mV/dec
1.	NCFET (spacer)	0.232	8.51×10^{-10}	0.000 599	1.035×10^6	148.28
2.	NCFET	0.224	01.187×10^{-9}	0.000 881	0.5048×10^6	157.29

6.4 Conclusion

In this work, the influence of a ferroelectric thick sheet (HfO_2FE) in NCFET is investigated in this paper using a Visual TCAD simulator. Meanwhile, NCFETs with spacer (NCFET (spacer)) replicate extreme performance than NCFETs, according to the theoretical result. The SS of the NCFET (spacer) is 5.7% lower than that of NCFETs. NCFET (spacer) has also decreased by 28% in DIBL compared to NCFETs. When compared to NCFETs, this device shows a 28% reduction in I_{off}. Linearity concert of the device is upgraded. The property of spacer offers numerous advantages of silicon nitride (Si_3N_4), including its ability to withstand high temperatures, minimal friction, great wear and chemical resistance, and high electrical resistivity. It has a low specific weight, a long service life, and is nonmagnetic. For the safety of people and the environment, nitride detection is essential. It was also observed that a better value of VIP3 and VIP2 is achieved for NCFET (spacer). Furthermore, it has been discovered that the NCFET (spacer) improves the IMD3, 1-dB compression point, and IIP3 points of the device. The suggested device's future research will focus on its analog, less distortion and RF performance. These results show that NCFET (spacer) is more effective than NCFETs for enhanced digital and wireless applications.

Acknowledgment

The authors thank to Nano Bioelectronics and Microelectronics Lab, Vinod Dham centre of excellence for semiconductor and Microelectronics (VDCoE4SM), Delhi Technological University for giving essential facilities.

References

1 Pathak, Y., Malhotra, B.D., and Chaujar, R. (2022). Analog/RF performance and effect of temperature on ferroelectric layer improved FET device with spacer. *Silicon* 14 (18): 1–12.
2 Pathak, Y., Malhotra, B.D., and Chaujar, R. (2022). Detection of biomolecules in dielectric modulated double metal below ferroelectric layer FET with improved sensitivity. *Journal of Materials Science: Materials in Electronics* 33 (17): 1–10.
3 Moore, G.E. (1965). *Cramming More Components onto Integrated Circuits*. New York: McGraw-Hill.
4 Mann, R. and Chaujar, R. (2021). TCAD investigation of ferroelectric based substrate MOSFET for digital application. *Silicon* 14 (9): 1–10.
5 Waldrop, M.M. (2016). More than Moore. *Nature* 530 (7589): 144–148.
6 Kobayashi, M. and Hiramoto, T. (2016). On device design for steep-slope negative-capacitance field-effect-transistor operating at sub-0.2 V supply voltage with ferroelectric HfO_2 thin film. *AIP Advances* 6 (2): 025113.
7 Pathak, Y., Malhotra, B.D., and Chaujar, R. (2021). TCAD analysis and simulation of double metal negative capacitance FET (DM NCFET). *2021 Devices for Integrated Circuit (DevIC)*, Kalyani, India (19–20 May 2021). IEEE. pp. 224–228
8 Kish, L.B. (2002). End of Moore's law: thermal (noise) death of integration in micro and nano electronics. *Physics Letters A* 305 (3–4): 144–149.
9 Moselund, K., Bouvet, D., Pott, V. et al. (2008). Punch-through impact ionization MOSFET (PIMOS): from device principle to applications. *Solid-State Electronics* 52 (9): 1336–1344.
10 Salahuddin, S. and Datta, S. (2008). Use of negative capacitance to provide voltage amplification for low power nanoscale devices. *Nano Letters* 8 (2): 405–410.

11 Khan, A.I., Yeung, C.W., Hu, C., and Salahuddin, S. (2011). Ferroelectric negative capacitance MOSFET: capacitance tuning & antiferroelectric operation. *2011 International Electron Devices Meeting*, Washington, DC, USA (5–7 December 2011). IEEE. pp. 11–3.

12 Li, X., George, S., Ma, K. et al. (2017). Advancing nonvolatile computing with nonvolatile NCFET latches and flip-flops. *IEEE Transactions on Circuits and Systems I: Regular Papers* 64 (11): 2907–2919.

13 George, S., Aziz, A., Li, X., et al. (2016). Device circuit co design of FEFET based logic for low voltage processors. *2016 IEEE Computer Society Annual Symposium on VLSI (ISVLSI)*, Pittsburgh, PA, USA (11–13 July 2016). IEEE. pp. 649–654.

14 Khan, A.I., Chatterjee, K., Duarte, J.P. et al. (2015). Negative capacitance in short-channel FinFETs externally connected to an epitaxial ferroelectric capacitor. *IEEE Electron Device Letters* 37 (1): 111–114.

15 Salvatore, G.A., Lattanzio, L., Bouvet, D. et al. (2010). Ferroelectric transistors with improved characteristics at high temperature. *Applied Physics Letters* 97 (5): 053503.

16 Gupta, N. and Chaujar, R. (2016). Optimization of high-k and gate metal workfunction for improved analog and intermodulation performance of gate stack (GS)-GEWE-SiNW MOSFET. *Superlattices and Microstructures* 97: 630–641.

17 Pradhan, K.P., Mohapatra, S.K., Sahu, P.K., and Behera, D. (2014). Impact of high-k gate dielectric on analog and RF performance of nanoscale DG-MOSFET. *Microelectronics Journal* 45 (2): 144–151.

18 Narendar, V. and Girdhardas, K.A. (2018). Surface potential modeling of graded-channel gate-stack (GCGS) high-k dielectric dual-material double-gate (DMDG) MOSFET and analog/RF performance study. *Silicon* 10 (6): 2865–2875.

19 Pathak, Y., Malhotra, B.D., and Chaujar, R. (2021). A numerical study of analog parameter of negative capacitance field effect transistor with spacer. *2021 7th International Conference on Signal Processing and Communication (ICSC)*, Noida, India (25–27 November 2021). IEEE. pp. 277–281.

20 Pathak, Y., Malhotra, B.D., and Chaujar, R. (2023). DFT based atomic modeling and Analog/RF analysis of ferroelectric HfO2 based improved FET device. *Physica Scripta* 98 (8): 085933.

21 Sharma, M. and Chaujar, R. (2022). Ultrascaled 10 nm T-gate E-mode InAlN/AlN HEMT with polarized doped buffer for high power microwave applications. *International Journal of RF and Microwave Computer-Aided Engineering* 32 (4): 23057.

22 Pahwa, G., Dutta, T., Agarwal, A., and Chauhan, Y.S. (2017). Compact model for ferroelectric negative capacitance transistor with MFIS structure. *IEEE Transactions on Electron Devices* 64 (3): 1366–1374.

23 Pathak, Y., Mishra, P., Sharma, M. et al. (2024). Experimental circuit design and TCAD analysis of ion sensitive field effect transistor (ISFET) for pH sensing. *Materials Science and Engineering: B* 299: 116951.

24 Sharma, S. and Chaujar, R. (2022). RF, linearity and intermodulation distortion analysis with small-signal parameters extraction of tunable bandgap arsenide/antimonide tunneling interfaced JLTET. *Microsystem Technologies* 28: 1–9.

25 Sharma, M. and Chaujar, R. (2021). Design and investigation of recessed-T-gate double channel HEMT with InGaN back barrier for enhanced performance. *Arabian Journal for Science and Engineering* 47: 1–8.

26 Gupta, N. and Chaujar, R. (2016). Investigation of temperature variations on analog/RF and linearity performance of stacked gate GEWE-SiNW MOSFET for improved device reliability. *Microelectronics Reliability* 64: 235–241.

27 Chaujar, R., Kaur, R., Saxena, M. et al. (2009). TCAD assessment of gate electrode workfunction engineered recessed channel (GEWE-RC) MOSFET and its multi-layered gate architecture, part II: analog and large signal performance evaluation. *Superlattices and Microstructures* 46 (4): 645–655.

7

Electrically Doped Nano Devices: A First Principle Paradigm

Debarati D. Roy[1,3], Pradipta Roy[2], and Debashis De[3,4]

[1] *Department of Electronics and Communication Engineering, B. P. Poddar Institute of Management and Technology, Kolkata, West Bengal, India*
[2] *Department of Computer Application, Dr. B. C. Roy Academy of Professional Courses, Durgapur, West Bengal, India*
[3] *Department of Computer Science and Engineering, Maulana Abul Kalam Azad University of Technology, Kolkata, India*
[4] *Department of Physics, University of Western Australia, Perth, Western Australia, Australia*

7.1 Introduction

Recent trends in nanodevice fabrication and related investigations into the characteristics of the molecular level nanoscale devices are the blooming trends toward the theoretical movement of nanoscale device fabrication. Density functional theory (DFT) and non-equilibrium Greens' function (NEGF)-based first principle approach are the key methods to investigate the nanoscale device designing method at the molecular level. This analytical approach is again flourishing with the electrical doping procedure. This method of doping helps to make a fundamental approach to design empirical models of nanoscale devices. Ultra-low device stress, satisfactory high operating frequency, and high device density make these electrically doped nanodevices as a raising research area in the nano semi-conducting field to suppress the conventionally doped semiconductor circuit. Molecular devices based on either organic or inorganic materials make this approach successful for the researchers. Moreover, using this doping approach, fundamental electronic devices can be theoretically designed, and their characteristics are also investigated. For example, diode, FET, or the switch can be designed using first-principle formalisms. Furthermore, the various quantum-ballistic transport phenomena and quantum-mechanical characteristics are also investigated for these nanoscale devices. Besides, these devices' bio-molecular quantum cellular automata (QCA) and bio-molecular

Advanced Nanoscale MOSFET Architectures: Current Trends and Future Perspectives,
First Edition. Edited by Kalyan Biswas and Angsuman Sarkar.

nanotubes are also efficiently designed at room temperature. The room temperature operation and high circuit density of these nanoscale devices pledge the energy-efficient design of the logic circuit at a nanoscale level. On the other hand, the requirement of a large number of logic gates and disadvantageous garbage outputs may limit the functionality of this electrically doped analytical model approach to some extent. In this chapter, a brief overview of some related works based on the first principle approach, the electrical doping process, DFT, NEGF, and their formalisms, along with computational methodology, is outlined.

Since the foundation of DFT and NEGF-based first principle theory, the design of analytical nano-scale devices has become easier. This first principle approach has made a theoretical study and characterization of nanoscale device designing as a strong alternative to conventional ones. The electrically doped analytical nano-scale molecular electronic device design method has attracted significant attention as a first principle approach-based powerful future substitute for conventional semiconductor electronic devices. When the size of molecular devices becomes close to the order of de Broglie wavelength of electrons, the quantum-confinement effect and quantum-ballistic transmission affect the electronic characteristics of these nanoscale molecular devices. This raises expectations for many noble applications in nano-electronics technology. As the nanotechnology advances, more and more versatile, noble molecular devices are reported using both theoretical and practical approaches. This provides a good platform to study and investigate various electronic properties of low-dimensional devices and systems. Among these nanoscale devices, we choose to design, study, and investigate the characteristics of organic (adenine, thymine, cytosine, and guanine) and in-organic (GaAs) molecular devices (Diode, FET, Switch, BMQCA, Logic gates).

Miniaturization is a key topic in the field of semiconductor technology. Traditional electronic devices have been manufactured using in-organic semi-conductors. But nanoengineering has made a smooth bridge between bio-molecules and in-organic materials. Conjugation of in-organic materials with bio-molecules attracts researchers towards the formation of nanobiotechnology. Most of the researchers as well as physicists are interested in this collaboration of bio-molecules and in-organic materials. This approach plays a crucial role in various fields, for example, drug delivery system, design of mathematical tools, device diagnosis at the molecular level, physical therapy, and contrast of reagents [1]. NEGF and DFT together have made a wide platform for the researchers to design plenty of molecular devices. Downscaling of nanodevices has been done to achieve a higher degree of performance and faster-operating speed. This downscaling has been carried out using molecular device designing. Nowadays, bio-molecules show great performance in nanodevice designing. Even when these bio-molecules combine with in-organic molecules then also they exhibit great performance in the semi-conductor nanotechnology domain. These bio-molecules are integrated according to minimum stress, optimum atomic-level force, minimization of stress-induced leakage current, etc. [2].

The main concern of these molecular devices is to achieve maximum tunneling current with minimum stress and optimum force level. The tunneling current is mainly considered as quantum-ballistic tunneling current. This work is a theoretical framework for the fundamental atomic-scale theory of molecules or atoms, especially in relation to current flow. These theoretical investigations are supported by the most advanced concepts of DFT and NEGF-based first principle approach. These formalisms mainly provide information on quantum transport mechanisms at the atomic scale. This conceptual framework emphasizes electronic properties, especially the tunneling current flow through the molecular channel of the nano dimension devices [3]. In this simulation work, we have proposed an electrical doping concept to avoid the basic problems related to the conventional doping process. This process has been implemented by creating potential differences between the two ends of the device. These two ends are generally described as the electrodes. This approach is generally known as the two-probe experiments. By introducing electrical doping, the possibilities for structural defects and physical ionization heating effect can be partially removed. The main objective of the proposed work is to theoretically design and investigate quantum electronic characteristics of various electronic molecular devices using bio-molecules and GaAs semiconductors. At the same time, we must investigate whether the molecular device is sustainable at room temperature without losing its electronic properties. Another aim of this molecular device design is to get a high operating frequency.

This chapter presents the theoretical background for the analysis of nanoscale molecular electronic devices using the electrical doping process. The first principle approach is generally a conjugated form of DFT and NEGF. The major areas of discussion in this chapter are DFT, NEGF, electrical doping process, etc. Several molecular electronics nano-devices can be designed using the first principle approach. Though these devices are based on the analytical or theoretical study, this can be a future aspect of designing various molecular electronic devices. These molecules are either organic or inorganic types. Moreover, in this approach, we used a master-equation-based approach to incorporate electronic correlations beyond the mean-field-approximation of DFT. Here we demonstrate and survey the important theoretical tools that are needed for the master equation, which is derived from DFT. It is emphasized that these molecular layers of these devices are interacting in nature and thus help to be considered for this type of master-equation approach. Firstly, we will discuss a brief overview of the electrical doping process.

Secondly, a general discussion will be held on the first principle approach comprises of DFT and NEGF formalisms along with electrical doping process. Finally, the computational methods and molecular simulation process are discussed briefly.

7.2 Electrical Doping

Electrical doping property is one of the unique properties that help the researchers to dope the organic–metal or organic–semiconductor or organic–organic interfaces at the molecular level. Implementation of the electrical doping process generated by carrier injection into the thin film can be improved for the molecular device. By introducing this doping process, numerous glitches that are related to diminishing tunneling current can be avoided [4]. The tunnel current is affected due to carrier-back scattering effect at the molecular level.

Nowadays, nano dimension device designing is a thought-provoking facet for researchers. Diodes, transistors, logic gates, and many other nanoscale devices are implemented at the molecular level. There is additional choice for researchers for the implementation of nano-bio-inspired semiconductor devices at the molecular level. Few of these bio-molecular devices have already been introduced in the field of bio-medicine. The theoretical strategy of these nanodevices can be implemented using the Atomistix-tool Kit and Virtual Nano Laboratory (ATK-VNL)-based quantum-wise software simulator version 12.8.0 [5–11]. Even more, QCA logic can be theoretically implemented using DFT and NEGF-based first principle approach [12]. Several logic gates can be made designed using bio-molecules, and the results obtained from these theoretical implications are also validated using Multi-Sim, SPICE, or other simulation techniques [5]. The electrical doping process is a significant amendment that is introduced to obtain optimal current. Tunnel current through the molecular channel is affected by various factors, like the back-scattering effect. By implementing this doping process, we can avoid the problems related to the conventional doping process. The Schottky barrier tuning of dipole combination model is also suggested at the metal–semiconductor interface at the molecular level [13]. The first principle approach is also applicable for magnetic tunnel junction and their quantum electronic properties have been analyzed [14]. To calculate leakage current through SiO2 and Si$O_x N_y$-based MOSFET, researchers used DFT and NEGF-based first principle approach [15]. This ab initio modeling is applied for representation of Schottky barrier height tuning using the Yttrium and Nickel Silicide atomic-scale interfaces [16]. Direct band-to-band tunneling in reverse-biased MOS_2 p-n junction nanoribbons can be described using DFT and NEGF [17]. The effect of the incorporation of opposite polarities dopant

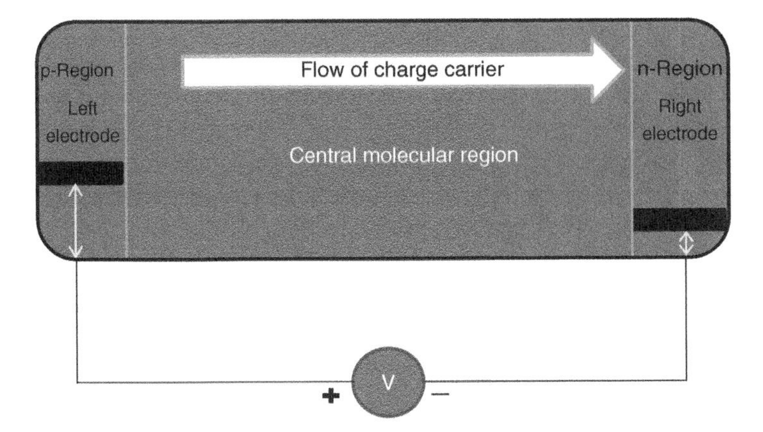

Figure 7.1 Schematic diagram of the conceptual electrical doping process.

atoms into the nanowire exhibits electrical properties like Zener diode [18]. The dual-spin filtering effect can be seen in the half-metallic Yattrium nitrite YN2 [19]. The detection process of various gases using a bio-molecular heterojunction chain through multilayer GaAs nanopores has been implemented using DFT and NEGF [20]. Using this theoretical approach, an electrically doped bio-molecular switch is designed when using single-wall carbon nanotube (SWCNT) as electrodes [21]. NEGF formalisms help to design graphene-based anti-dot resonant tunnel diodes [22]. Atomistic characteristics of two-dimensional silicon p-n junctions have been demonstrated using the first principle approach [23]. The graphical representation of the electrical doping process is shown in Figure 7.1 [24]. Diodes and transistors are the basic building blocks of any electronic circuit. Logic gates can also be implemented using diodes and transistors. Therefore, any logic can be implemented using first principle formalisms.

The various quantum-electronic properties are therefore investigated. The device has been investigated using two probe experiments. The two electrodes are used to find out the results. The simulation parameters have been chosen according to the desired outcome. The geometrically optimized nanoscale devices are generally designed using adenine, guanine, cytosine, thymine bio-molecules, and their combinations. Another important crystallographic atomic device has been made using GaAs crystal at the atomic scale. The representation of this analytical model is derived from a combination of DFT and NEGF formalisms. The quantum-ballistic transmission properties have been investigated for the devices. Depending on the nature of this transmission, various electronic properties have been studied. This ATK is a powerful software simulation tool that is used to implement various nanoscale devices analytically. The calculations for DFT have been performed using generalized gradient approximation (GGA) along with Perdew–Burke–Emzerhof single or double-zeta polarization approaches. The mesh cut-off density has been chosen at 150 Ry (maximum). The k-point sampling is taken as $1 \times 1 \times 100$. That means the maximum sampling point has been taken along the z-direction. The Brillouin zone integration is performed using the Monkhorst-Pack k-point grid. The quantum-ballistic transmission has occurred along the z-direction for these 2-D analytical devices, so the maximum numbers of samples have been taken along the z-direction.

7.3 First Principle

In recent trends in nanotechnology, the investigation of quantum transport through mesoscopic and molecular-scale systems is one of the most challenging and interesting aspects for scientists. This quantum transport at the molecular level is generally divided into two approaches. One is stationary, and another is a time-dependent phenomenon. For the last few decades, DFT and NEGF formalisms have been the dominant approaches for the quantum mechanical simulation process. The quantum chemical scientists adopt these approaches to simulate the energy surfaces in molecules. This chapter introduces the basic concepts of DFT and NEGF theory and outlines the basic mathematical concepts of DFT and NEGF formalisms. A short overview will be given of the electrical doping process and the various analytical design approaches to molecular-level nanodevices. Electrical oping is one of the most fabulous approaches where no external impurity is provided to the main molecular nanosheet or wire. This doping method can be adopted for low-dimension nanoscale device design at ultra-low temperatures. DFT is one of the major tools to calculate molecular structure efficiently. DFT- and NEGF-based first principle approach is used to calculate analytical model representation of the nanodevices at the molecular level. In-organic crystalline, and amorphous structures of the materials, as well as their bio-molecular structure, can be analyzed with the help of the first principle approach. The quantum-ballistic transport phenomenon by the active electrical carriers has been illustrated using the first principle method. Real-space, NEGF formalisms, and spin-polarization method-based DFT methods can be illustrated using the ATK-VNL software simulation package. By introducing these formalisms, a molecular device has been divided into three main portions, such as left and right electrodes and central molecular region. The analytical calculations for two electrodes are performed using the sampling of the Brillouin zone integration method along with Monkhorst-pack parameters with regular k-point sampling, i.e. $1 \times 1 \times 100$. Double-zeta or single-zeta polarization methods have been taken into account. Self-consistent calculations play an important role in DFT. High mesh cut-off density gives more accurate results for this first principle approach [25].

7.3.1 DFT

Since the last 30 years of downscaling in the semiconductor industry, DFT has played an important and obvious role in observing quantum-ballistic transport phenomenon. These formalisms are widely adopted by the researchers to discover the energies in molecules. The main associated theorem with this formalism is Hohenberg–Kohn theorems demonstrated in 1964. The first theorem states that "*The electron density establishes the exterior potential (to within an additive constant)*". If this declaration is accurate, then it follows the rule of "*electron density uniquely determines the Hamiltonian operator*". Hohenberg and Kohn directly prove this theorem and give a generalized solution for the inclusion of systems along with de-generate states [26]. There are several steps associated with DFT formalisms. These steps are as follows:

- By solving and finding the solutions for the Schrödinger Equation.
- The Hohenberg–Kohn Theorems.
- By finding functional energy.
- The local density approximation.

The generalized gradient approximation

- Meta-GGA functional.
- Hybrid exchange functional.
- The performance of various functional [26].

DFT is an extremely useful tool to design nanoscale electronic devices. For the last 50 years, DFT has dominated the quantum-mechanical and quantum-ballistic simulation of the periodic systems. It can also be used to calculate the surface energy of the nanoscale molecular devices. In this chapter, we introduce and outline the basic concepts and features of DFT. Therefore, modern signs of progress in exchange-correlation functionals are introduced in this chapter. The main formula for DFT is derived from the solution of the Schrödinger equation. The energy can be calculated by solving Schrödinger equation on a time-independent, non-relativistic Born–Oppenheimer approximation platform in Eq. (7.1).

$$H \dagger (r1r2 \cdots rn) = E \dagger (r1r2 \cdots rn) \tag{7.1}$$

The Hamiltonian operator (H) is a summation of three individuals, for example, the kinetic energy, the interactive part of the external potential (Vext), and the interaction between two electrons (Vee). The Vext which is the force that is acting between atomic nuclei and electron plays an important role in the simulation of materials which is shown in Eq. (7.2).

$$V_{\text{ext}} = \sum_{\alpha}^{\text{Nat}} |Z_\alpha / r_i - R_\alpha| \tag{7.2}$$

In Eq. (7.2), *ri* is the electron coordinate for *i* and *Rα* is the charge on the nucleus at *Zα*. Equation (7.1) is solved for a set of Ψ subject to the restraint that the Ψ are anti-symmetric and the sign of electrons is opposite if their coordinates are exchanged. The lowest energy Eigen-value, *E*0, is the ground state energy, and the probability density of detecting an electron with any specific set of coordinates {*ri*} is $|\Psi 0|^2$. The average energy can be formulated as shown in Eq. (7.3).

$$E(\dagger) = \int \dagger^* H \dagger \, \mathrm{d}r \tag{7.3}$$

Ψ describes the wave function of the functionals. Ψ0 is the ground state wave function which is related to energy as stated in Eq. (7.4)

$$E(\dagger) \geq E0 \tag{7.4}$$

Hartree–Fock theory can also be implemented for Ψ which is assumed to be a product of *I* which is obtained for a single electron coordinate system, which is described in Eq. (7.5).

$$I_{\text{HF}} = \frac{1}{\sqrt{N!}} \det [\emptyset_1 \emptyset_2 \cdots \emptyset_N] \tag{7.5}$$

The Hartree–Fock equations explain electrons that are non-interacting under the pressure of an average field potential in the circumstances of classical Coulomb potential and a non-local exchange potential [26–29]. The year-wise evolution of DFT is shown in Figure 7.2 [30].

Moreover, it can be formulated for many-body complex calculations of electronic structure where nuclei of the treated molecules are assumed to be fixed with an external static potential *V* and the wave function is Ψ. Therefore, after solving the time-independent Schrodinger equation, the formalisms of DFT are implemented. DFT is often described as the ab initio method by the researchers. Therefore, for *N*-electron system, DFT can be formulated as shown in Eq. (7.6), where *H* is Hamiltonian, *U* is the energy between two electrons, *E* is the total energy for kinetic energy (*T*) and potential energy (*V*), which are received from the exterior field because of positively charged nuclei.

$$\text{HT} = [T + V + U]T \tag{7.6}$$

Now *T* can be represented as the equation which is given in Eq. (7.7)

$$T \Xi^1 \int T^*(r) \cdot T(r) \mathrm{d}r \tag{7.7}$$

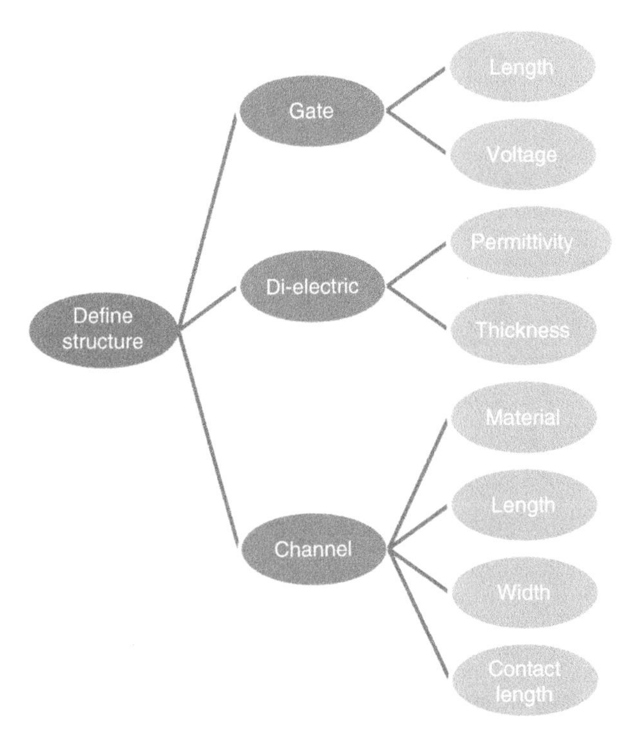

Figure 7.2 Working flow chart diagram of Quantumwise ATK-VNL step-1 (pre-simulation).

V is shown in Eq. (7.8).

$$V \Xi \int v(r)T^*(r)T(r)\mathrm{d}r \tag{7.8}$$

U can be formulated as shown in Eq. (7.9)

$$U \Xi^1 \int^1 T^*(r)T^*(r')T(r')T(r)\mathrm{d}r\mathrm{d}r' \tag{7.9}$$

In the context of computational material science and chemistry, DFT plays a crucial role in allowing the prediction and computation of material behavior based on quantum mechanical considerations. It does not require higher-order fundamental material properties as its parameters. Whereas it is the technique to calculate the many-body electronic system and also evaluate potentials that act on the system. The origin of DFT is Hohenberg–Kohn theorem, which was formulated by Walter Kohn and Pierre Hohenberg in the era of the 1970s [31]. Generally, DFT has its application in the fields of molecular chemistry and materials science to determine and predict the complex system activities on the atomic scale. Particularly, DFT is a computational method that is implemented to

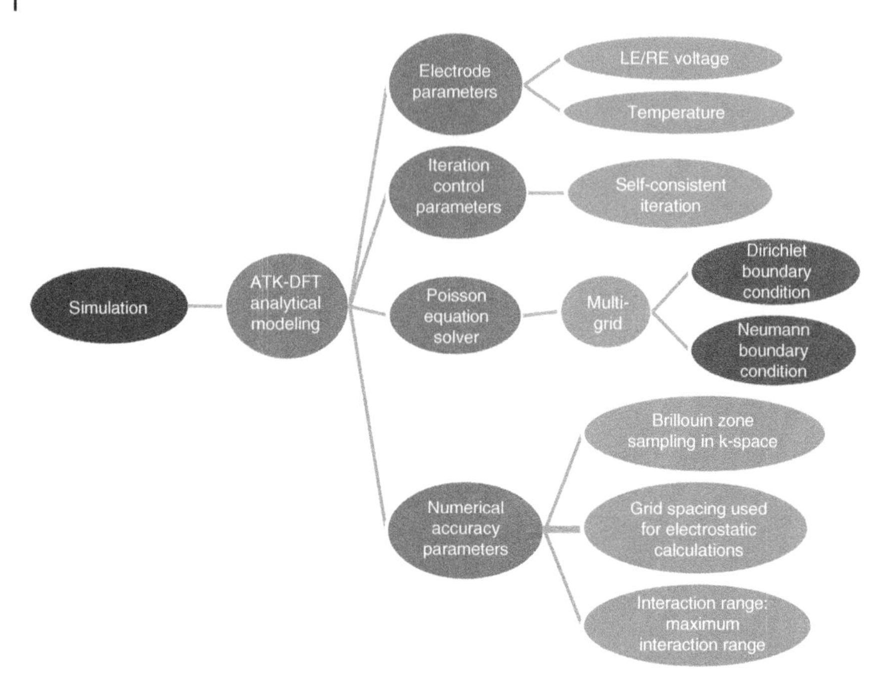

Figure 7.3 Working flow chart diagram of Quantumwise ATK-VNL step-2 (simulation).

synthesize many-body complex systems. Figure 7.3 gives a clear view of solving the path for the Schrodinger equation. For example, to emphasize the effect of dopants on phase alteration performance in oxides and magnetic actions in dilute magnetic semiconductor materials, study and investigate the magnetic and electro-mechanical properties of semiconductor materials [32–45]. Figure 7.4 gives a glance to DFT [37].

7.3.2 NEGF

Investigation and design of the mesoscopic and nano-scale systems are some of the interesting as well as challenging topics for researchers nowadays. The quantum-ballistic transport phenomenon is the key feature for molecular modeling. NEGF formalism is extensively used to demonstrate quantum transmission. This formalism was again used to solve the time-dependent Schrödinger equation. This is used to study static and time-dependent transport phenomena for nanoscale devices. This theory is also known as Keldysh formalism. The other equilibrium theory is different from the NEGF formalism. In the case of NEGF, all the time-dependent functions are described for time arguments as a contour plot, which is called a Keldysh contour. In this formalism, perturbation theory

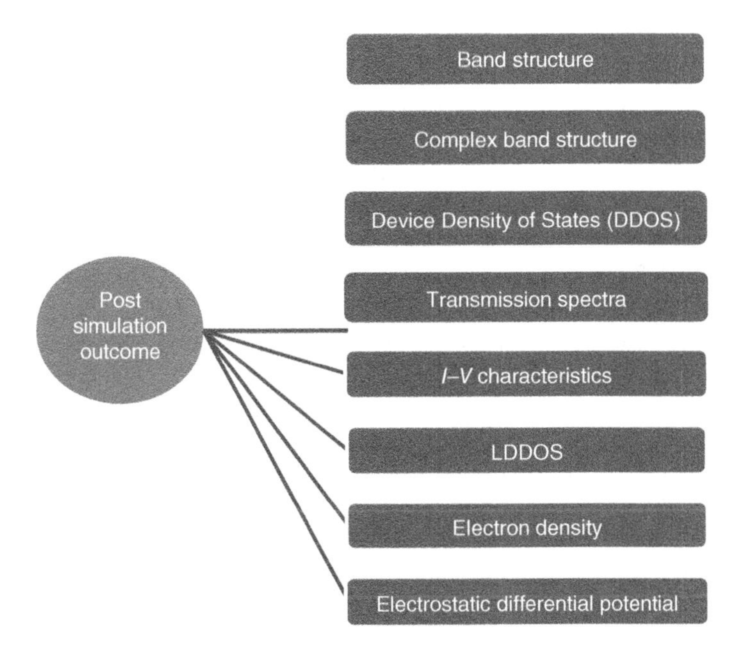

Figure 7.4 Working flow chart diagram of Quantumwise ATK-VNL step-3 (post-simulation).

also plays an important role. There are several parts that are associated to form NEGF formalism, as mentioned below:

- Contour-ordered Green's function
- Keldysh contour
- Analytic continuation: Langreth theorem
- Keldysh formulation
- Application to steady-state transport
- Time-dependent transport [38]

The quantum-ballistic transport phenomenon can be satisfactorily explained with the help of NEGF. It is well described for the mesoscopic and nanoscale systems. These nanoscale and mesoscopic systems are the thrust area for the researchers nowadays. This low-dimension quantum-ballistic transmission phenomenon is to be divided into two parts, such as stationery and time-dependent occurrences. NEGF is also known as Keldysh formalism, which is extensively used to describe the quantum transport phenomenon in the nanoscale regime. NEGF helps to investigate and predict the interaction of particles within a time-dependent many-body system. Time-dependent current–voltage (I–V) characteristics can be eventually solved by using NEGF formalisms, which are divided into two basic phases like (i) static and (ii) time-dependent electronic transport in mesoscopic systems. The main difference between ordinary equilibrium theory and NEGF formalisms is that all time-dependent functions are determined for time-arguments on a contour, which is called the Keldysh contour. To create equilibrium perturbation theory, it is to be considered that the complex system should return to its first state as $t \to +\infty$, but in NEGF theory, this may not work as the initial state at $t = -\infty$ can be dissimilar from the final state $t = +\infty$. Therefore, in the NEGF formalisms, any reference to large times should be avoided.

In the case of NEGF formalisms, a nanodevice is theoretically divided into three parts, namely, right electrode, left electrode, and central molecular region. Figure 7.1 shows the fundamentals of the molecular modeling approach [39]. The Fermi levels for right and left electrodes are $EF(R)$ and $EF(L)$, respectively. A potential drop is generated when an equal and opposite bias voltage (Vd) applied at the two ends of the electrodes. The amount of potential drop into the central molecular region is calculated using Eq. (7.10).

$$E_F(R) - E_F(L) = qV_d \tag{7.10}$$

The left and right electrodes and the central molecular region are modeled in Figure 7.1, which approximates the bulk characteristics of the material used in them, while the central region is modeled by its density of states $D(E)$, dependent on dimensions, position x, and self-consistent effective potential Veff [11, 24].

7.4 Molecular Simulation

The complete algorithm for the nanoscale device simulation is as follows:

1. Start with an initial guess for $V^{\text{eff}}(r)$.
2. For each Eigen energy, calculate G for retarded Green's function is shown in Eq. (7.11) and spectral function in Eq. (7.12), where $\sum 1$, $\sum 2$ are the left and right self-energies and $\Gamma 1$ is the broadening function.

$$G = 1/\left[\text{EI} - H - \sum \Gamma_1 - \sum \Gamma_2\right] \tag{7.11}$$

$$A1 = G\Gamma_1 G^{\text{H}} A1 = G\Gamma_1 G^{\text{H}} \tag{7.12}$$

3. Determine the density matrix $\rho 1(x)$ using Eq. (7.13).

$$\rho_1(r) = \int 2 \times 2\text{d}EA_1 f(E_{\text{F}} - E) \tag{7.13}$$

4. Solve Poisson's equation for $V^{\text{eff}}(r)$ based on $\rho 1(r)$. If $V^{\text{eff}}(r)$ is no longer varying (aka. $V^{\text{eff}}(r) - V^{\text{eff}}(r) < V^{\text{eff}}(r)$), the calculation is self-consistent, proceed to 5. *old*

Otherwise, go to 2.

$$T(E) = \text{trace}\,[\Gamma_1 G \Gamma_2 G^{\mathrm{H}}] \tag{7.14}$$

$$I = \int_{h}^{2q} T(E)[f(E - E_{\mathrm{F}}(R)) - f(E - E(L))]\mathrm{d}E \tag{7.15}$$

Calculate various properties of the system, such as transmission co-efficient, and current using Eqs. (7.14) and (7.15) respectively. The formula for current is identical to the one that is derived from the Landauer formula [24].

The nanodevices are analytically designed and simulated using quantum-wise software simulator version 13.8.0. The various quantum-electronic properties are therefore investigated. The device has been investigated using two probe experiments. The two electrodes are used to find out the results.

The simulation parameters have been chosen according to the desired outcome. The geometrically optimized nanoscale devices are generally designed using adenine, guanine, cytosine, thymine bio-molecules, and their combinations. Another important crystallographic atomic device has been made using GaAs crystals at the molecular level. To design this analytical design, a combination of DFT and NEGF formalisms has been implemented. The quantum-ballistic transmission properties have been investigated for the devices.

Depending on the nature of this transmission, various electronic properties have been studied. This ATK is a powerful software simulation tool that is used to implement various nanoscale devices analytically. The calculations for DFT have been performed using GGA along with Perdew–Burke–Emzerhof single or double-zeta polarization approaches. The mesh cut-off density has been chosen at 150 Ry (maximum). The k-point sampling is taken as $1 \times 1 \times 100$. That means the maximum sampling point has been taken along the z-direction. The Brillouin zone integration is performed using a Monkhorst–Pack k-point grid. The quantum-ballistic transmission has occurred along z-direction for these 2-D analytical devices, so the maximum numbers of samples have been taken along the z-direction. The flow chart with different modules that are used to design these nanoscale devices is shown in Figures 7.2–7.4 respectively [11, 24].

7.5 Conclusion

The major plus point of theoretical nanoscale device designing is that it works based on the computational methodology. Therefore, no such conventional doping process needs to implement this type of design methodology. The electrical doping process, using the first principle approach, abolishes the adverse effects of the conventional doping phenomenon. Furthermore, the quantum-ballistic transport phenomenon helps to characterize the devices. Besides this, depending on the characterization of these nanoscale devices, a comparative study can be made, which is important to implement further modifications of these devices. Moreover, geometrically stable devices can be modeled at a minimum stress level, which signifies further the stability of the devices. The atomistic design tools make it happened that various properties of these devices can be observed using different design rule platforms. For example, using DFT, extended Hückel theory, FFT, one can design the nanoscale devices. Even more, various properties of these nanoscale devices, for example, band structure, complex band structure; DDOS, transmission spectra, and *I*–*V* characteristics, can be investigated, and modification can also be imposed where necessary. The key features of these molecular structures can also be extracted with the help of post-simulation features. For example, like CNT, bio-molecular nanotubes can also be analytically designed and their key features extracted for further studies. Moreover, an approach is taken to design hetero-junction bio-molecular nanotubes, and their characteristics are also investigated. Lastly, the future aspect of this nanoscale device designing is to implement these molecular devices with their corresponding circuit-level simulation approach.

References

1 Nikfar, Z. and Shariatinia, Z. (2017). Phosphate functionalized (4,4)-armchair CNTs as novel drug delivery systems for alendronate and etidronate anti-osteoporosis drugs. *Journal of Molecular Graphics and Modelling (Elsevier)* 76: 86–105.

2 Nadimi, E., Planitz, P., Ottking, R. et al. (2010). Single and multiple oxygen vacancies in ultrathin SiO2 gate dielectric and their influence on the leakage current: an ab initio investigation. *IEEE Electron Device Letters* 31 (8): 881–883.

3 Datta, S. (2005). *Quantum Transport: Atom to Transistor*. Cambridge University Press.

4 Gao, W. and Kahn, A. (2003). Electrical doping: the impact on interfaces of π-conjugated molecular films. *Journal of Physics: Condensed Matter* 15 (38): S2757.

5 Dey, D. and De, D. (2018). A first principle approach toward circuit level modeling of electrically doped gated diode from single wall thymine nanotube-like structure. *Microsystem Technologies (Springer)* 24 (7): 3107–3121.

6 Dai, X., Zhang, L., Li, J., and Li, H. (2017). Metal–semiconductor transition of single-wall armchair boron nanotubes induced by atomic depression. *The Journal of Physical Chemistry C* 121 (46): 26096–26101.

7 Dey, D., Roy, P., and De, D. (2017). Atomic scale modeling of electrically doped pin FET from adenine based single wall nanotube. *Journal of Molecular Graphics and Modelling (Elsevier)* 76: 118–127.

8 Harada, N., Jippo, H., and Sato, S. (2017). Theoretical study on high-frequency graphene-nanoribbon heterojunction backward diode. *Applied Physics Express* 10 (7): 074001.

9 Wang, S., Wei, M.Z., Hu, G.C. et al. (2017). Mechanisms of the odd-even effect and its reversal in rectifying performance of ferrocenyl-*n*-alkanethiolate molecular diodes. *Organic Electronics* 49: 76–84.

10 Dey, D., Roy, P., Purkayastha, T., and De, D. (2016). A first principle approach to design gatedpin nanodiode. *Journal of Nano Research* 36: 16–30.

11 Dey, D., Roy, P., and De, D. (2016). Electronic characterisation of atomistic modelling based electrically doped nano bio pin FET. *IET Computers and Digital Techniques* 10 (5): 273–285.

12 Dey, D., Roy, P., and De, D. (2017). Design and electronic characterization of bio-molecular QCA: a first principle approach. *Journal of Nano Research* 49: 202–214.

13 Geng, L., Magyari-Kope, B., and Nishi, Y. (2009). Image charge and dipole combination model for the schottky barrier tuning at the dopant segregated metal/semiconductor interface. *IEEE Electron Device Letters* 30 (9): 963–965.

14 Chakraverty, M., Kittur, H.M., and Kumar, P.A. (2013). First principle simulations of various magnetic tunnel junctions for applications in magnetoresistive random access memories. *IEEE Transactions on Nanotechnology* 12 (6): 971–977.

15 Nadimi, E., Planitz, P., Ottking, R. et al. (2010). First principle calculation of the leakage current through SiO2 and Si*O*x*N*y gate dielectrics in MOSFETs. *IEEE Transactions on Electron Devices* 57 (3): 690–695.

16 Geng, L., Magyari-Kope, B., Zhang, Z., and Nishi, Y. (2008). Ab initio modeling of Schottky-barrier height tuning by yttrium at nickel silicide/silicon interface. *IEEE Electron Device Letters* 29 (7): 746–749.

17 Ghosh, R.K. and Mahapatra, S. (2012). Direct band-to-band tunneling in reverse biased MoS2 nanoribbon pn junctions. *IEEE Transactions on Electron Devices* 60 (1): 274–279.

18 Chakraverty, M., Harisankar, P.S., Gupta, K. et al. (2016). Simulation of electrical characteristics of silicon and germanium nanowires progressively doped to zener diode configuration using first principle calculations. *Microelectronics, Electromagnetics and Telecommunications (Springer)* 372 (Lecture Notes in Electrical Engineering),: 421–428.

19 Li, J., Gao, G., Min, Y., and Yao, K. (2016). Half-metallic YN_2 monolayer: dual spin filtering, dual spin diode and spin Seebeck effects. *Physical Chemistry Chemical Physics* 18 (40): 28018–28023.

20 Dey, D., Roy, P., and De, D. (2017). Detection of ammonia and phosphine gas using heterojunction biomolecular chain with multilayer GaAs nanopore electrode. *Journal of Nanostructures* 7 (1): 21–31.

21 Dey, D. and De, D. (2018). Electrically doped adenine based optical bio molecular pin switch with single walled carbon nanotube electrodes. *Journal of Active & Passive Electronic Devices* 13 (2–3): 107–118.

22 Palla, P., Ethiraj, A.S., and Raina, J.P. (2016). Resonant tunneling diode based on band gap engineered graphene antidot structures. *2nd International Conference on Emerging Technologies: Micro to Nano 2015. AIP Conference Proceedings*, Rajasthan, India (24–25 October 2015), vol. 1724. AIP Publishing. pp. 020069.

23 Tabe, M., Tan, H.N., Mizuno, T. et al. (2016). Atomistic nature in band-to-band tunneling in two-dimensional silicon pn tunnel diodes. *Applied Physics Letters* 108 (9): 093502.

24 Krotnev, I. (2013). Novel metallic field-effect transistors. Doctoral dissertation. University of Toronto.

25 Chauhan, S.S., Srivastava, P., and Shrivastava, A.K. (2014). Electronic and transport properties of boron and nitrogen doped graphene nanoribbons: an ab initio approach. *Applied Nanoscience* 4 (4): 461–467.

26 Harrison, N.M. (2003). An introduction to density functional theory. *Nato Science Series Sub Series III Computer and Systems Sciences* 187: 45–70.

27 Parr, R.G. (1980). Density functional theory of atoms and molecules. In: *Horizons of Quantum Chemistry. International Academy of Quantum Molecular Science*, vol. 3 (ed. K. Fukui and B. Pullman). Dordrecht: Springer https://doi.org/10.1007/978-94-009-9027-2_2.

28 March, N.H. (1999). *Electron Correlation in the Solid State*. London: Imperial College Press.

29 Callaway, J. and March, N.H. (1984). Density functional methods: theory and applications. *Solid State Physics* 38: 135–221.

30 https://www.eurekalert.org/pub_releases/2017-01/nioo-dft122916.php, the Date of Accession is 18th January, 2023.

31 Hohenberg, P. and Kohn, W. (1964). Inhomogeneous electron gas. *Physical Review* 136 (3B): B864–B871.

32 Roy, D.D. and De, D. (2023). Predicting model of I–V characteristics of quantum-confined GaAs nanotube: a machine learning and DFT-based combined framework. *Journal of Computational Electronics* 22 (4): 999–1009.

33 Roy, D.D., Roy, P., and De, D. (2023). Machine learning and DFT-based combined framework for predicting transmission spectra of quantum-confined bio-molecular nanotube. *Journal of Molecular Modeling* 29 (11): 338.

34 Rastegar, S.F., Hadipour, N.L., Tabar, M.B., and Soleymanabadi, H. (2013). DFT studies of acrolein molecule adsorption on pristine and Al-doped graphenes. *Journal of Molecular Modeling (Springer)* 19 (9): 3733–3740.

35 Dey Roy, D., Roy, P., and De, D. (2023). Bio-molecular nano scale devices using first principle paradigm: a comprehensive survey. *International Journal of Nano Dimension* 14 (2): 115–125.

36 https://www.bc.edu/content/dam/bc1/schools/mcas/physics/pdf/wien2k/DFT%20and%20APW.pdf, the Date of Accession is 18th January, 2023.

37 https://www.assignmentpoint.com/science/chemistry/dft-density-functional-theory.html, the Date of Accession is 18th January, 2023.

38 Datta, S. (2002). The non-equilibrium Green's function (NEGF) formalism: an elementary introduction. *Digest. International Electron Devices Meeting*, San Francisco, CA, USA (8–11 December 2002). IEEE. pp. 703–706.

39 Biswas, K., Sarkar, A., and Sarkar, C.K. (2016). Impact of Fin width scaling on RF/Analog performance of junctionless accumulation-mode bulk Fin-FET. *ACM Journal on Emerging Technologies in Computing Systems (JETC)* 12 (4): 1–12.

40 Baral, B., Das, A.K., De, D., and Sarkar, A. (2016). An analytical model of triple-material double-gate metal–oxide–semiconductor field-effect transistor to suppress short-channel effects. *International Journal of Numerical Modelling: Electronic Networks, Devices and Fields* 29 (1): 47–62.

41 Ghoshhajra, R., Biswas, K., and Sarkar, A. (2021). A review on machine learning approaches for predicting the effect of device parameters on performance of nanoscale MOSFETs. *2021 Devices for Integrated Circuit (DevIC)*, Kalyani, India (19–20 May 2021). IEEE. pp. 489–493.

42 Sarkar, A., De, S., and Sarkar, C.K. (2013). Asymmetric halo and symmetric single-halo dual-material gate and double-halo dual-material gate n-MOSFETs characteristic parameter modeling. *International Journal of Numerical Modelling: Electronic Networks, Devices and Fields* 26 (1): 41–55.

43 Sarkar, A. and Sarkar, C.K. (2013). RF and analogue performance investigation of DG tunnel FET. *International Journal of Electronics Letters* 1 (4): 210–217.

44 Dhar, R., Deyasi, A., and Sarkar, A. (2021). Generation of tunable low-noise millimeter-wave signal using optical frequency comb through electrical mixing at 94 GHz. *2021 Devices for Integrated Circuit (DevIC)*, Kalyani, India (19–20 May 2021). IEEE. pp. 13–18.

45 Biswas, K. and Sarkar, A. (2022). Chapter 6: MEMS-based optical switches. In: *Optical Switching: Device Technology and Applications in Networks* (ed. D. Nandi, S. Nandi, A. Sarkar, and C.K. Sarkar), 93–106. Wiley.

8

Tunnel FET: Principles and Operations

Zahra Ahangari

Department of Electronic, Yadegar -e- Imam Khomeini (RAH) Shahre Rey Branch, Islamic Azad University, Tehran, Iran

8.1 Introduction to Quantum Mechanics and Principles of Tunneling

Quantum tunneling is an incident that has no peer equivalent in classical physics, and it can be merely interpreted based on quantum mechanics concepts. In quantum mechanics, it is possible for a particle to tunnel through a sufficiently thin potential barrier that has a barrier height greater than the kinetic energy of the particle. This effect is due to the wave nature of the particle, which shows the probability of finding it in a certain location. Figure 8.1 illustrates the concept of quantum tunneling based on classical physics and quantum physics viewpoints. If the electron has enough kinetic energy (E), which is higher than the energy of potential barrier V ($E > V$), it can successfully surmount the barrier. However, if the electron does not have enough kinetic energy ($E < V$), it will never be able to get over the barrier. In principle, this is the classical viewpoint and is controlled by the law of conservation of energy. In contrast, when quantum mechanics effects are taken into consideration, the electron can tunnel through the potential barrier and appear on the other side, even if its kinetic energy is less than the energy of the potential barrier, which is due to the wave nature of the electron [1–3].

In principle, for effective tunneling of carriers through the potential barrier, the energy of carriers should be less than the energy of the potential barrier. Moreover, the barrier height should have a finite value of energy, and the barrier thickness should be thin enough. Quantum tunneling plays a fundamental role in physical phenomena, such as nuclear fusion, tunnel diodes, tunnel transistors, quantum computing, scanning tunneling microscopes, and nanoelectronics. The tunnel diode is an application of the p-n junction in a way that requires a quantum

Advanced Nanoscale MOSFET Architectures: Current Trends and Future Perspectives, First Edition. Edited by Kalyan Biswas and Angsuman Sarkar.

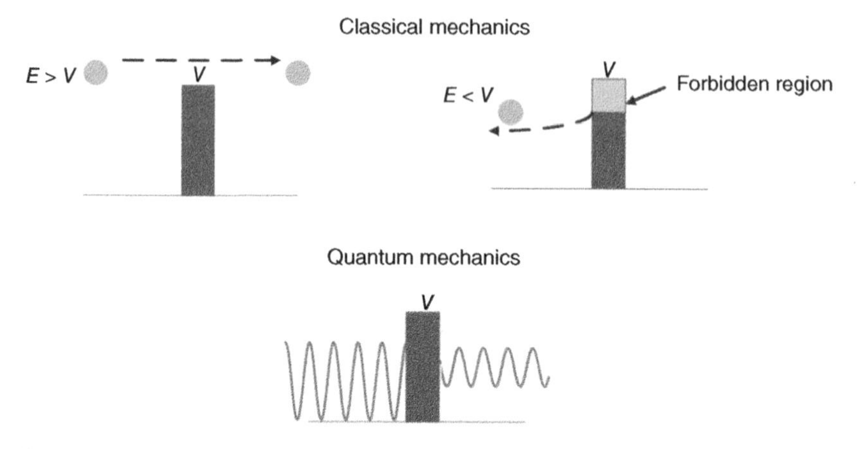

Figure 8.1 Quantum mechanics vs. classical mechanics.

tunneling phenomenon to operate, and carriers tunnel through a barrier that classically could not be surmounted [4, 5]. Basically, the tunnel diode or Esaki diode, named after the inventor, Leo Esaki, who discovered the effect in 1957, is a heavily doped p^+-n^+ junction (around 10^{19}–10^{20} cm^{-3}) that exploits quantum tunneling for carrier transport between the two sides of the diode. Unlike the conventional p-n junction, as a result of the very high doping, a tunnel diode possesses a very narrow space charge region, typically less than 10 nm (Figure 8.2). The main feature of the tunnel diode is that, unlike the conventional diode, it provides a region of negative resistance in the I–V transfer characteristics (Figure 8.3).

In Section 8.2, the operation principle of tunnel field-effect transistor (TFET) as an efficient high-speed low-power device that exploits quantum tunneling for carrier transport will be comprehensively discussed.

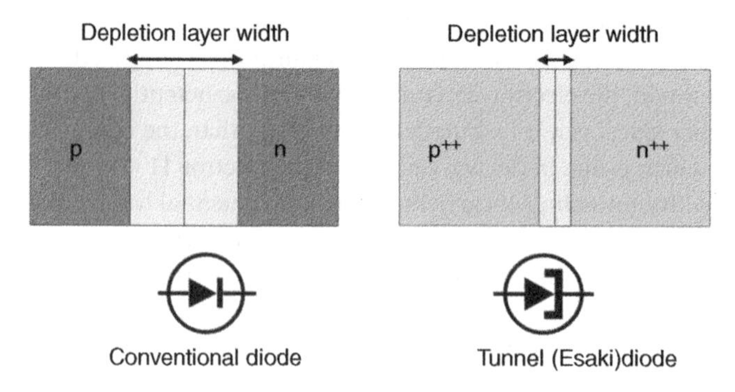

Figure 8.2 P-n junction schematics of a conventional diode and a tunnel diode.

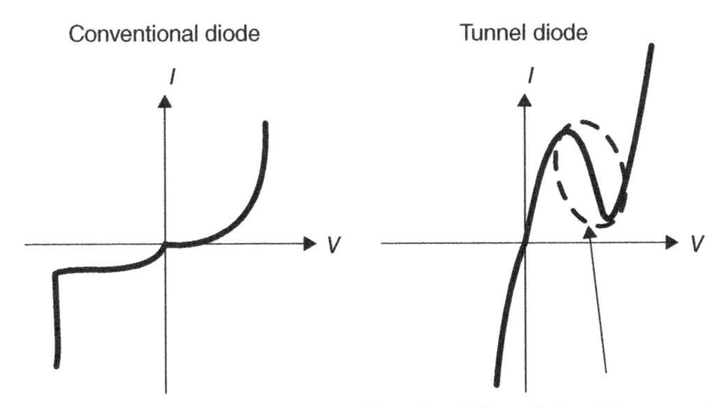

Figure 8.3 Transfer characteristics of a conventional diode and a tunnel diode, in which a negative differential resistance region is observed.

8.2 Tunnel Field-Effect Transistor

The downscaling of the dimensions of conventional metal–oxide–semiconductor field-effect transistor (MOSFET) continuously proceeds to meet the ever-increasing requirements of low-cost, low-power, efficient ICs with miscellaneous functionality. However, further downscaling of the device below 100 nm is seriously challenging due to the emergence of short-channel effects, which eventually degrade the devices performance [6–10]. The reduction of electrostatic gate controllability over the channel, besides the high value of supply voltage, results in high leakage current and high power dissipation. In principle, lowering the supply voltage degrades the switching speed of the device and reduces the drive current. Basically, in order to improve the switching speed and to maintain a high drive current while the supply voltage is decreased, the threshold voltage should be minimized. However, reduction of the threshold voltage results in the increment of the leakage current. Thus, the relationship between switching speed and power dissipation is critical in designing MOSFET with optimized electrical performance. By definition, the switching speed of the device is measured by a subthreshold swing figure of merit, which is defined as the amount of gate voltage that is required to increase the current by 1 order of magnitude at room temperature. Based on the Boltzmann limit, the 60 mV/dec limit for a subthreshold swing at room temperature is generally considered a fundamental limit for the conventional MOSFET. TFET with a different current mechanism has been introduced as a promising candidate for the conventional MOSFET, in which a properly designed TFET can provide a subthreshold swing of below 60 mV/dec at room temperature [11–13].

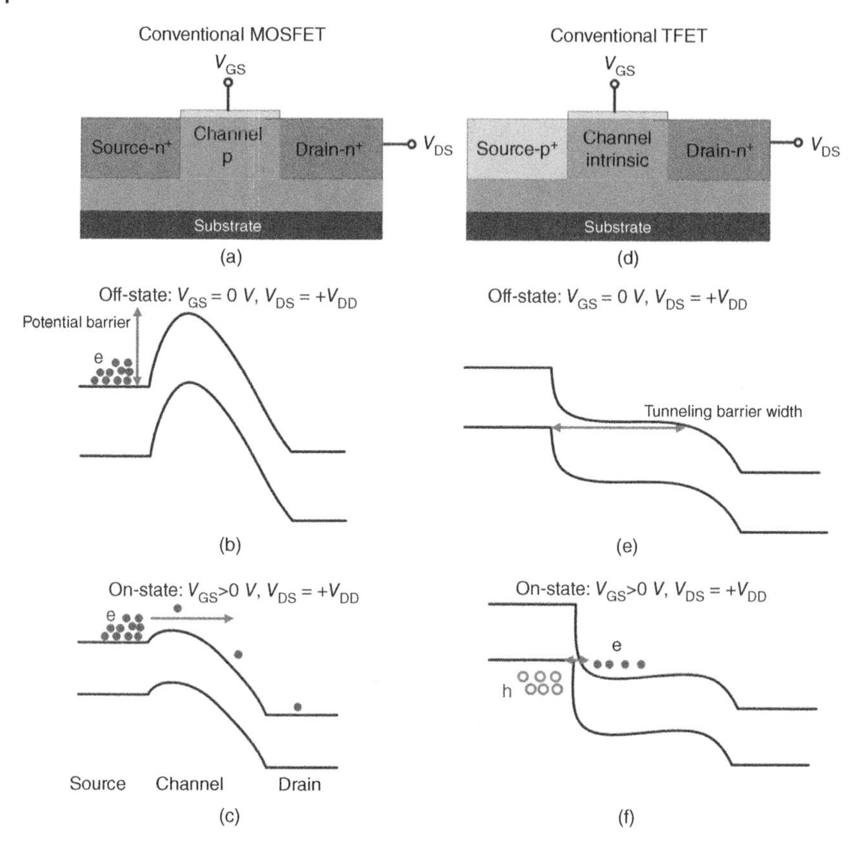

Figure 8.4 Schematic and energy band diagram of (a–c) conventional MOSFET and (d–f) conventional TFET in the off-state and on-state.

Figure 8.4 illustrates the schematics and energy band profiles of an n-channel conventional MOSFET and an n-TFET for comparison. In unipolar MOSFET devices, source and drain regions are n^+ heavily doped regions for efficient electron injection. However, the channel region is oppositely doped p region, and consequently, a potential barrier is created at the interface of source and channel regions. The current mechanism in MOSFET is dominated by the gate-modulated thermionic emission, in which carriers with sufficient energy can spill over the potential barrier. The gate voltage modulates the potential barrier height, and carriers transport from source to drain via drift and diffusion mechanisms under the influence of the electric field applied at the drain (Figure 8.4a–c). However, a TFET is a reverse-biased gated p-i-n diode. The source and drain regions of the device are heavily doped and of opposite types. In an n-type TFET, the source region is p^+ and the drain region is n^+, separated by an intrinsic channel. The gate voltage is responsible for modulating the tunneling barrier width at the interface of the source and channel regions.

Figure 8.4d–f illustrates the energy band diagram of a conventional TFET along the device from source to drain, in the off-state and on-state operations. Basically, due to the absence of any external electric field, the tunneling barrier width is not thin enough, and there is no possibility of quantum tunneling from source to drain. The TFET device operates by applying sufficient bias to the gate electrode so that the accumulation of electrons occurs in the intrinsic channel region. In this situation, the tunneling barrier becomes thin enough, and electrons from the valence band of the source region tunnel into the conduction band of the channel region, via band-to-band tunneling (BTBT) process.

Basically, the energy band bending at the interface of source and channel regions can be approximated by a triangular-like barrier. Based on the Wentzel–Kramers–Brillouin approximation (WKB), the tunneling probability of TFET can be described as [11, 14]:

$$T_{\mathrm{WKB}} \approx \exp\left(-\lambda\frac{4\sqrt{2m^{*}}E_{\mathrm{G}}^{3/2}}{3q\hbar(E_{\mathrm{G}}+\Delta\Phi)}\right) \tag{8.1}$$

where λ denotes the tunneling length and shows the extension of the transition region at the interface of source and channel regions, E_{G} is the band gap energy of the source region, and $\Delta\Phi$ represents the tunneling window, where the BTBT is allowed to occur (Figure 8.5). Basically, the smaller value of λ shows the greater band bending at the tunneling region. In principle, increment of the gate bias controls the tunneling probability, T_{WKB}, by reducing λ and simultaneously increasing the $\Delta\Phi$ window. To improve the working efficiency of TFET, the tunneling junction should be precisely designed to ensure a high BTBT rate and excellent gate controllability over the tunneling barrier. In other words, the tunneling barrier, W_{Tun}, should become thin enough for the onset of tunneling.

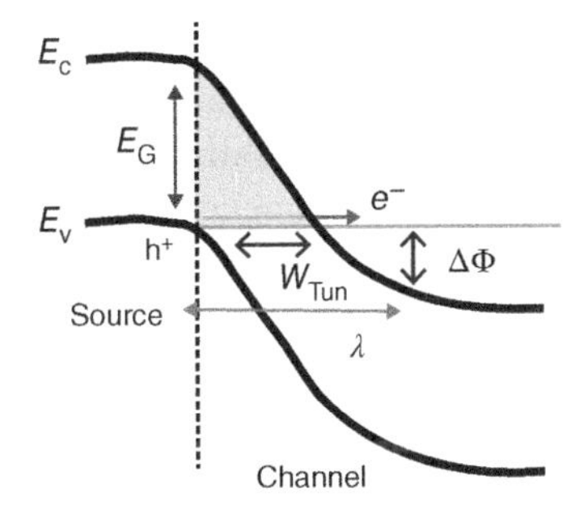

Figure 8.5 Energy band diagram of conventional TFET at the interface of the source and channel regions.

8.3 Challenges of Tunnel Field-Effect Transistor

Basically, despite the excellent switching performance and reduced value of off-state current, TFETs suffer from critical limitations for efficient operation of the device, including low on-state current, ambipolarity, drain-induced barrier thinning (DIBT) effect, and trap-assisted tunneling (TAT), which limit their commercial practical applications. These inherent disadvantages will be discussed in the following sections.

8.3.1 Low On-state Current

In principle, BTBT as the main current transport mechanism of tunnel FET causes low on-state current, which restricts the devices performance for competitive electrical features compared with ITRS requirements. The limited tunneling area at the interface of the source and channel is one of the main factors that reduces the BTBT current [15]. Moreover, a steep tunneling junction is fundamental for efficient tunneling and increment of the drive current. However, fabrication of abrupt p^+-n^+ junction seems challenging due to the diffusion of dopant atoms from heavily doped source/drain regions into the adjacent channel, resulting from high-temperature fabrication process. In addition, the required voltage for the band alignment and trigger of BTBT is a fundamental factor that may affect the on-state current. Basically, employment of different materials providing different band alignments at the tunneling junction has been considered an impressive solution for the amplification of the tunneling rate.

8.3.2 Drain-Induced Barrier Thinning Effect

Basically, scaled TFET devices are fundamental to be competitive with conventional MOSFETs with channel lengths less than 10 nm for enabling high integration density of transistors in a chip. Similar to conventional MOSFETs, short-channel effects are one of the challenging difficulties that restrict the scaling of traditional TFETs into sub-30 nm regimes due to the occurrence of tunneling in the absence of the gate bias. In conventional TFETs, if the channel length is short enough, the drain electric field can penetrate deeply into the interface of source and channel regions and, consequently, reduce the tunneling barrier. This effect is called drain-induced barrier thinning (DIBT) or drain-induced source tunneling (DIST), which effectively deteriorates the subthreshold swing [16–18]. Doping engineering, employment of wide band gap materials in the channel, and multi-gate structures are feasible techniques that can effectively reduce this effect and provide better short-channel effect immunity. Since these techniques can modify other problems that are associated with the electrical performance of conventional TFET, their practicality will be discussed in the following section.

8.3.3 Ambipolarity

It is observed that ambipolarity can be considered one of the main limitations of TFET devices [19–21]. Ambipolarity, by definition, refers to the conduction of carriers in two directions, both for negative and positive gate bias. This effect mainly occurs due to the transfer of tunneling junction from source to drain when the gate voltage is negative for an n-TFET operation. The energy band diagram of a n-TFET is illustrated in Figure 8.6 in the case of off-state, on-state, and ambipolar state. It is clearly observed that in the case of negative gate bias, holes are accumulated in the channel and a tunneling pass is created at the drain junction. This is an undesired effect for complementary logic circuit applications and thus degrades the feasibility of the device for digital circuit purposes.

To optimize the TFET performance and mitigate the risk of ambipolarity, different techniques and structures are introduced, including gate workfunction engineering [19, 22], material engineering [23, 24], gate oxide (material and thickness) engineering [25, 26], gate-drain underlap engineering [27, 28], spacer engineering [29–31], and doping engineering [32, 33].

In gate workfunction engineering, multi-materials are employed for the gate electrodes, possessing different workfunctions. In this case, even if the same gate voltage is applied, the energy band bending is different along the channel from source to drain. For an n-TFET, the gate electrode contains two different materials. The gate electrode that controls the tunneling at the interface of the source and channel should be adjusted to facilitate carrier tunneling and the gate electrode at the drain side should be tuned to prevent ambipolar tunneling. Material engineering can be considered another useful technique to attenuate the ambipolarity effect. It is observed that the employment of heterostructures with large band gap materials at the drain side can enhance the tunneling barrier

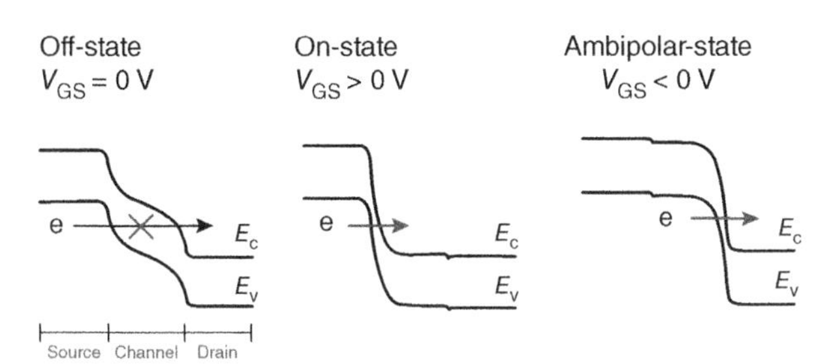

Figure 8.6 Energy band diagram of conventional TFET in the off-state, on-state, and ambipolar-state.

width and as a consequence can degrade the ambipolar current. To degrade the ambipolar tunneling, the electric field at the drain side should be reduced either by gate oxide material engineering or gate oxide thickness engineering. Basically, by employing high-k and low-k materials at the source and drain sides, respectively, the electric field distribution at both ends of the channel can be controlled. This method can also be accomplished by employing asymmetric oxide thickness at the source and drain sides. It is evident that gate oxide thickness at the source should be thin enough to boost the electric field at the tunneling junction. However, to weaken the electric field at the drain side, thicker gate oxide similar to the material of the source region is applied. In gate–drain underlap engineering technique, the gate electrode does not cover all the channel regions, and a distance is created between the gate and drain electrodes. In this case, the electric field at drain side is reduced, and suppression of ambipolar current is expected. Spacer engineering is another effective method for diminishing the ambipolarity effect. In this method, the coupling between drain and gate electric fields is suppressed by the presence of a low-k spacer. In the case of doping engineering technique, the drain doping density is reduced to increase the extent of space charge region in the drain.

8.3.4 Trap-Assisted Tunneling

In principle, TAT can be defined as the process of BTBT current generation via the contribution of phonons. Basically, TAT provides a leakage current before the onset of BTBT, which causes a considerable degradation of subthreshold swing in TFETs [31, 34, 35]. Figure 8.7 shows the TAT process that occurs due to the presence of the defects in the forbidden gap. In the first step, electrons are transferred from the valence band to the traps via absorbing phonons. In the next step, the electrons are moved in the traps and eventually tunnel into the conduction band. This process occurs even in the absence of the gate bias and also before the complete band alignment, and it highly depends upon the temperature. Similarly, hole emission and tunneling from a trap are also possible. It is worth noting that this leakage interband carrier transition is also possible when phonon scattering is considered alone. Basically, traps at the gate oxide and channel interface, defects, and traps at the semiconductor/semiconductor interface in the case of

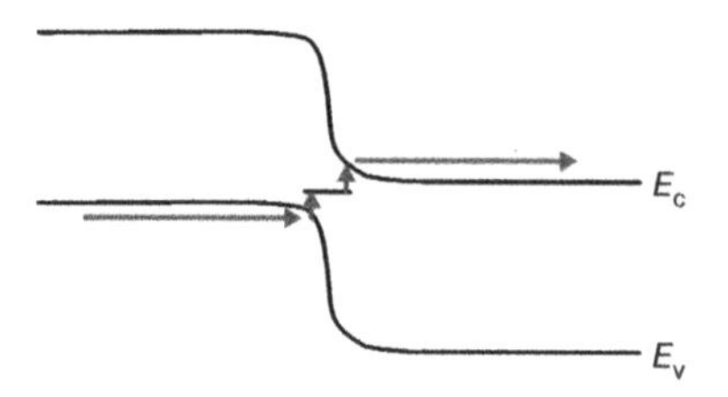

Figure 8.7 Trap-assisted tunneling via the presence of traps in the forbidden gap.

heterojunction TFETs, and defects from doping and implantation process are the limiting factors for achieving a steep slope in the subthreshold region. Depending on the materials that are exploited at the tunneling junction, structure of the TFET, and the nature of traps, different techniques are employed to disseminate TAT current.

8.4 Techniques for Improving Electrical Performance of Tunnel Field-Effect Transistor

In this section, different engineering techniques for solving the problems that are associated with the conventional TFET are comprehensively discussed. These techniques include: Doping Engineering, Material Engineering, and Geometry/Structure Engineering.

8.4.1 Doping Engineering

Basically, the BTBT transmission mechanism of a p-i-n TFET, which differs from the current transport mechanism of conventional MOSFETs, leads to issues such as low drive current and high ambipolar current. Doping engineering is introduced as one of the effective methods to increase the tunneling rate and solve these problems, which will be discussed as follows:

- **Pocket Doping**: Pocket doing is a functional method to boost the electric field at the tunneling interface and, as a consequence, assists the increment of the tunneling rate. By definition, pocket doping is defined as a high density of dopants (different polarity in comparison with the source dopant) that is located at a confined region at the interface of the source–channel junction, depicted in Figure 8.8. Qualitatively, the width of the space charge region at

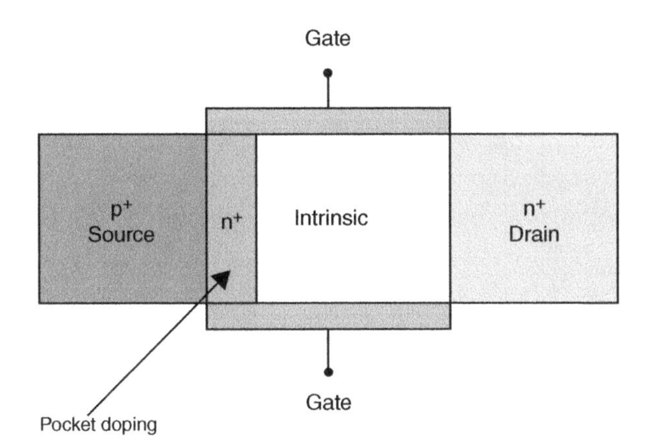

Figure 8.8 Pocket doping at the source interface for the TFET device.

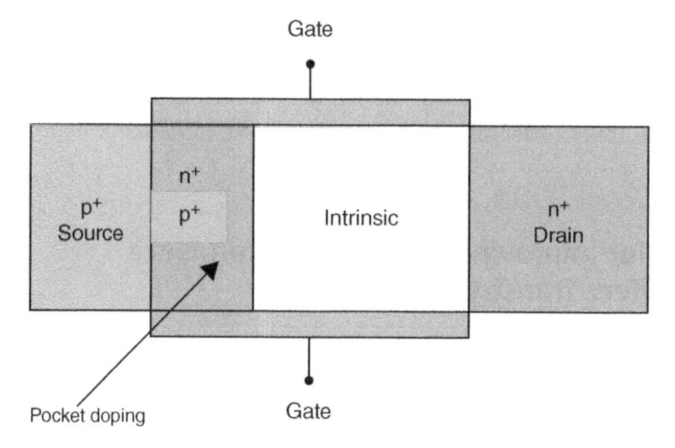

Figure 8.9 Square-shaped TFET with pocket doping.

the source–channel p-n junction is inversely proportional to the doping concentration of the n and p sides. It is evident that heavily doped pocket doping considerably reduces the tunneling barrier width, which provides an additional electric field to the primitive gate electric field [36–38]. In this case, the magnified electric field across the tunneling junction increases the tunneling current. The location, configuration, and dopant density of the pocket region are fundamental parameters that effectively modify the devices performance. Figure 8.9 illustrates the schematic of a TFET with square-shaped source configuration that is extended along the channel region [39]. This structure has both point and line tunneling, which provides additional improvement in the on-state tunneling current.

- **Semi-Junctionless TFET**: In principle, scaling the TFET dimensions into the nanoscale regime has become a substantial challenging problem that may result in the DIBT effect, which degrades the devices performance. Figure 8.10

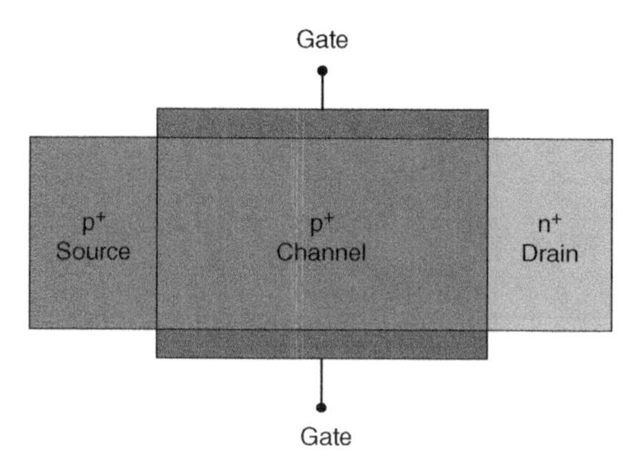

Figure 8.10 Schematic of a semi-junctionless TFET, which has an identical doping profile in the source and channel regions.

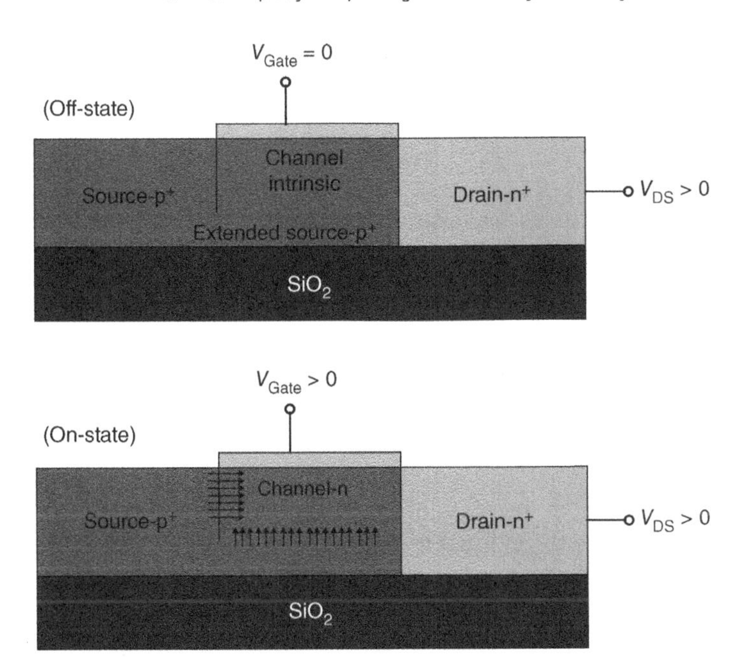

Figure 8.11 Extended source TFET in the off-state and on-state operations. The source region is extended in the channel region and horizontal and vertical tunneling current is expected. The arrows indicate the BTBT direction.

illustrates the schematics of a semi-junctionless n-TFET [40, 41]. It is observed that, unlike the conventional TFET with p-i-n doping profile, the proposed device has a similar doping profile in the source and channel regions, which has simpler fabrication process. Due to the creation of a p^+-n junction at the interface of channel and drain regions, the drain electric field is mainly confined at the lightly doped drain side and as a consequence, a considerable large tunneling barrier width is created in the off-state, which exhibits superior immunity over DIBT effect.

- **Extended Source TFET**: Figure 8.11 illustrates the structure of an extended source TFET in which the source region is extended along the channel, in the off-state and on-state operations. In this device, a heavily doped boron layer is implanted in the channel region [42, 43]. This structure is a useful solution for widening the tunneling area. Basically, when a positive gate voltage is applied with sufficient value, an inversion layer of electrons is created in the channel which facilitates the occurrence of BTBT both vertically (along the extended source and the top-layer electrons of the channel) and horizontally (at the interface of source and channel). The vertical component of the tunneling current significantly increases the drive current. The most critical part of

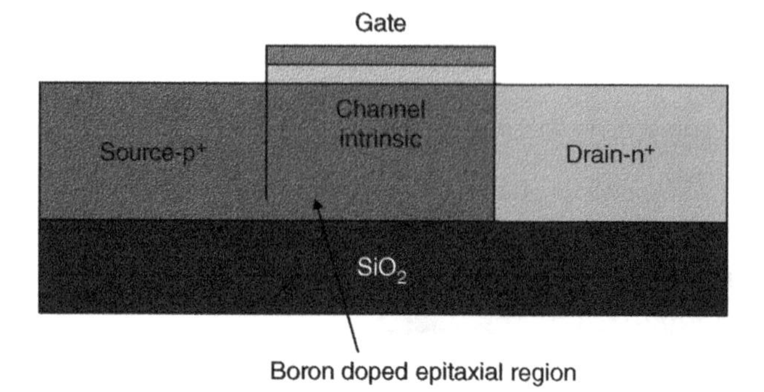

Figure 8.12 Schematic of a pure boron-doped TFET with extended epitaxial source region.

extended source TFET is the thickness of the intrinsic channel. This design parameter should be properly adjusted to provide a maximum rate of BTBT tunneling. Consequently, for large channel thicknesses, there is a degradation of gate control over the channel due to the lower electric field at the tunneling junction. In this situation, the gate cannot trigger the onset of tunneling, and as a result, degradation of the subthreshold swing is expected.

- **Pure Boron TFET**: The pure boron TFET structure is an improved architecture of extended source TFET devices. Basically, fabrication of thin, heavily doped extended source region with conventional implantation process is challenging and seems impractical. The created traps and defects, as a result of the ion implantation, will completely degrade the subthreshold swing of the device. The schematic of pure boron TFET is illustrated in Figure 8.12, in which a highly thin, heavily doped boron layer is created in the channel via epitaxial fabrication process [44]. Similar to extended source TFET, for an n-TFET operation, upon employment of adequate positive gate bias, an inversion layer of electrons is created in the channel, and vertical BTBT occurs along the channel thickness. The spacer between the channel and drain region is essential to eliminate lateral parasitic tunneling in this device.
- **Junctionless TFET**: In principle, the requirements to enhance the on-state drive current in TFETs highly depend upon the ability to create heavily doped abrupt tunneling junctions. Basically, the source doping density gradient is the fundamental physical design factor that leads to the threshold voltage variation in experimental TFET. This effect is more pronounced when the dimensions of the device are scaled and fabrication of ultra-steep p-n junction becomes more challenging.

  Junctionless TFET [45, 46] has been introduced as a potential candidate for the conventional TFET, which takes advantage of conventional junctionless transistors with identical doping profiles in the source, drain, and channel regions, as illustrated in Figure 8.13 for an n-TFET operation. In order to implement the structure of a junctionless TFET, two gates are employed on the source and channel regions, respectively, to create a p-i-n doping profile along the device. The gate contact over the channel is called the control gate (CG), and it is responsible for depleting carriers from the channel in the off-state. The program gate (PG) over the source region is accountable for electrically doping of the source region, possessing different workfunction in comparison with the workfunction of the control gate. It is worth notifying that no bias is applied to the PG. In the off-state, the channel is depleted from the electrons due to the workfunction difference between the gate material and channel region. This effect is more pronounced in the source region, in which due to the p-gate workfunction of the PG, the n^+ source region is inverted to a p^+ region via electrical doping process. Electrical doping is the process of electronic charge insertion or acceptance of them to the material via controlling the Fermi level, which reduces the risk of high-temperature contamination of extrinsic dopants. In the on-state, by applying sufficient bias to the control gate, electrons are accumulated in the channel and a steep p^+-n^+ junction is created at the interface of the source and channel regions. In this situation, BTBT occurs when the conduction band of the channel aligns with the valence band of the p^+ source region.

- **Electrically Doped TFET**: Generally, doping process is considered as an effective method to change the electrical properties of a semiconductor, by inserting the appropriate impurities in the material. However, conventional physical doping techniques have limitations, mainly in nanoscale regimes. It is observed that physical doping inserts various structural defects due to the interaction between the dopants and the host atoms changing the density of states of the material. Electrical doping of thin films has been introduced as an effective alternative to physical doping and has been previously employed in junctionless TFET. Basically, choosing the appropriate workfunction value for the control and PG metal electrodes is challenging. These values can effectively modify the steepness of p-n junction profile and, as a consequence, can dramatically affect the tunneling rate and subthreshold swing.

  Figure 8.14 illustrates the structure of an electrically doped TFET in which the workfunction of the control gate and PG have equal values. However, unlike junctionless TFET in which no bias is applied to the PG, in this device, sufficient voltage with an appropriate value is applied to the PG to alter the doping profile of the source region [47, 48]. For a p-TFET operation, the initial doping profile is p^+-p^+-p^+ in the source, channel, and drain regions, respectively. In the off-state, the control gate over the channel region is zero and the intrinsic channel is created in the channel due to the workfunction difference between the

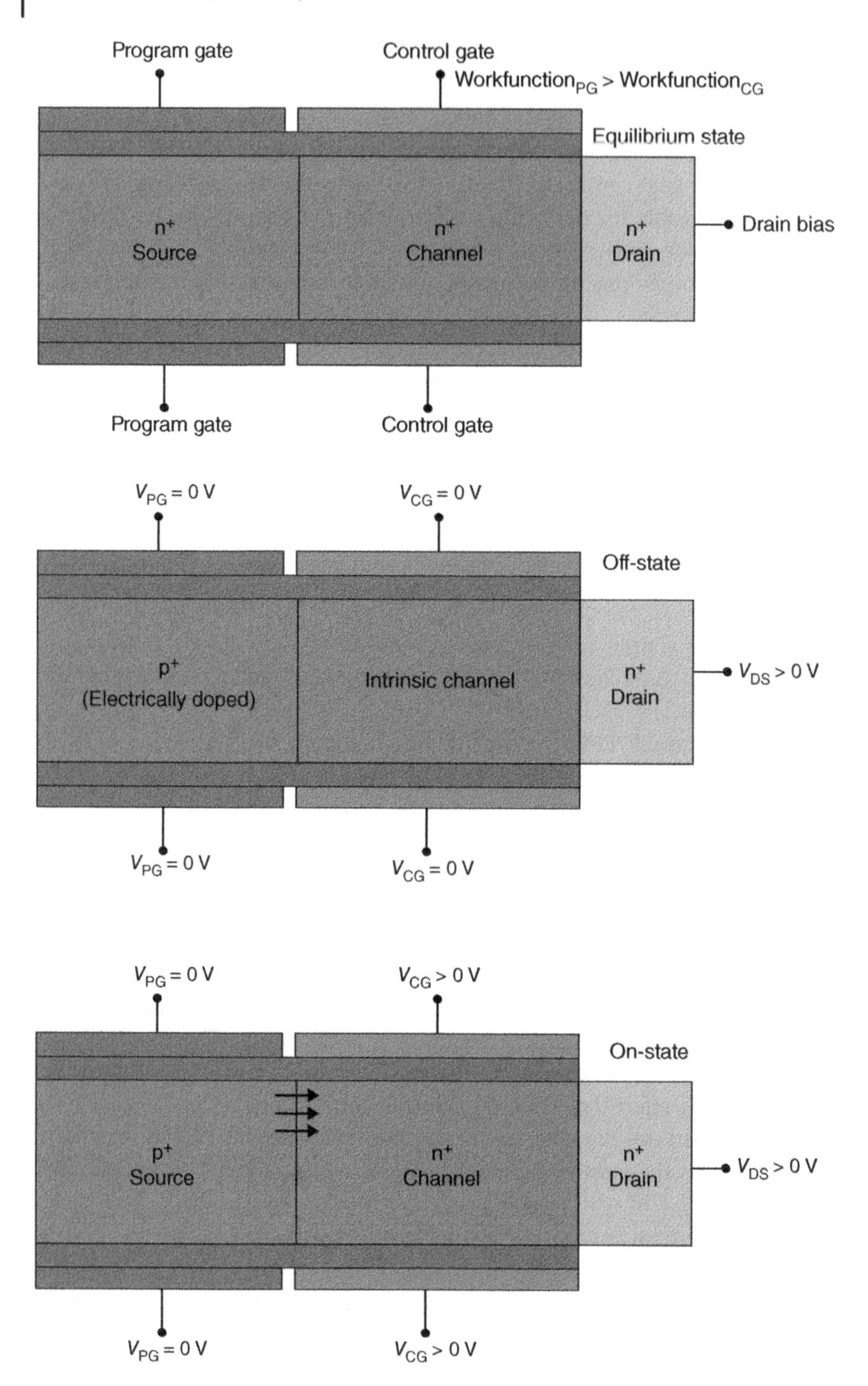

Figure 8.13 Schematic of an n-type junctionless TFET in the equilibrium state, off-state and on-state.

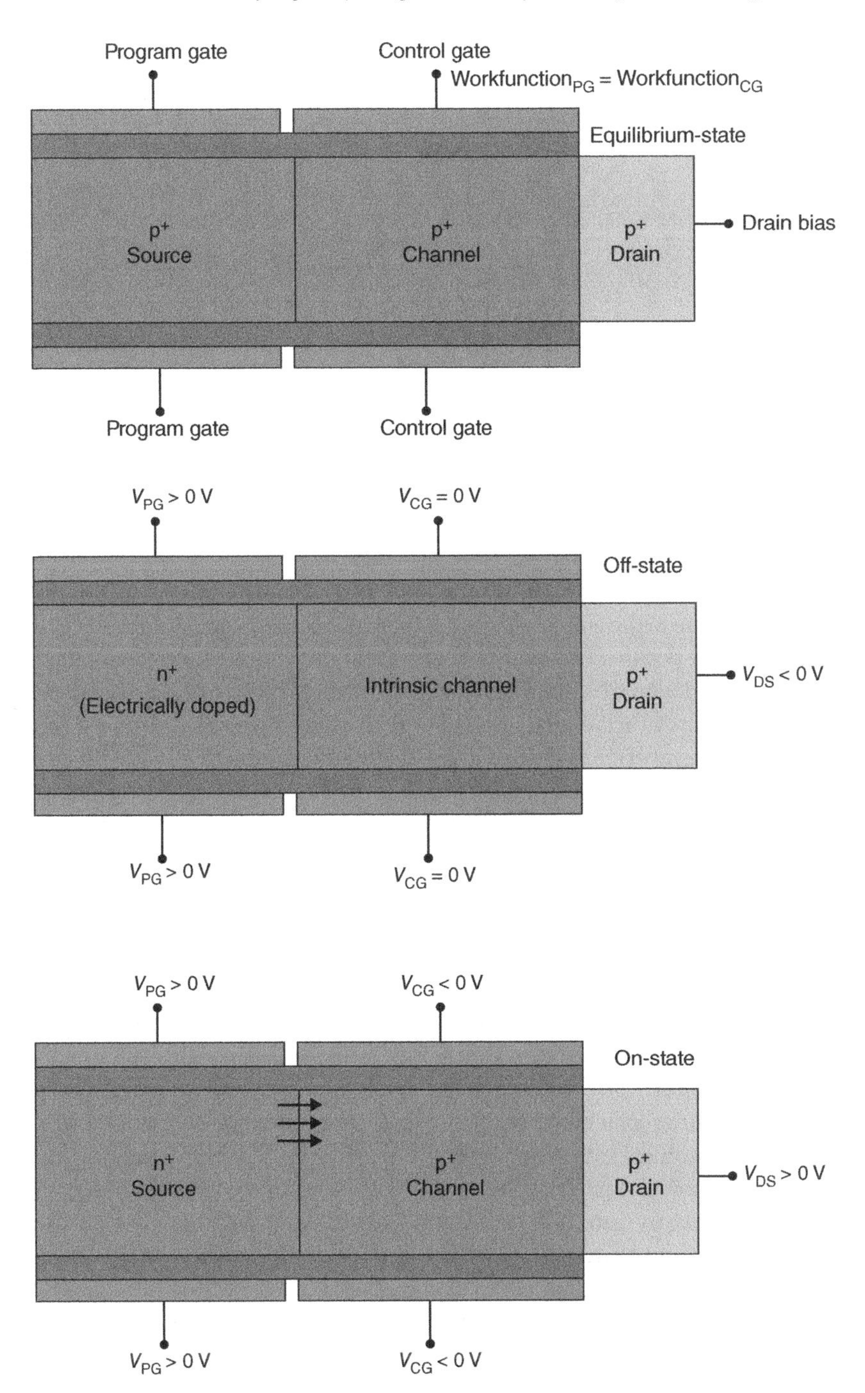

Figure 8.14 Schematic of a p-type electrically doped TFET in the equilibrium state, off-state and on-state.

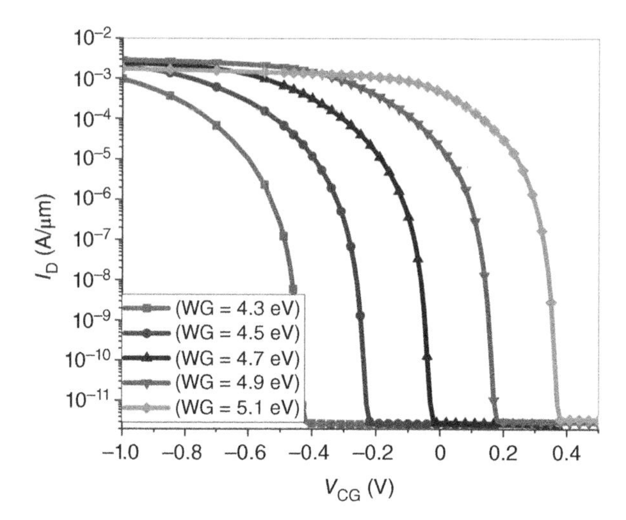

Figure 8.15 Transfer characteristics of an electrically doped TFET as a function of gate workfunction.

control gate and the channel. However, an n^+ inversion region is created in the source region due to the employment of sufficiently positive voltage to the PG. In the on-state, the positive PG bias is still applied to the source region to maintain the electrically doped n^+ source area and besides that, a negative bias is applied to the control gate to accumulate holes in the channel. In this situation, a sharp p^+-n^+ tunneling junction is created and BTBT starts to occur.

Figure 8.15 illustrates the transfer characteristics of an electrically doped p-TFET, as the gate workfunction (WG) is parametrized. The results demonstrate that the threshold voltage for the onset of tunneling varies with the workfunction. Basically, reduced value of the gate workfunction leads to the reduction of the hole density in the channel and shifts the transition voltage from off-to-on state to higher negative values. Moreover, it is worth noting that due to the symmetric gate workfunction values for the PG and control gate, there exists a competition between the PG workfunction and the polarity gate bias for the electrically induced charges in the source region. Figure 8.16 illustrates the electrically induced electron density in the source region as a function of polarity gate bias for different gate workfunctions, with drain bias of −0.05 V and control gate of zero voltage, to assess the sole impact of polarity gate workfunction and polarity gate bias on the electron density. It is observed that for low polarity gate bias, impact of polarity gate workfunction on the electron density is considerable. However, as the polarity gate bias increases, its workfunction value has a negligible effect on the induced electron density. The results manifest that proper combination of polarity gate bias and workfunction values is required for achieving steep tunneling junctions.

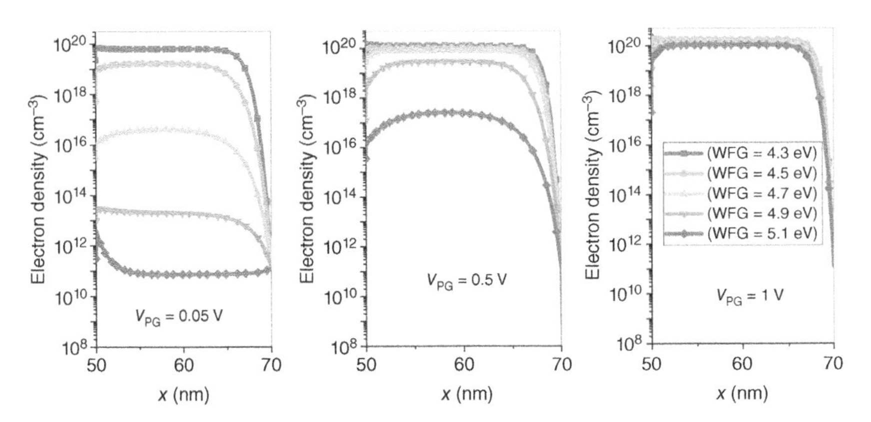

Figure 8.16 Electrically induced electron concentration in the source region of a p-type electrically doped TFET as the polarity gate bias and polarity gate workfunction are parameterized: (a) $V_{PG} = 0.05$ V, (b) $V_{PG} = 0.5$ V and (c) $V_{PG} = 1$ V.

8.4.2 Material Engineering

By definition, BTBT occurs when the electronic energy bands are aligned at the tunneling junction. It is evident that the materials in the source, channel, and drain regions can modify the band alignment configuration. Basically, semiconductor interfaces can be organized into one type of homojunction and three types of heterojunctions including straddling gap (type I), staggered gap (type II), and broken gap (type III), Figure 8.17. In heterojunction TFET, based on the source and channel material, different band alignments can be established, which can effectively modulate the tunneling rate and the devices performance. Heterojunction structures that provide narrower tunneling barrier width have been regarded as a powerful solution to address problems associated with TFET, mainly low on-state current and ambipolar behavior.

During recent years, the number of fabricated TFETs based on compound materials, mainly from III–V category has grown dramatically [49–51]. The main feature of these materials in comparison with silicon is their tunable direct-gap structure with small effective mass, which, based on WKB approximation, increases the probability of tunneling. The main feature of these materials is that the energy band value can be modulated by modifying the mole fraction of alloy compositions, making possible fabrication of different heterojunction configurations. The energy band diagrams of homojunction and staggered-type heterojunction TFET are illustrated in Figure 8.18. It is observed that heterojunction TFET is suitable for high-speed ultra-low applications as it requires a smaller gate bias for the onset of tunneling.

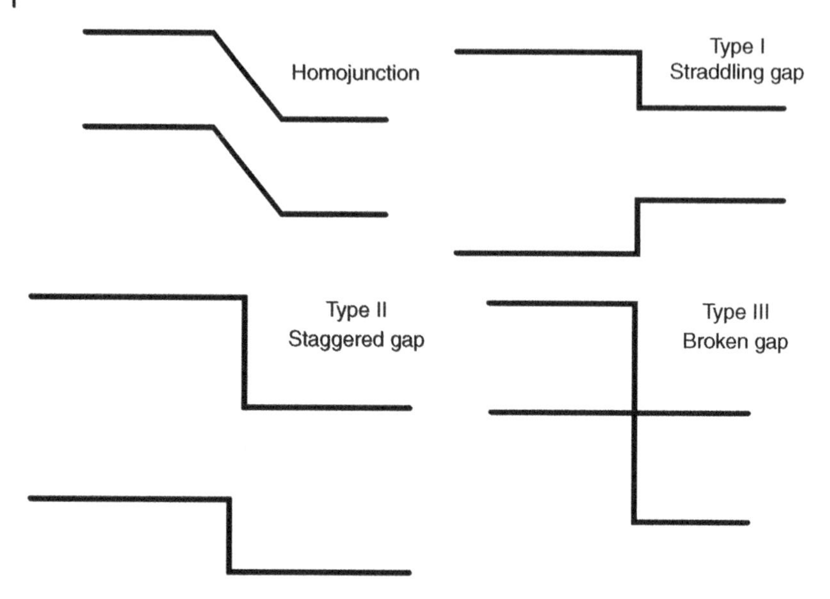

Figure 8.17 Energy band diagram of homojunction and different types of heterojunctions.

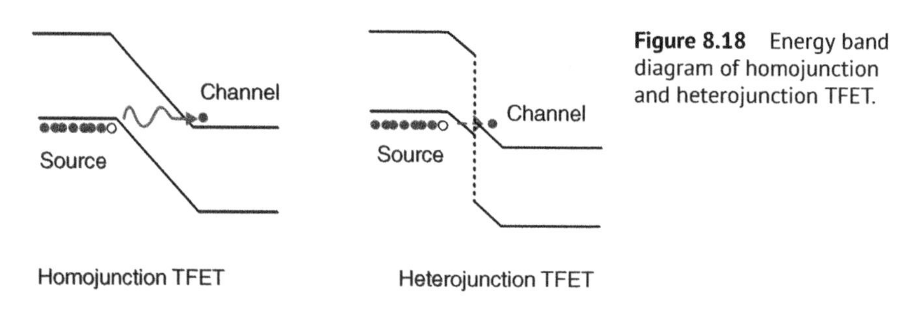

Figure 8.18 Energy band diagram of homojunction and heterojunction TFET.

Gallium nitride (GaN) is an efficient III–V direct band gap material with a wide band gap of 3.4 eV. This material can operate at high temperatures and its special properties make it perfect for applications in optoelectronic, high-power, and high-frequency devices. Basically, among III–V materials, only III–V nitrides have spontaneous polarization characteristics. This effect is due to the intrinsic asymmetry of the bonding in wurtzite crystal structure in equilibrium. In Ref. [52], a novel TFET based on GaN is proposed in which a thin layer of indium nitride (InN) is employed to exploit the polarization effect at the interface of GaN/InN (Figure 8.19). The intrinsic polarization of these materials provides accumulation of uncompensated sheet charge, leading to a high internal electric field. It is observed that due to this electric field, tunneling within a high band gap

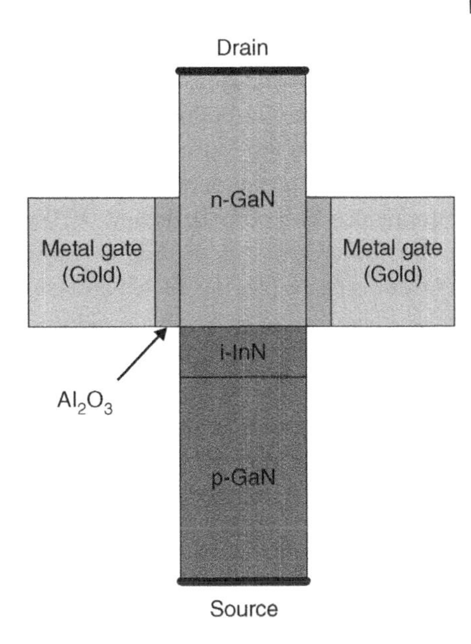

Figure 8.19 Schematic of a III–V InN/GaN heterojunction TFET.

material appears to be significant. Simulation results exhibit that the proposed heterojunction GaN/InN TFET is enabled to achieve subthreshold swing of 15 mV/dec and on/off current ratio of more than 10^6.

Recently, a special focus has been dedicated to III–V heterojunction TFET, and they are considered as building blocks for modern high-speed low-power semiconductor devices. Basically, forming a heterojunction lattice match (or a very small mismatch) between two adjacent semiconductor materials is critical to minimize the formation of defects and traps at the interface. Figure 8.20 illustrates

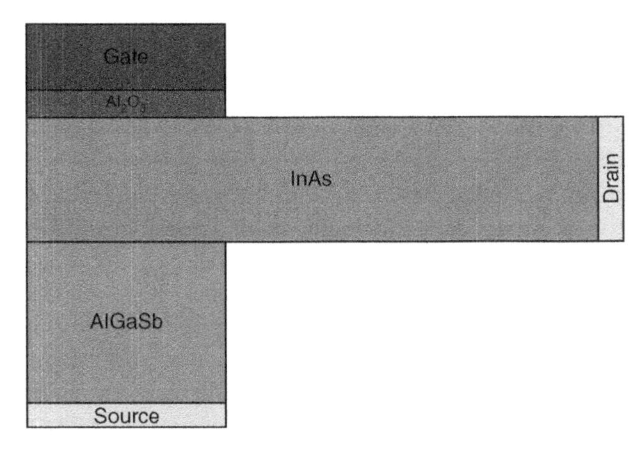

Figure 8.20 Schematic of heterojunction III–V TFET.

the schematics of AlGaSb/InAs heterojunction TFET [53]. The main feature of the proposed structure is that the gate electric field is designed to be in the same direction as the tunneling junction. Due to the occurrence of line tunneling at the interface of the source and channel high on-state current has been achieved. Moreover, the presence of the spacer and also the gate-drain underlap leads to the suppression of ambipolarity and DIBT effect.

The other efficient method to improve the tunneling rate and the on-state current is by employing a low band gap material in the source region. In order to take advantage of the high tunneling probability of narrow band gap materials in the source region and the high mobility of III–V materials in the channel region, a type-I heterostructure TFET based on Ge–$Al_xGa_{1-x}As$–Ge is introduced [54]. The optimum value for the Al mole fraction is about $x = 0.2$ (20%). The wide band gap channel material degrades the drain electric field at the source junction and as a consequence, reduces the short channel DIBT effect. Moreover, Ge–GaAs have been introduced as a lattice-matched heterostructure, which can provide improved digital and analog characteristics for high-performance TFET devices [55].

The successful emergence of graphene in 2004 has attracted great interest in exploiting this material for electronics, optoelectronics, spintronic, and other applications. By definition, the term single-layer materials or 2D materials refers to crystalline solids consisting of a single layer of atoms. Compared to graphene, the 2D transition metal dichalcogenides (TMDs) have the advantage of being semiconductors with feasible band gaps, which would allow their application for electronic devices. These materials can be a proper choice for TFETs as high on-state current may be achieved because of short tunneling distance in the range of van der Waals distance. The remarkable advantages of TFETs based on 2D materials include superior electrostatic gate control due to the significantly larger surface-to-volume ratio, high electrical conductivity, and high carrier mobility due to ballistic or quasi-ballistic transport [56–59]. Moreover, the rapid fabrication synthesis methods facilitate uniform deposition of atomically thin layers of 2D materials with high quality on various substrates, which disseminate the formation of dangling bonds and defects. However, excellent gate control over the tunneling junction is not the fundamental factor in improving the on-state current. Basically, other critical parameters including doping density, effective mass, and band gap play fundamental roles in the tunneling rate.

The monolayer of black phosphorus material is called phosphorene, which has a non-zero direct band gap at different layer thicknesses. The main feature of this material is that small effective mass in the transport direction without any degradation in the density of states. This effect is due to the high anisotropy in this material, which makes it a potential candidate for TFET device. Figure 8.21 illustrates the schematics of a phosphorene TFET [56]. Specifically, due to the

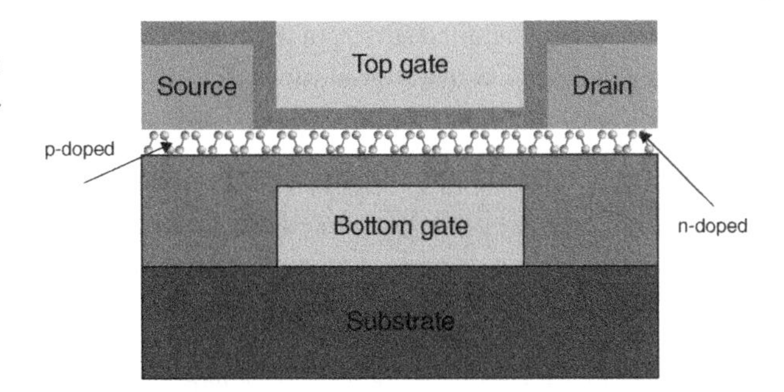

Figure 8.21 Schematic of a p-i-n TFET based on monolayer phosphorene.

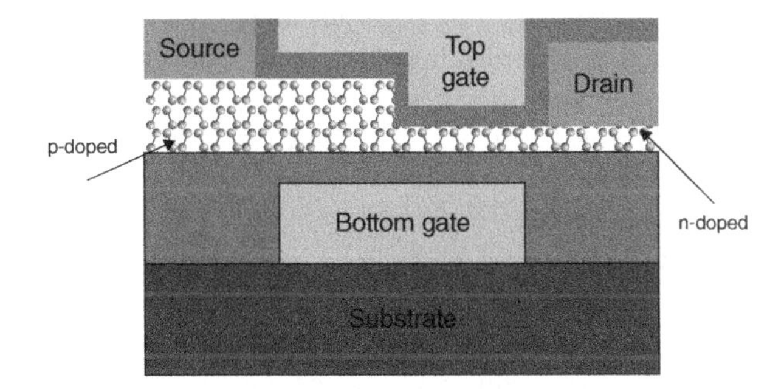

Figure 8.22 Heterojunction phosphorene TFET. The tri-layer has a lower band gap in comparison with the monolayer phosphorene.

small effective mass in the armchair direction, the tunneling current is much higher than that in the zigzag direction. The electrical performance of phosphorene can be modulated via appropriate uniaxial strain. Figure 8.22 depicts the structure of a heterojunction phosphorene TFET. The proposed device has a high immunity to ambipolar effect. In addition, the device provides low effective mass at the source region, which increases the on-state drive current without any enhancement of the off-state current. The $WSe_2/SnSe_2$ heterostructure material system [60] is introduced as an efficient 2D-2D TFET, which due to the beneficial band alignment and also due to the defect-free van der Waals atomic junctions, subthreshold swing of 35 mV/dec and on/off current ratio of more than 10^5 can be achieved.

8.4.3 Geometry and Structure Engineering

It is evident that the BTBT phenomenon is mainly administered by the point tunneling in a narrow region at the interface of the source and channel region, which makes the drain current not competitive. Therefore, it is essential to enhance the tunneling current by expanding the BTBT generation area that is governed by line tunneling mechanism. In addition, exploiting multi-gate structures provides better gate controllability over the channel which may amplify the electric field at the tunneling junction and, as a consequence, improve the tunneling current. This section presents a detailed survey of the various techniques for geometry and structure engineering of conventional tunnel FETs to boost the on-state current along with improvement of the subthreshold swing.

- **F-Shaped Tunnel FET**: Figure 8.23 illustrates the schematic of an F-shaped TFET in which parallel extended source regions resembling fingers are distributed in the channel [61]. The proposed structure can reduce the required gate voltage for the occurrence of BTBT due to the effect of amplified electric field crowding. The device can be fabricated in a self-aligned process and the number of finger-like source regions can be increased to enhance the BTBT current. This structure takes advantage of enhanced tunneling window as well as electric field crowding effect which reduces the required voltage for the onset of tunneling. However, the length of fingers, the distance between fingers, and the thickness of fingers are critical design parameters that may fundamentally affect the tunneling rate and should be adjusted for the efficient performance of the device.
- **L-Shaped Tunnel FET**: The L-shaped TFET is a vertical structure that has been proposed to amplify the tunneling rate and suppress ambipolarity [62]. In this structure, as shown in Figure 8.24, the source and channel regions are overlapped providing an L-shaped configuration, and are an experimentally fabricated line-tunneling TFET. The BTBT tunneling of carriers occurs in two vertical and horizontal directions one from the bottom source and the other from the side source, respectively, and eventually results in a considerable high on-state current.
- **Vertical Dual-Channel TFET**: The schematic of vertical dual-channel [63] TEFT is illustrated in Figure 8.25, in which tunneling junction at the source–channel interface is extended vertically which considerably improves

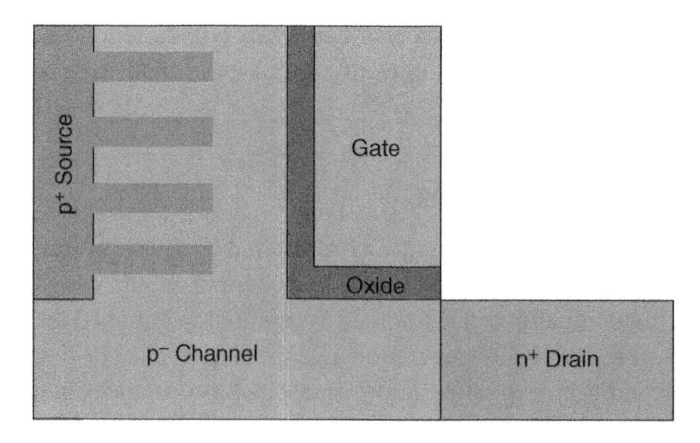

Figure 8.23 Schematic of a F-shaped TFET with multiple fingers as the source region.

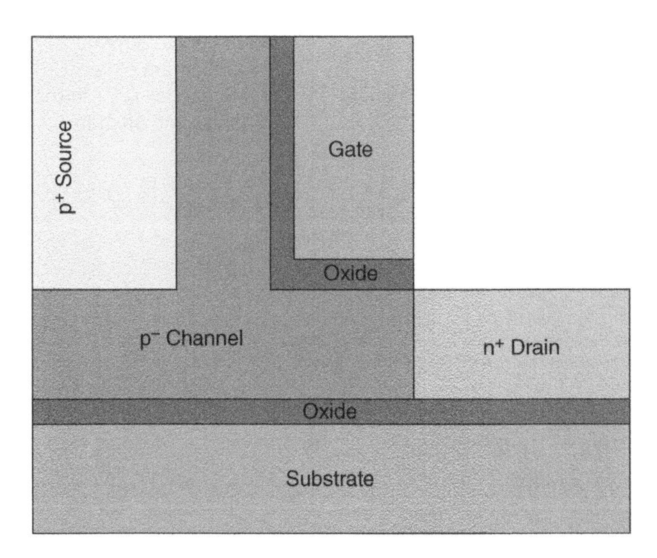

Figure 8.24 Schematic of an L-shaped TFET.

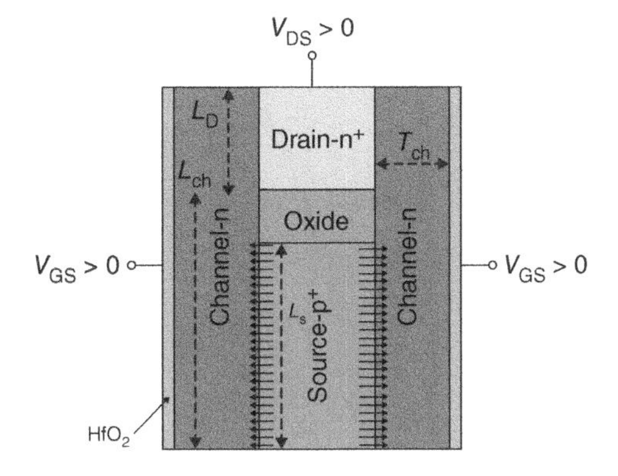

Figure 8.25 Schematic of a vertical dual-channel TFET with two sidewall channels.

the devices performance. In addition, two parallel channels are created at the source sidewalls that can fundamentally improve the on-state current. The oxide region between the top side of the source and drain considerably provides high immunity to short-channel effects such as DIST effect. Figure 8.26 illustrates the conduction and valence band of vertical TFET in the off-state ($V_{GS} = 0$ V, $V_{DS} = 1$ V) and on-state ($V_{GS} = 1$ V, $V_{DS} = 1$ V). In the off-state, the tunneling barrier width is not thin enough for the carriers to tunnel through. However, as the positive gate bias goes beyond the threshold voltage, the space charge region at the source–channel junction shrinks considerably and a considerable amount of charged carrier tunnel through the channel.

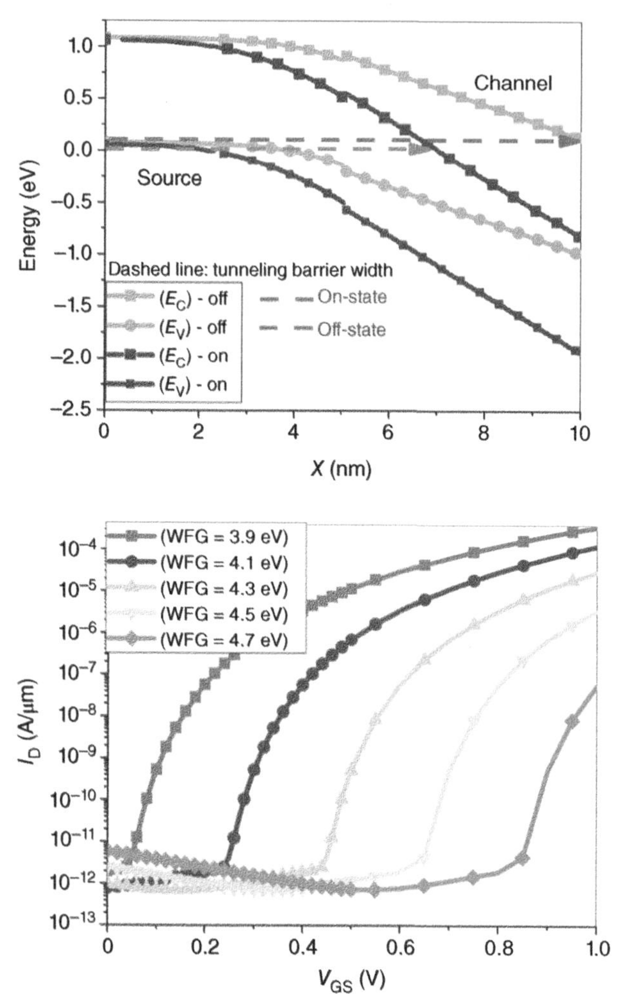

Figure 8.26 Energy band diagram of vertical dual-channel TFET in the off-state and on-state.

Figure 8.27 Transfer characteristics of vertical dual-channel TFET as the sidewall gates' workfunction are parametrized.

Gate workfunction (WFG) is an important design parameter that can modulate electron concentration in the channel. Clearly, a heavily doped p^+-n^+ region is essential for the onset of carrier tunneling. Figure 8.27 illustrates the transfer characteristics of the device as a function of gate workfunction. The threshold voltage, which is defined as the required gate voltage for the onset of tunneling considerably increases as the metal gate workfunction value enhances, which confirms decrement of electron density in the channel. Basically, it is essential to find optimum value for the gate workfunction, which

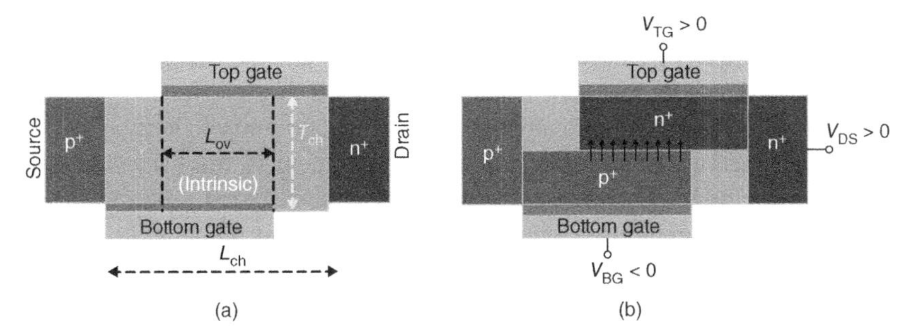

Figure 8.28 Schematic of an n-type EHBTFET in (a) equilibrium state and (b) on-state. Line tunneling occurs vertically along the channel thickness. The p^+-n^+ tunneling junction is electrically created in the intrinsic channel.

apart from the compatibility with the CMOS fabrication process, provides low positive threshold voltage, high on/off current ratio, and steep subthreshold slope. Basically, source doping, channel thickness, and gate workfunction are critical design parameters that affect the tunneling rate, which manifests the employment of gate workfunction engineering technique as well as heavily doped source region for optimizing the devices performance, for each channel thickness.

- **Electron–Hole Bilayer TFET**: Figure 8.28 shows the schematics of electron–hole bilayer tunnel field-effect transistor (EHBTFET). This device exploits the carrier vertical line tunneling through an electrically doped electron–hole bilayer in order to achieve improved switching speed and higher drive current when compared to a conventional lateral p–i–n junction TFET. In the proposed structure, the tunneling window is extended along the channel thickness in the intrinsic channel. The bias-induced p^+-n^+ tunneling region is created via the top and bottom gate polarity. In the initial device structure that has been introduced by Ionescu and coworkers in Ref. [64], in which the top (V_{TG}) and bottom gate (V_{BG}) biases are different. For n-TFET configuration, the negative bias of the bottom gate accumulates holes in the bottom layer of the channel. In the off-state, the electron density in the top layer of the channel is not sufficient to provide a steep p^+-n^+ tunneling junction. Therefore, no BTBT is expected. However, by applying adequate positive bias to the top gate, electrons are accumulated in the top layer and the tunneling occurs when the first subbands of electrons and holes overlap each other.

  Different structures and materials are introduced to improve the performance of the EBTFET device, with symmetric workfunction for the top and bottom gate. However, in [65], asymmetric workfunction engineering technique is employed in which different workfunctions are exploited for the top and bottom gates. In this configuration opposite bias of the bottom gate is eliminated and the carriers are electrically induced in bottom layer of the channel via workfunction difference between the bottom gate and the channel.

- **Multi-gate Structures**: Basically, increasing the gate controllability over the tunneling junction can provide superior improvement in the tunneling rate besides reducing short channel effects. Clearly, the required voltage for the onset of tunneling and the on-state current sensitively depend upon the electric field at the tunneling junction. Consequently, exploiting multi-gate TFET architectures amplifies electric field crowding at the interface of the source and channel and makes a considerable improvement in device electrical measures. The essential analog and digital parameters of the nanowire gate all around TFET are investigated and a low-power application two-stage operational amplifier is implemented based on SiGe nanowire TFET which exhibits a high gain, high power supply rejection ratio (PSRR) value, and gain bandwidth (GBW) value of 90.5 dB, −47.4 dB, and 78.61 MHz, respectively. The proposed structure has very low off-state current and it consumes a very low power of 4.45 μW which manifests device feasibility for low-power applications [66].
  Nanotube core–shell structure with external shell gate and internal core gate surrounding the channel is an electrostatically optimized device that can effectively improve the tunneling current [67, 68]. Figure 8.29 illustrates the schematics of an electrically doped core–shell heterojunction p-TFET with low band gap InAs material in the source region. The core–shell structure and

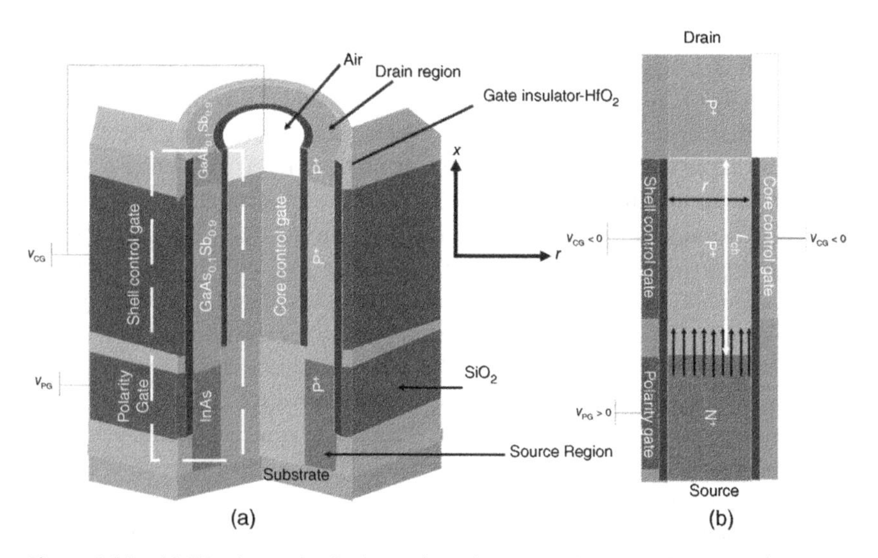

Figure 8.29 (a) 3D schematic of a heterojunction nanotube core–shell electrically doped p-TFET, (b) 2D cut-plane of the TFET device in the on-state.

employment of InAs in the source channel with a low band gap of 0.385 eV provides inherent high drive current.

The electrical characteristics of an electron–hole bilayer n-TFET in a core–shell configuration, with InN as the channel material, are considered based on a simulation study [69]. The device combines the advantages of core–shell gates including line tunneling bilayer TFET. The device possesses different workfunction values for the core and shell gates. The inner core gate induces holes in half of the channel diameter, without employing any bias to it. The electrically hole accumulated area is created by the workfunction difference between the channel and the core gate. The outer shell gate is responsible for switching the device from off to on-state. Clearly, by applying positive bias electrons are electrically induced in half of the channel diameter, and consequently, line tunneling occurs vertically along the channel length.

8.5 Conclusion

To summarize, the present chapter discussed operation principle of TFETs and comprehensively presented different techniques and architectures to satisfactorily establish these devices as one of the great promising candidates for conventional CMOS devices providing sub-thermal subthreshold swing. It is demonstrated that TFET electrical features such as drive current, switching speed, and power consumption can be further improved by exploiting line-tunneling architecture combined with heterostructure material systems. However, experimental results still lag simulation predictions, which fundamentally affects the prospects of TFET technology. TAT, a phonon-assisted BTBT current generation process aided by defects and trap states, emerges from a complex fabrication process of different materials and is found to impose significant subthreshold slope degradation in TFETs. Finally, a comprehensive research effort in terms of fabrication of nanomaterials, besides the development of compact models for TFETs based on emerging innovative materials, is still demanded to effectively integrate such devices into commercially feasible nanoelectronic products and circuits.

References

1 Schwabl, F. (2007). *Quantum Mechanics*. Springer Science & Business Media.

2 Mercier de Lépinay, L., Ockeloen-Korppi, C.F., Woolley, M.J., and Sillanpää, M.A. (2021). Quantum mechanics–free subsystem with mechanical oscillators. *Science* 372 (6542): 625–629.

3 Dodd, J.E. and Gripaios, B. (2020). *The Ideas of Particle Physics*. Cambridge University Press.

4 Goto, E., Murata, K., Nakazawa, K. et al. (1960). Esaki diode high-speed logical circuits. *IRE Transactions on Electronic Computers* 1: 25–29.

5 Oehme, M., Hähnel, D., Werner, J. et al. (2009). Si Esaki diodes with high peak to valley current ratios. *Applied Physics Letters* 95 (24): 242109.

6 Xie, Q., Lee, C.J., Xu, J. et al. (2013). Comprehensive analysis of short-channel effects in ultrathin SOI MOSFETs. *IEEE Transactions on Electron Devices* 60 (6): 1814–1819.

7 Pretet, J., Monfray, S., Cristoloveanu, S., and Skotnicki, T. (2004). Silicon-on-nothing MOSFETs: performance, short-channel effects, and back-gate coupling. *IEEE Transactions on Electron Devices* 51 (2): 240–245.

8 Narendar, V. and Mishra, R.A. (2015). Analytical modeling and simulation of multigate FinFET devices and the impact of high-k dielectrics on short channel effects (SCEs). *Superlattices and Microstructures* 85: 357–369.

9 Kwon, D., Chatterjee, K., Tan, A.J. et al. (2017). Improved subthreshold swing and short channel effect in FDSOI n-channel negative capacitance field effect transistors. *IEEE Electron Device Letters* 39 (2): 300–303.

10 Dhanaselvam, P.S. and Balamurugan, N.B. (2013). Analytical approach of a nanoscale triple-material surrounding gate (TMSG) MOSFETs for reduced short-channel effects. *Microelectronics Journal* 44 (5): 400–404.

11 Ionescu, A.M. and Riel, H. (2011). Tunnel field-effect transistors as energy-efficient electronic switches. *Nature* 479 (7373): 329–337.

12 Boucart, K. and Ionescu, A.M. (2007). Double-gate tunnel FET with high-κ gate dielectric. *IEEE Transactions on Electron Devices* 54 (7): 1725–1733.

13 Avci, U.E., Morris, D.H., and Young, I.A. (2015). Tunnel field-effect transistors: prospects and challenges. *IEEE Journal of the Electron Devices Society* 3 (3): 88–95.

14 Shen, C., Yang, L.T., Samudra, G., and Yeo, Y.C. (2011). A new robust non-local algorithm for band-to-band tunneling simulation and its application to tunnel-FET. *Solid State Electronics* 57 (1): 23–30.

15 Lin, Z., Chen, P., Ye, L. et al. (2020). Challenges and solutions of the TFET circuit design. *IEEE Transactions on Circuits and Systems I: Regular Papers* 67 (12): 4918–4931.

16 Wu, J., Min, J., and Taur, Y. (2015). Short-channel effects in tunnel FETs. *IEEE Transactions on Electron Devices* 62 (9): 3019–3024.

17 Xu, W., Wong, H., Iwai, H. et al. (2017). Analytical modeling on the drain current characteristics of gate-all-around TFET with the incorporation of short-channel effects. *Solid State Electronics* 138: 24–29.

18 Chien, N.D. and Shih, C.H. (2016). Short channel effects in tunnel field-effect transistors with different configurations of abrupt and graded Si/SiGe heterojunctions. *Superlattices and Microstructures* 100: 857–866.

19 Raad, B., Nigam, K., Sharma, D., and Kondekar, P. (2016). Dielectric and work function engineered TFET for ambipolar suppression and RF performance enhancement. *Electronics Letters* 52 (9): 770–772.

20 Tiwari, S. and Saha, R. (2022). Methods to reduce ambipolar current of various TFET structures: a review. *Silicon* 14 (12): 6507–6515.

21 Shaikh, M.R. and Loan, S.A. (2019). Drain-engineered TFET with fully suppressed ambipolarity for high-frequency application. *IEEE Transactions on Electron Devices* 66 (4): 1628–1634.

22 Nigam, K., Kondekar, P., and Sharma, D. (2016). Approach for ambipolar behaviour suppression in tunnel FET by workfunction engineering. *Micro & Nano Letters* 11 (8): 460–464.

23 Shaw, N., Sen, G., and Mukhopadhyay, B. (2020). An analytical approach of elimination of ambipolarity of DPDG-TFET using strained type II staggered SiGeSn heterostructure. *Superlattices and Microstructures* 141: 106488.

24 Kumar, P. and Bhowmick, B. (2019). Comparative analysis of hetero gate dielectric hetero structure tunnel FET and Schottky barrier FET with n^+ pocket doping for suppression of ambipolar conduction and improved RF/linearity. *Journal of Nanoelectronics and Optoelectronics* 14 (2): 261–271.

25 Madan, J. and Chaujar, R. (2016). Gate drain-overlapped-asymmetric gate dielectric-GAA-TFET: a solution for suppressed ambipolarity and enhanced ON state behavior. *Applied Physics A* 122: 1–9.

26 Kumar, S., Singh, K.S., Nigam, K. et al. (2019). Dual-material dual-oxide double-gate TFET for improvement in DC characteristics, analog/RF and linearity performance. *Applied Physics A* 125 (5): 353.

27 Madan, J. and Chaujar, R. (2017). Gate drain underlapped-PNIN-GAA-TFET for comprehensively upgraded analog/RF performance. *Superlattices and Microstructures* 102: 17–26.

28 Gracia, D., Nirmal, D., and Moni, D.J. (2018). Impact of leakage current in germanium channel based DMDG TFET using drain-gate underlap technique. *AEU – International Journal of Electronics and Communications* 96: 164–169.

29 Ranjan, R., Pradhan, K.P., Artola, L., and Sahu, P.K. (2016). Spacer engineered trigate SOI TFET: an investigation towards harsh temperature environment applications. *Superlattices and Microstructures* 97: 70–77.

30 Go, S., Kim, S., Park, J.Y. et al. (2022). Impact of sidewall spacer materials and gate underlap length on negative capacitance double-gate tunnel field-effect transistor (NCDG-TFET). *Solid State Electronics* 198: 108483.

31 Beohar, A., Yadav, N., and Vishvakarma, S.K. (2017). Analysis of trap-assisted tunnelling in asymmetrical underlap 3D-cylindrical GAA-TFET based on hetero-spacer engineering for improved device reliability. *Micro & Nano Letters* 12 (12): 982–986.

32 Bhattacharjee, D., Goswami, B., Dash, D.K. et al. (2019). Analytical modelling and simulation of drain doping engineered splitted drain structured TFET and its improved performance in subduing ambipolar effect. *IET Circuits, Devices and Systems* 13 (6): 888–895.

33 Saxena, M. and Gupta, M. (2022). Undoped drain graded doping (UDGD) based TFET design: an innovative concept. *Micro and Nanostructures* 163: 107147.

34 Sajjad, R.N., Chern, W., Hoyt, J.L., and Antoniadis, D.A. (2016). Trap assisted tunneling and its effect on subthreshold swing of tunnel FETs. *IEEE Transactions on Electron Devices* 63 (11): 4380–4387.

35 Joseph, H.B., Singh, S.K., Hariharan, R.M. et al. (2019). Simulation study of gated nanowire InAs/Si hetero p channel TFET and effects of interface trap. *Materials Science in Semiconductor Processing* 103: 104605.

36 Devi, W.V. and Bhowmick, B. (2019). Optimisation of pocket doped junctionless TFET and its application in digital inverter. *Micro & Nano Letters* 14 (1): 69–73.

37 Li, W. and Woo, J.C. (2018). Optimization and scaling of Ge-pocket TFET. *IEEE Transactions on Electron Devices* 65 (12): 5289–5294.

38 Devi, W.V., Bhowmick, B., and Pukhrambam, P.D. (2020). N^+ pocket-doped vertical TFET for enhanced sensitivity in biosensing applications: modeling and simulation. *IEEE Transactions on Electron Devices* 67 (5): 2133–2139.

39 Sabaghi, M. (2020). Stability performance of a quantum square-shaped extended source PNPN TFET with silicon carbide substrate by means of a physics-based analytical model. *Tecciencia* 15 (28): 28–35.

40 Ahangari, Z. (2019). Design and analysis of energy efficient semi-junctionless n^+ n^+ p heterojunction p-channel tunnel field effect transistor. *Materials Research Express* 6 (6): 065901.

41 Ahangari, Z. (2021). Performance optimization of a nanotube core–shell semi-junctionless p^+ p^+ n heterojunction tunnel field effect transistor. *Indian Journal of Physics* 95: 1091–1099.

42 Joshi, T., Singh, Y., and Singh, B. (2020). Extended-source double-gate tunnel FET with improved DC and analog/RF performance. *IEEE Transactions on Electron Devices* 67 (4): 1873–1879.

43 Karmakar, P., Patil, P., and Sahu, P.K. (2021). Triple metal extended source double gate vertical TFET with boosted DC and analog/RF performance for low power applications. *Silicon* 14: 6403–6413. https://doi.org/10.1007/s12633-021-01425-5.

44 Nanver, L.K., Scholtes, T.L., Sarubbi, F., et al. Pure-boron chemical-vapor-deposited layers: a new material for silicon device processing. *18th International Conference on Advanced Thermal Processing of Semiconductors (RTP)*, Gainesville, FL, USA (28 September 2010). IEEE. pp. 136–139.

45 Sharma, S. and Chaujar, R. (2021). Performance enhancement in a novel amalgamation of arsenide/antimonide tunneling interface with charge plasma junctionless-TFET. *AEU – International Journal of Electronics and Communications* 133: 153669.

46 Xie, H., Liu, H., Wang, S. et al. (2019). Improvement of electrical performance in heterostructure junctionless TFET based on dual material gate. *Applied Sciences* 10 (1): 126.

47 Biswas, A., Rajan, C., and Samajdar, D.P. (2021). Sensitivity analysis of physically doped, charge plasma and electrically doped TFET biosensors. *Silicon* 14: 6895–6908. https://doi.org/10.1007/s12633-021-01461-1.

48 Li, J., Liu, Y., Wei, S.F., and Shan, C. (2020). In-built N^+ pocket electrically doped tunnel FET with improved DC and analog/RF performance. *Micromachines* 11 (11): 960.

49 Shikha, U.S., Krishna, B., Harikumar, H. et al. (2023). OFF current reduction in negative capacitance heterojunction TFET. *Journal of Electronic Materials* 25: 1–3.

50 Joseph, J.A., Adilakshmi, G., Robin, C.R. et al. (2023). Simulation based investigation of triple heterojunction TFET (THJ-TFET) for low power applications. *Silicon* 15 (1): 127–131.

51 Boggarapu, L. and Lakshmi, B. (2023). Design of universal logic gates using homo and hetero-junction double gate TFETs with pseudo-derived logic. *International Journal of Electronics* 110 (3): 442–462.

52 Mao, W., Peng, Z., Yang, C. et al. (2019). A polarization-induced InN-based tunnel FET without physical doping. *Semiconductor Science and Technology* 34 (6): 065015.

53 Spano, C.E., Mo, F., Claudino, R.A. et al. (2022). Tunnel field-effect transistor: impact of the asymmetric and symmetric ambipolarity on fault and performance in digital circuits. *Journal of Low Power Electronics and Applications.* 12 (4): 58.

54 Cho, S.J., Sun, M.C., Kim, G.R. et al. (2011). Design optimization of a type-I heterojunction tunneling field-effect transistor (I-HTFET) for high performance logic technology. *JSTS: Journal of Semiconductor Technology and Science.* 11 (3): 182–189.

55 Yoon, Y.J., Cho, S., Seo, J.H. et al. (2013). Compound semiconductor tunneling field-effect transistor based on Ge/GaAs heterojunction with tunneling-boost layer for high-performance operation. *Japanese Journal of Applied Physics* 52 (4S): 04CC04.

56 Kanungo, S., Ahmad, G., Sahatiya, P. et al. (2022). 2D materials-based nanoscale tunneling field effect transistors: current developments and future prospects. *npj 2D Materials and Applications* 6 (1): 83.

57 Xu, J., Jia, J., Lai, S. et al. (2017). Tunneling field effect transistor integrated with black phosphorus-MoS_2 junction and ion gel dielectric. *Applied Physics Letters* 110 (3): 033103.

58 Jiang, X., Shi, X., Zhang, M. et al. (2019). A symmetric tunnel field-effect transistor based on MoS_2/black phosphorus/MoS_2 nanolayered heterostructures. *ACS Applied Nano Materials* 2 (9): 5674–5680.

59 Afzalian, A., Akhoundi, E., Gaddemane, G. et al. (2021). Advanced DFT–NEGF transport techniques for novel 2-D material and device exploration including HfS_2/WSe_2 van der Waals heterojunction TFET and WTe_2/WS_2 metal/semiconductor contact. *IEEE Transactions on Electron Devices* 68 (11): 5372–5379.

60 Oliva, N., Backman, J., Capua, L. et al. (2020). $WSe_2/SnSe_2$ vdW heterojunction tunnel FET with subthermionic characteristic and MOSFET co-integrated on same WSe_2 flake. *npj 2D Materials and Applications* 4 (1): 5.

61 Yun, S., Oh, J., Kang, S. et al. (2019). F-shaped tunnel field-effect transistor (TFET) for the low-power application. *Micromachines* 10 (11): 760.

62 Kim, S.W., Kim, J.H., Liu, T.J. et al. (2015). Demonstration of L-shaped tunnel field-effect transistors. *IEEE Transactions on Electron Devices* 63 (4): 1774–1778.

63 Kim, J.H., Kim, S., and Park, B.G. (2019). Double-gate TFET with vertical channel sandwiched by lightly doped Si. *IEEE Transactions on Electron Devices* 66 (4): 1656–1661.

64 Lattanzio, L., De Michielis, L., and Ionescu, A.M. (2012). The electron–hole bilayer tunnel FET. *Solid State Electronics* 74: 85–90.

65 Masoudi, A., Ahangari, Z., and Fathipour, M. (2019). Performance optimization of a nanoscale GaSb P-channel electron-hole bilayer tunnel field effect transistor using metal gate workfunction engineering. *Materials Research Express* 6 (9): 096311.

66 Patel, J., Sharma, D., Yadav, S. et al. (2019). Performance improvement of nano wire TFET by hetero-dielectric and hetero-material: at device and circuit level. *Microelectronics Journal* 85: 72–82.

67 Kumar, N., Amin, S.I., and Anand, S. (2020). Design and performance optimization of novel core–shell dopingless GAA-nanotube TFET with $Si_{0.5}Ge_{0.5}$-based source. *IEEE Transactions on Electron Devices* 67 (3): 789–795.

68 Gedam, A., Acharya, B., and Mishra, G.P. (2021). Design and performance assessment of dielectrically modulated nanotube TFET biosensor. *IEEE Sensors Journal* 21 (15): 16761–16769.

69 Ahangari, Z. (2019). Performance investigation of steep-slope core–shell nanotube indium nitride electron–hole bilayer tunnel field effect transistor. *Applied Physics A* 125 (6): 405.

9

GaN Devices for Optoelectronics Applications

Nagarajan Mohankumar[1] *and Girish S. Mishra*[2]

[1] *Symbiosis Institute of Technology, Nagpur Campus, Symbiosis International (Deemed University), Pune, India*
[2] *EECE, School of Technology, GITAM, Bengaluru, India*

9.1 Introduction

Since 1985, the introduction of Silicon (Si)-based devices by Gibbons' Stanford lab, the Si-based devices have been widely explored for various applications in semiconductor industries [1]. Though conventional semiconductors like selenium and germanium were in existence, Si gathered significant research attention in the field of semiconductor devices owing to their reliability, ease of handling, and cost-effectiveness. Various Si-based devices like metal-oxide-semiconductor field effect transistors (MOSFET), and bipolar junction transistors (BJT), have been reported with high input impedance, high drain resistance, easy fabrication, and high speed of operation. However, Si-based MOSFETs have limitations in various device parameters, including mobility, power density, and efficiency [2–9]. In addition, it has been anticipated that the scalability of the Si-MOSFETs will soon reach its dusk owing to the inherent material properties leading to low saturation velocity, high device resistance, limited inversion layer mobility, and low breakdown voltage [10]. In the past few decades, various semiconducting materials, such as sulfide, phosphide, and arsenide, were reported to meet the existing needs for semiconductor devices with maximum output power at higher operating voltages and frequencies. Among the novel semiconducting materials, III-nitride semiconductors, namely gallium nitride (GaN), have attracted significant research interest owing to their higher power density with high-frequency switching and operational temperatures with high mechanical and thermal stability. GaN-based transistors offer improved performance, scalability, and cost-effectiveness for a wide range of power conversion devices. Due to the outstanding performance and inherited properties of the GaN, it is anticipated to replace the Si semiconductor

Advanced Nanoscale MOSFET Architectures: Current Trends and Future Perspectives, First Edition. Edited by Kalyan Biswas and Angsuman Sarkar.

devices market by the year 2030. The global market of GaN for the year 2021 was estimated at US$1.88 billion, and it is predicted to experience a growth of 24.4% (compound annual growth rate [CAGR]) in the next decade (i.e. 2022–2032).

Light-emitting diodes (LEDs) are semiconductor devices that emit photons via electron recombination and holes in an applied electric field. LEDs are used as a lighting source for solid-state lighting devices due to their inherent properties, including high energy efficiency, reliability, cost-effectiveness, environmental protection, and extended life span [11]. Recently, GaN-based LEDs have attracted significant research attention due to their ability to emit white (with yellow phosphors) and blue lights. To date, a considerable augmentation in the light efficiency of GaN-based LEDs has been a significant challenge in semiconductor devices. In addition, the light extraction efficiency (LEE) due to the internal reflections is another limitation of the GaN-based LEDs [12]. Various techniques dealing with chip shape, surface texturing, patterned substrate, etc., have been reported for improving the LEE of GaN LEDs. However, a wide scope of architectural and technological aspects should be considered for enhancing the efficacy of GaN-based semiconductor devices. Interestingly, the alloying of GaN with Al or In can facilitate turnability bandgaps to achieve emission of different wavelengths from deep ultraviolet to infrared (IR) spectrum [13]. Therefore, InGaN-based semiconductor devices have emerged as a promising platform for optoelectronics applications like blue laser diodes, visible-wavelength LEDs, and solar-blind detectors.

9.2 Properties of GaN-Based Material

Generally, GaN and its alloys with aluminum (Al) and indium (In) have been reported for various optoelectronics applications owing to their tunable bandgap. Noteworthy, the bandgap of InN and AlN are testified as 0.7 and 6.2 eV (theoretical), respectively, as shown in Figure 9.1. With a wide bandgap tunability, GaN-based alloys have been employed to develop highly efficacious production of LEDs with different wavelengths starting from UVs (UVA, UVB, and UVC) to visible to IR.

Though the conventional phosphorus and arsenide-based LEDs delivered high stability with improved environmental performance and device flexibility, the non-tunability of their bandgaps limited their applications as LED due to photon emission at the red and IR region of the spectra [14]. Hence, the GaN-based semiconductor is an alternative material for semiconductor-based optoelectronics applications. The wide bandgap variation of GaN-based semiconductors and their alloys results in high-frequency spectrum emitting photons in visible, IR, and UV regions producing green LEDs, blue LEDs, UV LEDs, Quantum-cascade lasers (QCLs), and blue-violet laser diodes (LDs) [15]. The real-time application

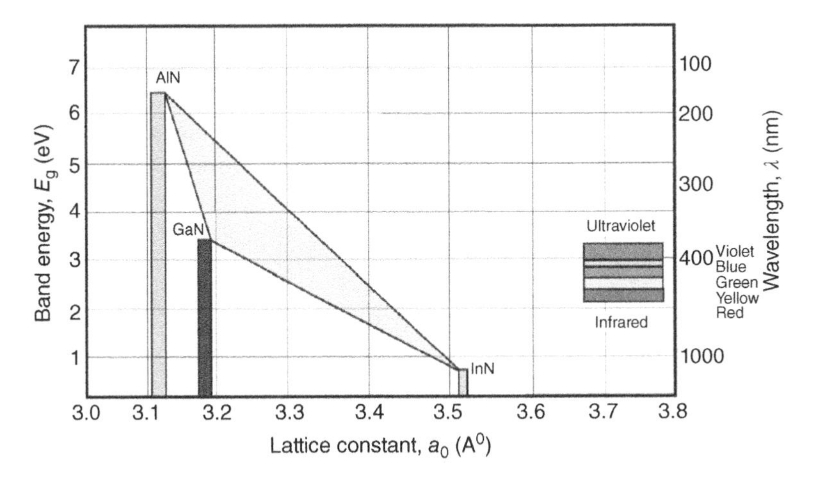

Figure 9.1 Bandgap energy and lattice constant for III-nitride semiconductors. Source: Adapted from Guo et al. [5].

of colored, IR, and UV LED has impacted our day-to-day life, from traffic lights, exterior/interior lighting for aircraft, indicators in domestic vehicles, etc., to blood gas measurement, security, sensing, and lithography [16]. Significant UV light has been reported with its germicidal properties, which have been extensively used in food safety industries and other viral prevention activities.

Table 9.1 compares semiconductors with various functional parameters like, bandgap (eV), breakdown field, electron mobility, maximum velocity, thermal conductivity, and dielectric constant. Herein, GaN has a wide energy bandgap with maximum saturation velocity, which can be utilized to fabricate power semiconductor devices. Besides, the figure of merits (FOM) is the globally accepted

Table 9.1 A material parameter of compound semiconductors with Si at 300 K.

Property	GaN	Si	GaAs	4H-SiC	Si
Bandgap energy, E_g (eV)	3.40	1.12	1.42	3.25	1.12
Saturation velocity, v_{sat} (10^7 cm/s)	3.0	1.0	2.0	2.0	1.0
Electron mobility, μ ($cm^2/(V s)$)	1300	1350	6000	800	1350
Dielectric constant, ε	9.0	11.8	12.8	9.7	11.8
Thermal conductivity, χ (W/(cm K))	1.3	1.5	0.5	4.9	1.5
Field breakdown, E_B (MV/cm)	4.0	0.25	0.4	3.0	0.25
CFOM	489	1	8	458	1

Source: Adapted from Nagakawa [2].

comparison parameter of semiconductors that clearly shows the efficacy of any semiconducting material. Johnson FOM and Baliga FOM are used to analyze high-speed and high-power semiconductor devices. The Johnson FOM states that the devices with high breakdown voltage are typically more sluggish than the lower voltage rating devices. The characteristic equation of Johnson FOM is given as Eq. (9.1)

$$\text{Johnson FOM} = f_{\mathrm{T}} \cdot V_{\mathrm{DS,max}} = \frac{E_{\mathrm{crit}} \cdot v_{\mathrm{Sat}}}{2\pi} \tag{9.1}$$

9.2.1 Bandgap of GaN

An energy band gap of any material is the primary factor in estimating its essential semiconductor characteristics and related applications. This energy bandgap exists in a material due to the chemical interaction between the atoms in the crystal lattice [17]. The higher bandgap semiconductor materials have dominated the market owing to less leakage current with the high-temperature operation and exceptional tunability of energy band. The bandgap of GaN and SiC are relatively high in comparison with Si.

9.2.2 Critical Electric Field of GaN

Impact ionization is influenced by the high electric field in GaN devices with a high chemical bonding nature triggering avalanche breakdown. The breakdown voltage of a device (V_{BR}) is directly proportional to the width of the drift region (w_{drift}) and is represented as follows:

$$V_{\mathrm{BR}} = \frac{1}{2} \cdot w_{\mathrm{drift}} \cdot E_{\mathrm{crit}} \tag{9.2}$$

Typically, for achieving the same breakdown voltage, the drift region of SiC and GaN is usually 10 times lower when compared with Si. Moreover, the drift region should possess depleted carriers in the saturation region to uphold the electric field [18]. Poisson's equation can be used to determine the number of electrons between the two terminals (assuming an N-type semiconductor) using the below formula:

$$q \cdot N_{\mathrm{D}} = \varepsilon_0 \cdot \varepsilon_{\mathrm{r}} \cdot E_{\mathrm{crit}}/w_{\mathrm{drift}} \tag{9.3}$$

where, E_{crit} = Critical field.

In Eq. (9.3), q represents the electron charge (1.6×10^{-19} C), N_{D} represents the total electrons in the given volume, ε_0 represent the permittivity/dielectric constant of the crystal (under DC conditions) of a vacuum measured in farads per meter (8.854×10^{-12} F/m), and ε_{r} represent the relative permittivity of the crystal compared to a vacuum.

9.2.3 ON-resistance of GaN

The ON-resistance (Ω) can be expressed as follows:

$$R_{\mathrm{ON}} = \frac{w_{\mathrm{drift}}}{q \cdot \mu_{\mathrm{n}} \cdot N_{\mathrm{D}}} \tag{9.4}$$

μ_n = Mobility of electron

By solving Eqs. (9.2)–(9.4), we get

$$R_{\mathrm{ON}} = \frac{4 \cdot V_{\mathrm{BR}}^2}{\varepsilon_0 \cdot \varepsilon_r \cdot E_{\mathrm{crit}}^3} \tag{9.5}$$

The temperature-dependent property of GaN w.r.t bandgap is projected via the Varshni relation given as Eq. (9.6):

$$E_{\mathrm{G}}(T) = E_{\mathrm{G,0}} - \frac{\alpha T^2}{T + \beta} \tag{9.6}$$

Table 9.2 summarizes the essential parameters for different compound semiconductors where the tunable alloy bandgap ranges from 0.7 to 6.2 eV [18]. These materials are used as nucleation and buffer regions in most AlGaN/GaN heterostructure-based GaN transistor devices.

The bandgap of ternary compound semiconductors differs from Vegard's rule, which follows the empirical expression shown as Eq. (9.7).

$$E_{\mathrm{g}}(\mathrm{A}_x\mathrm{B}_{1-x}\mathrm{N}) = xE_{\mathrm{g}}(\mathrm{AN}) + (1-x)E_{\mathrm{g}}(\mathrm{BN}) - x(1-x)b \tag{9.7}$$

where, b is a bowing parameter, x is the mole fraction of A, $E_{\mathrm{g}}(\mathrm{AN})$ is the bandgap of AN, and $E_{\mathrm{g}}(\mathrm{BN})$ is the bandgap BN.

$$E_{\mathrm{g}}(\mathrm{Al}_x\mathrm{Ga}_{1-x}\mathrm{N}) = 6.0x + 3.42(1-x) - 1.0x(1-x) \tag{9.8}$$

For GaN, effective density states N_{C} (conduction band) = 2.24×10^{18} cm^{-3}, N_{V} (valence band) = 4.56×10^{19} cm^{-3} at room temperature as well as the effective mass of electrons and holes are 0.20 and 1.49, respectively.

Table 9.2 Comparison of bandgap and Varshni parameters for different compound semiconductors.

Material	∝ (meV/K)	β (K)	$E_{G,0}$ (eV)
Gallium nitride	0.909	830	3.507
Aluminum nitride	1.799	1462	6.23
Indium nitride	0.414	454	0.69

Source: Adapted from [19–22].

9.2.4 Two-dimensional Electron Gas Formation at AlGaN/GaN Interface

The GaN material is dipole in nature with a hexagonal wurtzite crystal structure. The electronegativity difference is responsible for the shift of the electron cloud asymmetrically between the bond related to gallium (Ga) and nitrogen (N), forming a dipole structure on either end of the unit cell [23]. Charge carriers in a semiconductor are improved via doping techniques to enhance the conductivity effectively, but on the other hand, the mobility of the carriers is degraded. Interestingly, only donors like Si (where $E_C - E_D = 0.015\,\text{eV}$) and acceptors like magnesium (0.16 eV) are used as effective dopants for GaN semiconductor materials [24].

It is comparatively modest to accomplish high electron concentrations, while high hole densities require dopant levels beyond $10^{19}\,\text{cm}^{-3}$. It is vital to remember that even high-quality GaN exhibits residual n-type conductivity, leading to carrier densities that can range from 10^{15} to $10^{17}\,\text{cm}^{-3}$, depending on the characteristics of the material [25]. GaN, in contrast to conventional semiconductors, has a unique property that makes it possible to achieve high sheet carrier densities without doping because of its strong polarization effects. The anion sublattice can move further away from the cation sublattice due to strain in the crystal, which can cause the III-Nitride semiconductors to become piezoelectrically polarized. Figure 9.2 depicts the ball and stick illustration of the tetrahedral bond between gallium and nitrogen; the polarization is (simplified) as represented [26]. The Ga-polar

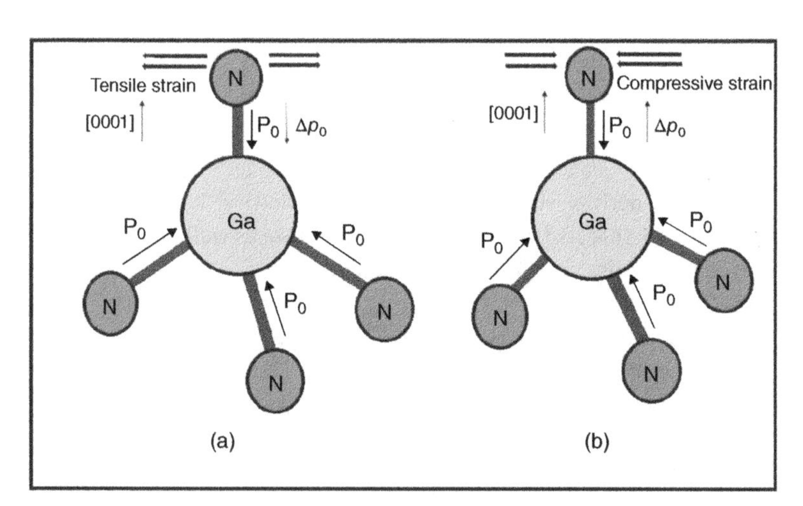

Figure 9.2 A GaN tetrahedron in a schematic ball-and-stick form with in-plane (a) tensile and (b) compressive strain and a net polarization effect.

arrangement is depicted in Figure 9.2. The polarization vectors are generated due to the closeness between the electron cloud and the nitrogen atoms. The vertical and in-plane polarization mechanisms cancel each other in a perfect tetrahedron. The polarization produced by the triple bonds diminishes when an in-plane tensile strain is applied, as revealed in Figure 9.2a. This fallout in augmentation of net polarization along the $(000\bar{1})$ direction [27–29]. When a compressive strain is applied, the triple bond polarization increases, leading to augmentation of net polarization along the (0001) direction.

As shown in Figure 9.2a, the electric field and the polarization are in opposite directions of motion in the field. An AlGaN/GaN interface is created by epitaxially growing AlGaN on GaN. The lattice mismatch of AlGaN/GaN is less because AlN has smaller atom spacing than GaN. Therefore, the AlGaN atoms in this layer are drawn in the direction of the GaN channel, creating tensile strain (piezoelectric polarization). The compressive strain in the GaN layer is produced by the pseudomorphic lattice connection between the relaxed AlN and the overlying GaN as shown in Figure 9.2b. A positive charge is produced at the AlGaN/GaN interface by piezoelectric polarization (P_{pz}) and net spontaneous polarization (P_{sp}), and it will form the two-dimensional electron gas (2DEG) at AlGaN/GaN interface. The electrons in the 2DEG will conduct electric current by applying a variable potential across it [30–33] (Figure 9.3).

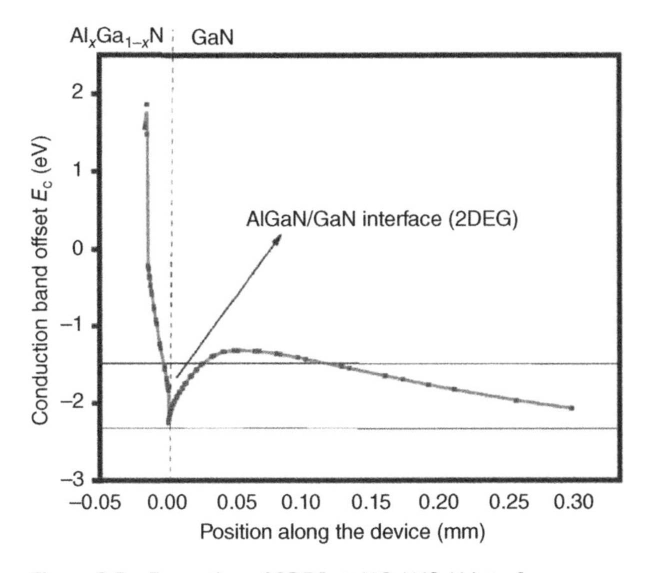

Figure 9.3 Formation of 2DEG at AlGaN/GaN interface.

9.3 GaN LEDs

Our daily lives depend heavily on light. While biological processes like photosynthesis, cellular reactions, and other crucial life-sustaining activities are made possible by natural light sources, artificial lighting has dramatically influenced human existence by enabling us to work into the hours of sunrise and sunset [34–36]. Moreover, various uses for these artificial light sources in lighting, communications, transportation, healthcare, and other fields have significantly changed our civilization. The energy efficiency of artificial light sources has increased tremendously, starting with fire as a source of light to the development of the incandescent bulb and modern solid-state lighting devices. The PN junction diode-based LEDs are well-known semiconductor light sources. Near bandgap, radiation is produced herein as the electrons and holes at the P-N junction recombine at a high forward voltage. Most of the early LED structures used homojunction semiconductors, as shown in Figure 9.4 [37, 38]. Although homojunction semiconductor devices are far simpler to fabricate, most researchers continue to use heterojunction semiconductor structures as LED architectures.

The two vital advantages of heterojunction structures are as follows:

(1) The carrier injection efficiency of the device can be enhanced by using the quantum well (QW) structure, as shown in Figure 9.5. Herein, to create distinct energy levels, the electrons and holes in the QW are confined to a one-dimensional or quasi-two-dimensional space [39]. Due to the thin QW layers, the quantum confinement effect produces an extremely high injection efficiency in LEDs.
(2) The quantum barriers with wide-bandgap, which are translucent to the photons produced in the narrow-bandgap QW layers, decrease the photon reabsorption. This light generated by QW layers can travel in any direction

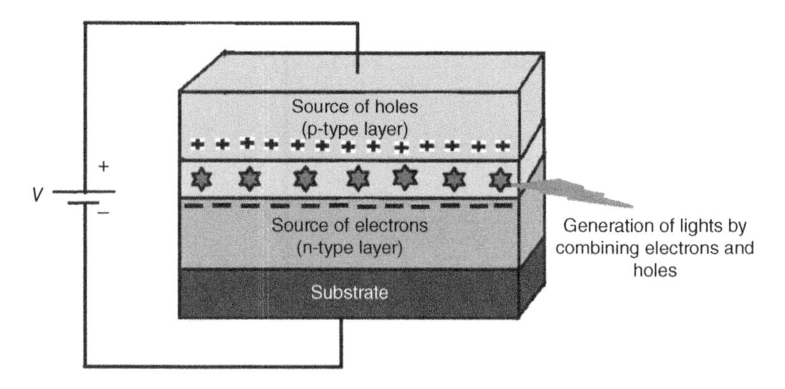

Figure 9.4 Diagrammatic representation of homojunction semiconductor LED structure.

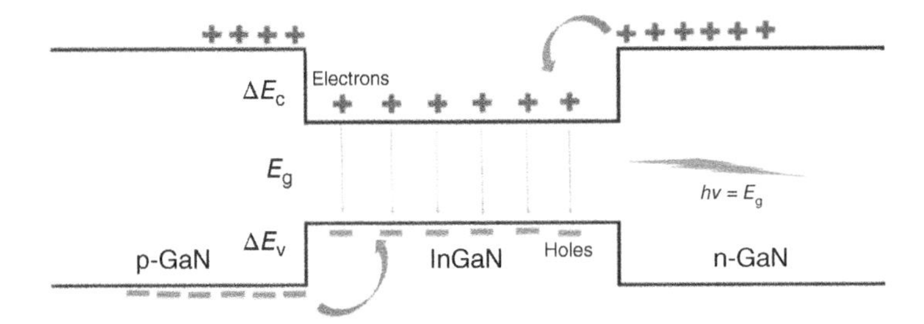

Figure 9.5 Diagrammatic representation of homojunction semiconductor LED structure.

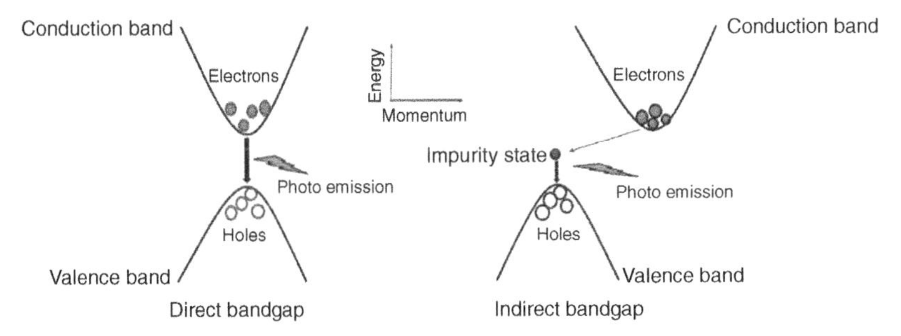

Figure 9.6 Diagrammatic representation of radiative recombination in direct and indirect bandgap semiconductors.

that is possible. In order to focus the light upwards, a standard hemispherical dome-shaped separate LED is used.

The electron–hole recombination process in semiconductor devices can be broadly classified into two types (i) direct and (ii) indirect, as illustrated in Figure 9.6.

9.3.1 Different Colors LEDs

The wavelength of the light emitted depends on the energy bandgap of the semiconductor material. The wavelength of an LED with a homojunction structure is determined by:

$$\lambda_0 = \frac{1.24}{E_{\text{gap}}}\ \mu\text{m} \tag{9.9}$$

When electrons and holes recombine from sub-bands in a QW structure with the same emitting material, the emission is blue-shifted, which means the produced light has a shorter wavelength than the λ_0. We require various materials with different energy bandgaps as emitting materials to achieve diverse colors [40, 41].

Table 9.3 Different colors of LEDs for different semiconductor materials.

Semiconductor material	Wavelength range	Color
GaAs AlGaAs	>760 nm	Infrared
AlGaAs GaAsP AlGaInP	610–760 nm	Red
AlGaInP GaAsP GaP	570–610 nm	Amber/Yellow
AlGaInP InGaN AlGaP GaP	500–570 nm	Green
ZnSe InGaN SiC	450–500 nm	Blue
InGaN	380–450 nm	Violet
C (diamond) AlN AlGaN AlGaInN	<380 nm	Ultraviolet

Table 9.3 displays the lists of inorganic semiconductor materials and their corresponding LED's luminous color.

In addition to these hues, white LEDs have dominated the market in lighting applications and LED displays. There are two ways to create white LEDs because white light is not monochromatic. Combining three monochromatic LEDs is one approach. The conversion of monochromatic emission of blue or violet light LEDs to red, green, or yellow is achieved via a simple covering of the LEDs with one or more layers of fluorescent materials [42]. When these hues are combined with any of the blue or violet light of the LEDs, they appear white to the naked eye.

9.3.2 μ-LEDs

OLEDs face major challenges, including poor efficiency, average performance, and short lifespan. So, there is an urgent need to possess a technological transformation from OLEDs which are widely used in electronic appliances like mobile

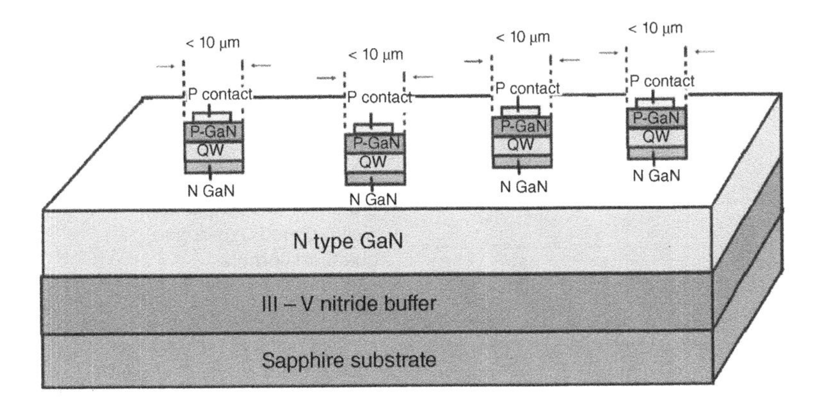

Figure 9.7 Diagrammatic representation of GaN/InGaN μ-LEDs in microns dimension.

phones, and televisions, in the near future [43]. Presently, μ-LEDs displays are widely used for new-generation display screens. These displays combine the RGB color LEDs on the same substrate as a single display panel and are generally measured in a geometrical scale of microns. Interestingly, OLED and LCDs are outperformed by μ-LED displays in terms of luminosity, speed, and contrast. Still, the efficiency performance of μ-LEDs, which results from the surface recombination effect, is a setback [44]. Early in the twenty-first century, some researchers created micron-sized GaN-based μ-LEDs (Figure 9.7), although the efficiency was around 1.5%. The excessive surface recombination of the μ-LEDs is the cause of the low efficiency [45].

9.3.3 Micro-LEDs with GaN-based N-doped Quantum Barriers

Figure 9.8 depicts the GaN/InGaN-LEDs with doped quantum barriers. Generally, the GaN quantum barriers are intrinsic or undoped, leading to lower efficiency. However, recently the GaN has been doped as either p-type or n-type as the quantum barrier layer widely in LEDs. It can be observed that GaN/InGaN μ-LEDs with doped quantum barriers perform more efficiently than μ-LEDs with un-doped quantum barriers at some working current densities as depicted in [45–47].

9.3.4 Blue Light Emission in GaN-based LEDs

Direct bandgap semiconductors in which the light emission is blue ($E_g = 2.6\,\text{eV}$) are suitable candidates for future Blue light LEDs. The development of blue LEDs is crucial because they are required for manufacturing LEDs in all three primary hues (R, G, and B) (Red-Green-Blue) [48]. These color combination sources can be mixed for different wavelengths in the visible spectrum. The process of generating

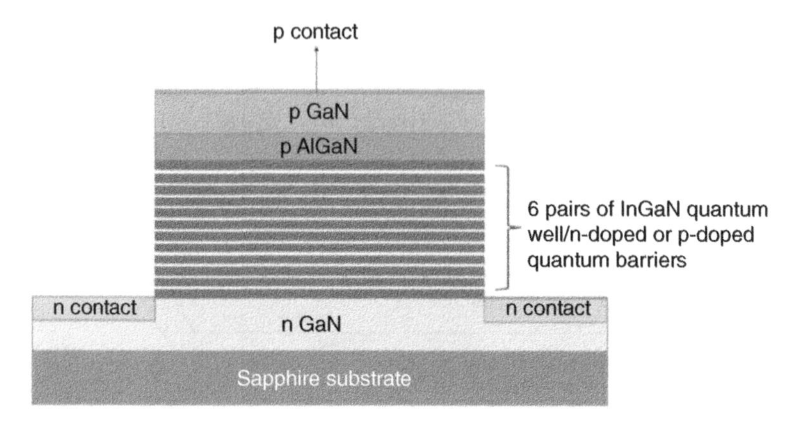

Figure 9.8 Diagrammatic representation of Micro QW GaN/InGaN μ-LEDs undoped quantum barriers.

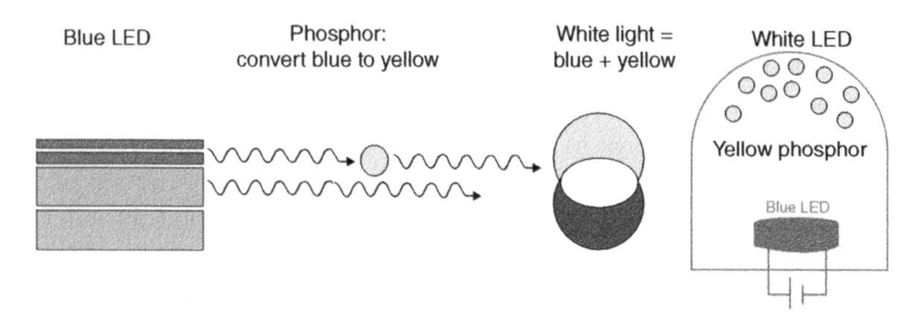

Figure 9.9 Diagrammatic representation of the white light generation process utilizing a blue LED and a yellow phosphor.

white light utilizing a blue LED and a yellow phosphor is shown in Figure 9.9. In the past, several materials, such as ZnSe, and GaN, were considered for blue light emission. On well-known substrates, crystalline GaN with regulated surface roughness was difficult to develop [49–52]. Moreover, it was very challenging to dope this material, particularly to create a p-type GaN, because introducing dopants induces flaws in the host material.

9.3.5 Characteristics

LED characteristics are not only defined by their emission wavelength. The material and epitaxial choices during their fabrication process can also define the device format, electrical characteristics, emission pattern, and optical performance properties. In this chapter, the LED devices utilized throughout are micro-pixelated LED arrays (μ-LEDs). The difference between these μ-LEDs and conventional broad-area devices is that each emitter element in a μ-LED array will have a diameter of 100 μm or less. In contrast, more conventional broad-area devices typically have active areas of 0.1–1 mm^2. Broad areas of other such devices are typically used for high-powered general-purpose illumination applications [53]. On the other hand, μ-LED devices are the basis of new types of high-resolution display technology and have other advantageous characteristics, such as high modulation bandwidths.

9.4 GaN Lasers

A laser is defined as Light amplification through stimulated emission of radiation. This is achieved via population inversion, where the system has the maximum number of electrons at the excited state due to some applied energy [11]. Later, the excited electrons release energy from the photon or phonon to reach the ground state to retain the stability or state of equilibrium. Various potential transitions could happen between the valance and conduction bands and raise their level to the conduction band. These are (a) absorption, (b) spontaneous emission, and (c) stimulated emission in a semiconductor (Figure 9.10). Herein, the lower energy levels dominate the emission energy as the majority of carriers relax into the lowest energy level. The energy of the emitted photon is calculated using Eq. (9.10), as shown below,

$$\text{The energy of the emitted photon, } \Delta E = E_C - E_V = \frac{hc}{\lambda} \tag{9.10}$$

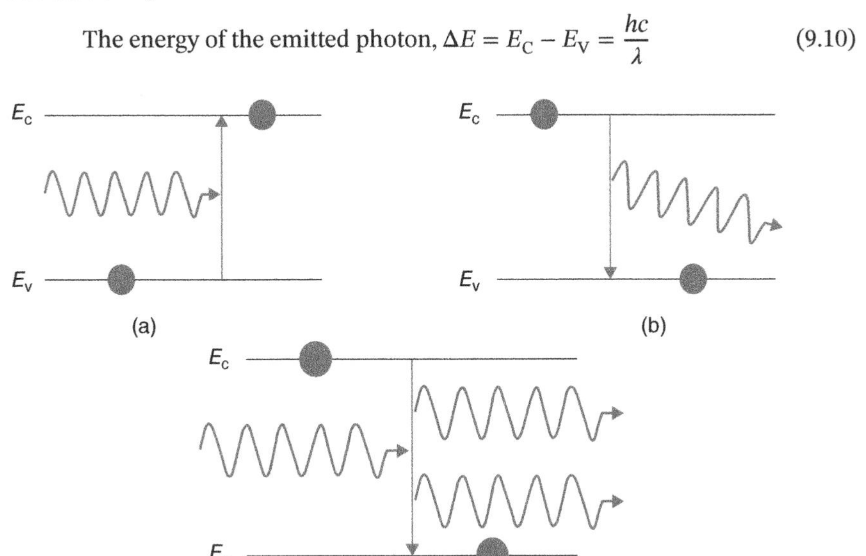

Figure 9.10 Diagrammatic representation of (a) absorption, (b) spontaneous emission, and (c) stimulated emission in a semiconductor.

where

λ	–	Electromagnetic wavelength of the emitted wave
h	–	Planck's constant
c	–	Velocity of light in vacuum
E_C	–	The electrons ground-state energy
E_V	–	Energy of the electron in an excited state

9.4.1 Blue Laser Diodes

The Blue Laser diodes are known for their emission wavelength ranging from 440 to 485 nm. These are widely used in the data storage applications like read and write CDs, and DVDs. In addition, they are employed in high-resolution printing applications. The bandgap of the material needed to emit photons in the blue area is 2.64 eV [30]. Interestingly, SiC with Group III–V and II–VI are the materials under evaluation owing to their adequate energy gap for emission in the blue area. For the active area of the laser diode, SiC was excluded because of its low internal quantum efficiency and indirect bandgap. The direct bandgap material systems ZnO (3.3 eV), GaN (3.4 eV), and ZnSe (2.7 eV) were also taken into account. Due to their short lifespan, ZnSe-based lasers were never commercially viable, and the development of low-resistivity p-contact ZnO light emitters is essential for future lighting applications [39]. Although GaN has a high dislocation density, it has a high internal quantum efficiency and a longer lifetime. Due to these outcomes, blue laser diodes based on GaN became predominant in the market depicted in Figure 9.11.

To date, adapting AlGaN as the cladding and GaN as the optical waveguide, a wide spectrum of design layouts for laser diodes have been reported. To minimize the optical losses, the refractive index difference between core and cladding should be high and subsequently achieved with AlGaN as the cladding layer. This optimizes the device gain due to the electric field's effective confinement. Surprisingly, many researchers have analyzed SiC devices adopting multiple quantum wells in the active layers [30]. Moreover, a p-AlGaN layer that blocks electrons was present preventing electron overflow. By changing the characteristics of each of these layers, these devices exhibit various wavelengths and efficiency with high output power. ELOG (Epitaxially laterally overgrown GaN) on sapphire has been reported with a reduced defective propagation at the active region during growth of GaN. A p-AlGaN electron breaking layer (EBL) is created as the barrier to inject sufficient holes into the quantum well.

Under low bias conditions, the major challenge of blue laser diode operation is doping n and p-type AlGaN cladding layers with high Al mole fraction. The laser diode external quantum efficiency is lowered due to the low conductivity of

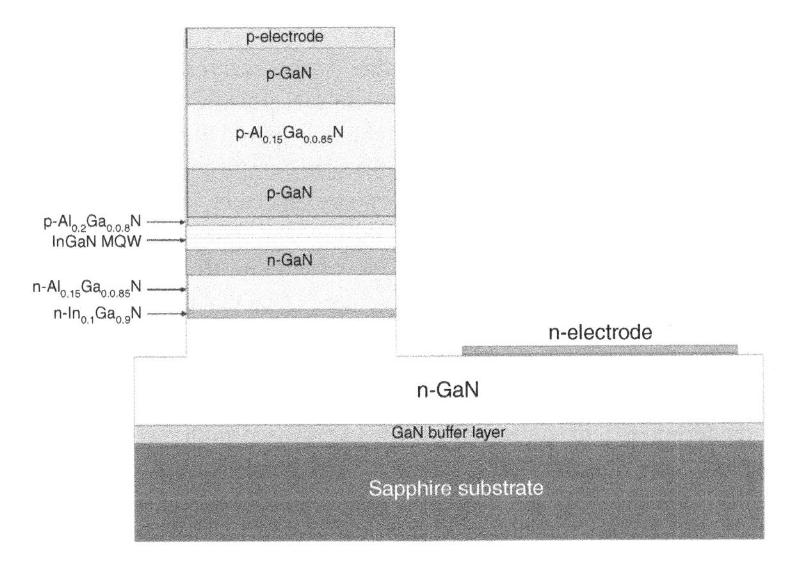

Figure 9.11 Diagrammatic representation of GaN Blue Laser.

AlGaN cladding layers and the increased series resistance. The InGaN quantum well-containing active region is crucial for efficient laser diode operation [39]. The active region is in charge of trapping the carriers through finite potential InGaN quantum wells. The poor confinement of carriers in the quantum well is brought on by the short conduction band offset, which results in electron current overflow. As a result, lasing requires a higher carrier density.

9.5 GaN HEMTs for Optoelectronics

The tunable direct bandgap energies of the III-nitride material system vary between 2 and 6.2 eV. This bandgap energy includes the spectrum regions of near-infrared, visible, and deep ultraviolet and corresponds to a wavelength range from about 650 to around 300 nm [54]. GaN has excellent material characteristics, including its high electron saturation drift velocity, acceptable thermal conductivity, and chemical stability, giving it great potential in power electronic applications and ensuring that it will be a promising material in optoelectronic devices. AlGaAs, InAlGaP, GaAsP, and GaN are frequently employed in LED materials [55]. The most extensive visible spectrum range, from violet to red, is covered by the GaN material system. GaN and its alloys are the only materials that can emit light with a wavelength of less than 500 nm. In 1971, Pankove et al. [56] developed the first GaN LED with a metal-insulated semiconductor structure, generating blue light with a peak wavelength of 475 nm. GaN LEDs based on PN junction topologies and high-brightness blue double heterostructures were presented in 1989 and 1993 due to advancements in GaN growing techniques and GaN epitaxial layers [57, 58]. GaN blue LED technology paved the way for white solid-state lighting, which was predicted to outperform current conventional and fluorescent lighting technologies in terms of energy efficiency, longer lifetime, and size reduction [58]. A phosphor can be used in a GaN-based LED package to produce a white emission by converting and combining the blue light, or white solid-state lighting can be created by integrating LEDs that emit red, green, and blue light individually into a single device [59–61]. The two methods for producing white light are shown in Figure 9.12. Compared to the white LED technique using phosphors, the color-mixing method is more expensive and has better color rendering [61]. In contrast, the white LEDs based on the color-mixing approach perform less well than expected because of the poor quality of green LEDs [62]. Moreover, the phosphor technique involved low cost and small package size attracted a lot of interest. In one of the most widely used phosphor-based white LED approaches, blue emission from InGaN/GaN multiple-quantum-well (MQW) LEDs is converted to yellow light to produce white light [63]. Another method is to pump a wide-spectrum phosphor for the creation of white light using a UV LED [54]. Due to its good characteristics of a precise absorption spectrum,

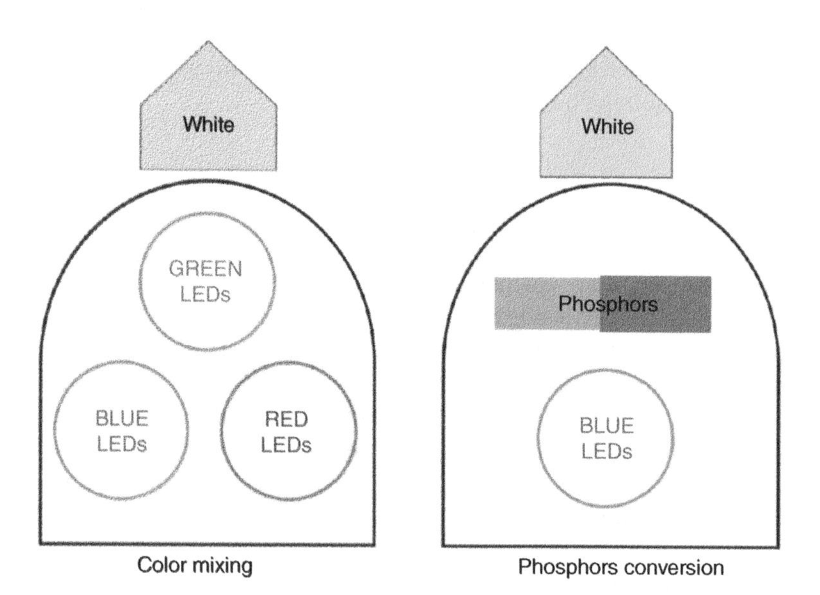

Figure 9.12 Diagrammatic representation of an illustration for white solid-state lighting.

excellent photoluminescence efficiency, and a wide emission wavelength range, quantum dots (QDs) have recently been used as phosphors in solid-state white light technology [64]. Further advancing the study of QD phosphors in white LEDs is the discovery of nonradiative energy transfer between InGaN/GaN QWs and QDs, which is thought to be a high-efficiency color conversion by avoiding multiple color conversion steps [65].

9.6 GaN Sensors

High-sensitivity pH sensing has become a significant area of research because the changes in pH can be constructively used as a sign for malignant tumors and acidosis-related diseases. Humans have a nominal pH between 7.35 and 7.45. They are modeling the sensing response as a change in the channel potential caused by a change in surface charge at the top AlGaN surface due to the production of a hydroxyl group that is dependent on the pH of the solution under test. A large diversity of sensing methods has evolved for the same, including assay methods along with ELISA (Enzyme-Linked Immunosorbent Assay) [66], electrochemical methods [67], mass spectroscopy [68], and fluorescent biosensing methods [69]. Moreover, to achieve high sensitivity, label-free detection, rapid response, and high resolution, ISFET (Ion Sensitive Field Effect Transistors) based sensors have been created [70, 71]. Modern intelligent healthcare applications can readily integrate nanoscale Si-based devices like Si nanowires, carbon nanotubes, and graphene, as biosensors. However, they are restricted by solution instability, material degradation, and low reliability due to long-term performance shifts [72–75]. GaN-based devices, which have the inherent qualities of biocompatibility and high stability in aqueous solutions, can be used to overcome the above limitations [76]. Due to the presence of spontaneous and piezoelectric polarization fields, GaN HEMTs are capable of attaining 2DEG sheet carrier densities (n_s) of above $10^{13}/cm^2$ (without any form of active doping). The quantum well generated at the interface can be modified by adjusting the Al mole fraction in the barrier AlGaN layer. The quantum well can be further improved by adding a thin AlN layer between the AlGaN and GaN layers. AlGaN/GaN HEMT, due to the presence of 2DEG at the heterointerface, shows the potential in bio/chemical sensing applications. Each change in the environment affects the surface charge of the HEMT created by this 2DEG channel at the interface, which in turn increases the channel density and potential at the device interface. In order to improve device sensitivity, inter-digitated sensing areas, and numerous sensing segments have been employed with a high sensitivity of 1.35 mA/pH [77]. Meanwhile, the device has the limitation of its complicated design, which makes it challenging in the fabrication process. The advantages of the proposed device architecture include

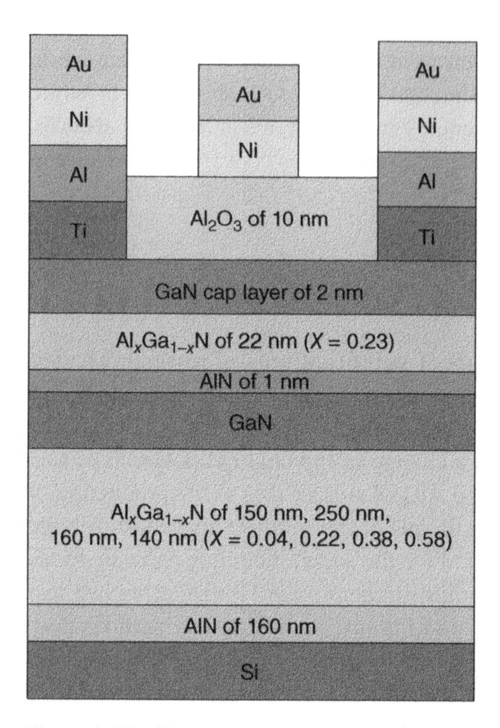

Figure 9.13 Diagrammatic representation of testing for pH sensitivity of dielectric modulated MOS-HEMT for biosensing applications.

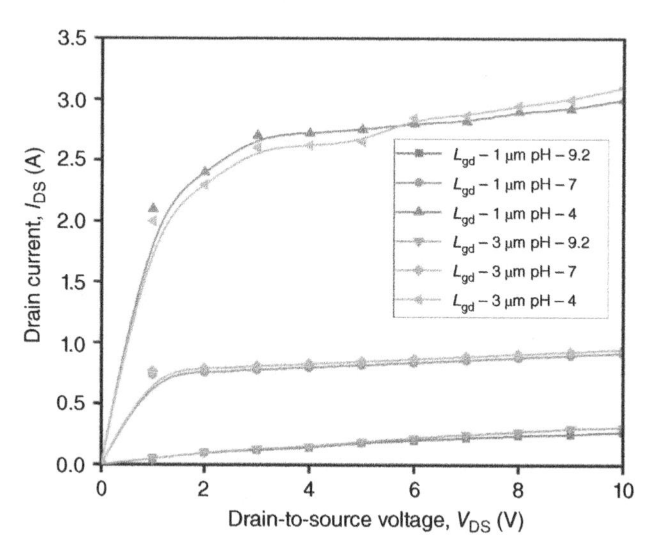

Figure 9.14 Diagrammatic representation of $I_{DS}-V_{DS}$ characteristics by changing the pH (4, 7, 9.2).

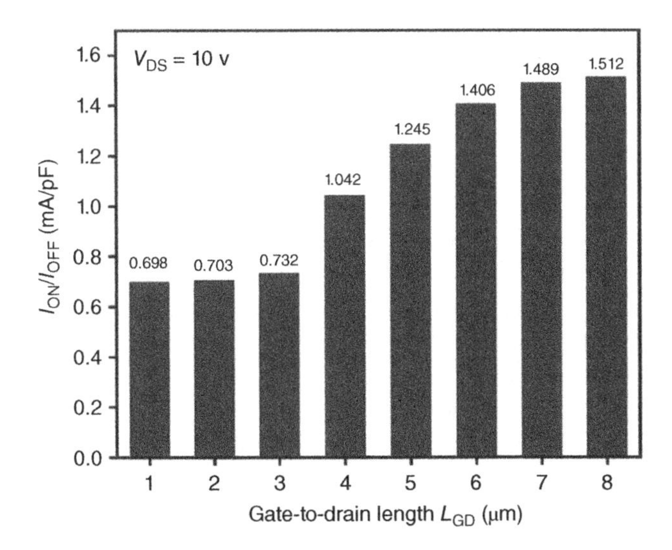

Figure 9.15 Diagrammatic representation of Sensitivity variation for device with at $V_{DS} = 10\,V$.

the high sensitivity and resolution that may be achieved without intentional doping or gate functionalization, making them appropriate for integration into bioelectronic systems (Figure 9.13).

The I_{DS}–V_{DS} characteristics for varying pH (4, 7, 9.2) are depicted in Figure 9.14. The figure visualizes that the device with $L_{GD} = 1\,\mu m$ displays a greater ON current than the device with $L_{GD} = 3\,\mu m$ at pH = 7. The device sensitivities show the opposite behavior, increasing by 5% with an increase in L_{GD} from 1 to 3 μm, producing less sensitivities for 1 μm as compared to 3 μm, respectively, as shown in Figure 9.15. This suggests that although the ON current drops with increasing L_{GD} in a real-life setting, sensitivity improves with increasing spacing, and lower fabrication complexity can be ensured by utilizing greater L_{GD} values. The sensitivity is observed to be 1.042 mA/pH when L_{GD} is made double L_{SG}. Moreover, it has been observed that devices become more sensitive when the L_{GD} is doubled or increased.

References

1 Lécuyer, C. (2005). What do universities really owe industry? The case of solid-state electronics at Stanford. *Minerva* 43 (1): 51–71.

2 Nakagawa, A. (2008). Recent advancement in high voltage power devices and ICs; Challenges to achieve silicon limit characteristics. *2008 International*

Symposium on VLSI Technology, Systems and Applications (VLSI-TSA), Hsinchu, Taiwan (21–23 April 2008). *IEEE*. pp. 103–104.

3 Pan, A. and Chui, C.O. (2014). RF performance limits of ballistic Si field-effect transistors. *2014 IEEE 14th Topical Meeting on Silicon Monolithic Integrated Circuits in RF Systems*, Newport Beach, CA, USA (19–23 January 2014). *IEEE*. pp. 68–70.

4 Frank, D.J. (2002). Power-constrained CMOS scaling limits. *IBM Journal of Research and Development* 46 (2): 235–244.

5 Guo, J., Datta, S., Lundstrom, M., et al. (2002). Assessment of silicon MOS and carbon nanotube FET performance limits using a general theory of ballistic transistors. *Digest. International Electron Devices Meeting* (vol. 2), San Francisco, CA, USA (8–11 December 2002). *IEEE*. pp. 711–714.

6 Dhar, R., Deyasi, A., and Sarkar, A. Analysis of optical performance of dual-order RAMAN amplifier beyond 100 THz spectrum. In: *Advanced Materials for Future Terahertz Devices, Circuits and Systems*, vol. 727 (ed. A. Acharyya and P. Das), 193. Springer.

7 Kawaguchi, Y., Yamaguchi, Y., Kanie, S., Baba, A., and Nakagawa, A. (2008). Proposal of the method for high efficiency DC-DC converters and the efficiency limit restricted by silicon properties. *2008 IEEE Power Electronics Specialists Conference*, Rhodes, Greece (15–19 June 2008). *IEEE*. pp. 147–152.

8 Assad, F., Ren, Z., Vasileska, D. et al. (2000). On the performance limits for Si MOSFET's: a theoretical study. *IEEE Transactions on Electron Devices* 47 (1): 232–240.

9 Meindl, J.D. (1983). Theoretical, practical and analogical limits in ULSI. *1983 International Electron Devices Meeting*, Washington, DC, USA (5–7 December 1983). *IEEE*. pp. 8–13.

10 Adler, M., Owyang, K., Baliga, B., and Kokosa, R. (1984). The evolution of power device technology. *IEEE Transactions on Electron Devices* 31: 1570.

11 DenBaars, S.P., Feezell, D., Kelchner, K. et al. (2013). Development of gallium-nitride-based light-emitting diodes (LEDs) and laser diodes for energy-efficient lighting and displays. *Acta Materialia* 61 (3): 945–951, ISSN 1359-6454, https://doi.org/10.1016/j.actamat.2012.10.042.

12 Kaiser, S., Jakob, M., Zweck, J. et al. (2000). Structural properties of AlGaN/GaN heterostructures on Si(111) substrates suitable for high-electron mobility transistors. *Journal of Vacuum Science & Technology, B: Microelectronics and Nanometer Structures–Processing, Measurement, and Phenomena* 18 (2): 733–740.

13 Aktas, O., Kim, W., Fan, Z., et al. (1995). High-transconductance GaN MODFETs. *Proceedings of International Electron Devices Meeting*, Washington, DC, USA (10–13 December 1995). *IEEE*. pp. 205–208.

14 Nakamura, S., Senoh, M., and Mukai, T. (1993). P-GaN/N-InGaN/N-GaN double-heterostructure blue-light-emitting diodes. *Japanese Journal of Applied Physics* 32: L8.

15 Nakamura, S., Mukai, T., Senoh, M. et al. (1993). $In_xGa_{(1-x)}N/In_yGa_{(1-y)}N$ superlattices grown on GaN films. *Journal of Applied Physics* 74: 3911.

16 Nakamura, S., Mukai, T., and Senoh, M. (1994). Candela-class high-brightness InGaN/AlGaN double-heterostructure blue-light-emitting diodes. *Applied Physics Letters* 64: 1687.

17 Vampola, K.J., Iza, M., Keller, S. et al. (2009). Measurement of electron overflow in 450 nm InGaN light-emitting diode structures. *Applied Physics Letters* 94 (6): 061116.

18 Efficient Power Conversion Corporation (2013). EPC2010 – Enhancement-mode power transistor. EPC2010 datasheet, March 2011 (Revised February 2013). https://epc-co.com/epc/Portals/0/epc/documents/datasheets/EPC2010_datasheet.pdf.

19 Vurgaftman, I., Meyer, J.R., and Ram-Mohan, L.R. (2001). Band parameters for III–V compound semiconductors and their alloys. *Journal of Applied Physics* 89 (11): 5815–5875.

20 Basak, A. and Sarkar, A. (2020). Drain current modelling of asymmetric junctionless dual material double gate MOSFET with high K gate stack for analog and RF performance. *Silicon* 14: 75–86.

21 Bouchkour, Z., Orlianges, J.C., Thune, E. (2012). Quantum confinement effect in nanostructured AlN films. *The First International Conference on Materials Energy and Environment (ICMEE 2012),* 2012 2nd International Conference on Future Environment and Energy (ICFEE 2012), Conference, in-person, 26th to 28th February 2012, Singapore.

22 Walukiewicz, W., Li, S.X., Wu, J. et al. (2004). Optical properties and electronic structure of InN and In-rich group III-nitride alloys. *Journal of Crystal Growth* 269 (1): 119–127.

23 Baliga, B.J. (1989). Power semiconductor device figure-of-merit for high frequency applications. *IEEE Electron Device Letters* 10: 455–457.

24 Beach, R. (2010). Master the fundamentals of your gallium-nitride power transistors. Electronic Design Europe, 29 April 2010.

25 Transphorm (2013). TPH3006PD – GaN power low-loss switch. Transform datasheet, 27 March 2013.

26 Reusch, D., Gilham, D., Su, Y., and Lee, F. (2012). Gallium nitride-based 3D integrated non-isolated point of load module. *2012 Twenty-Seventh Annual IEEE Applied Power Electronics Conference and Exposition (APEC), Twenty-Seventh Annual IEEE*, Orlando, FL, USA (5–9 February 2012). *IEEE*. pp. 38–45.

27 Deyasi, A. and Sarkar, A. (2022). Effect of material composition on noise performance of sub-micron high electron mobility transistor. *Microsystem Technologies* 28: 577–585.

28 Paul, S., Mondal, S., and Sarkar, A. Characterization and analysis of low-noise GaN-HEMT based inverter circuits. *Microsystem Technologies* 27: 3957–3965.

29 Efficient Power Conversion Corporation (2013). EPC2001 – Enhancement-mode power transistor. EPC2001 datasheet, March 2011 (Revised January 2013). http://epc-co.com/epc/documents/datasheets/EPC2001_datasheet.pdf.

30 Wen, B., Xu, C., Wang, S. et al. (2018). Dual-lasing channel quantum cascade laser based on scattering-assisted injection design. *Optics Express* 26 (7): 9194–9204.

31 Jin, S.X., Li, J., Lin, J.Y., and Jiang, H.X. (2000). InGaN/GaN quantum well interconnected microdisk light emitting diodes. *Applied Physics Letters* 77 (20): 3236–3238.

32 Pong, B.J., Chen, C.H., Hsu, J.F., Tun, C.J., and Chi, G.C. (2005). Electro-ridge for large injection current of micro-size InGaN light emitting diode. *Fifth International Conference on Solid State Lighting* (vol. 5941). International Society for Optics and Photonics, OPTICS AND PHOTONICS 2005, 31 July–4 August 2005, San Diego, California, United States. p. 59410Y.

33 Tian, P., McKendry, J.J., Gong, Z. et al. (2012). Size-dependent efficiency and efficiency droop of blue InGaN micro-light emitting diodes. *Applied Physics Letters* 101 (23): 231110.

34 Grimmeiss, H.G. and Koelmans, H. (1961). Analysis of p-n luminescence in Zn-Doped GaP. *Physics Review* 123: 1939.

35 Starkiewicz, J. and Allen, J.W. (1962). Injection electroluminescence at p-n junctions in zinc-doped gallium phosphide. *Journal of Physics and Chemistry of Solids* 23: 881.

36 Gershenzon, M. and Mikulyak, R.M. (1961). Electroluminescence at p-n junctions in gallium phosphide. *Journal of Applied Physics* 32: 1338.

37 Holonyak, N. Jr., and Bevacqua, S.F. (1962). Coherent (visible) light emission from $Ga(As_{1-x}P_x)$ junctions. *Applied Physics Letters* 1: 82.

38 Neamen, D.A. (2011). *Semiconductor Physics and Devices: Basic Principles.* McGraw-Hill.

39 Sun, G. (2010). The intersubband approach to Si-based lasers. In: *Advances in Lasers and Electro Optics* (ed. N. Costa and A. Cartaxo). InTech, https://doi.org/10.5772/193.

40 Biswas, K., Sarkar, A., and Sarkar, C.K. Effect of varying indium concentration of InGaAs channel on device and circuit performance of nanoscale double gate heterostructure MOSFET. *Micro & Nano Letters* 13 (5): 690–694.

41 Nakamura, S., Mukai, T., and Senoh, M. (1994). Candela-class high-brightness InGaN/AlGaN double-heterostructure blue-light-emitting diodes. *Applied Physics Letters* 64 (13): 1687–1689.

42 Nakamura, S., Mukai, T., and Senoh, M. (1994). High-brightness InGaN/AlGaN double-heterostructure blue-green-light-emitting diodes. *Journal of Applied Physics* 76 (12): 8189–8191.

43 Pong, B.J., Chen, C.H., Hsu, J.F. et al. (2005). Electro-ridge for large injection current of micro-size InGaN light emitting diode. *5th International Conference on Solid State Lighting*, Volume 5941, 59410Y (13 September 2005). San Diego, CA, USA: International Society for Optics and Photonics. https://www.spiedigitallibrary.org/ (accessed 24 April 2024).

44 Jiang, H.X., Jin, S.X., Li, J. et al. (2001). III-nitride blue microdisplays. *Applied Physics Letters* 78 (9): 1303–1305.

45 Choi, H.W., Jeon, C.W., Dawson, M.D. et al. (2003). Fabrication and performance of parallel-addressed InGaN micro-LED arrays. *IEEE Photonics Technology Letters* 15 (4): 510–512.

46 Choi, H.W., Jeon, C.W., Dawson, M.D. et al. (2003). Mechanism of enhanced light output efficiency in InGaN-based microlight emitting diodes. *Journal of Applied Physics* 93 (10): 5978–5982.

47 Jin, S.X., Li, J., Lin, J.Y., and Jiang, H.X. (2000). InGaN/GaN quantum well interconnected microdisk light emitting diodes. *Applied Physics Letters* 77 (20): 3236–3238.

48 Pankove, J.I. (1973). Luminescence in GaN. *Journal of Luminescence* 7: 114.

49 Pimputkar, S., Speck, J.S., DenBaars, S.P., and Nakamura, S. (2009). Prospects for LED lighting. *Nature Photonics* 3: 180.

50 Maruska, H.P. and Tietjen, J.J. (1969). The preparation and properties of vapor-deposited single crystalline GaN. *Applied Physics Letters* 15: 327.

51 Chakraborty, A. and Sarkar, A. Analytical modeling and sensitivity analysis of dielectric-modulated junctionless gate stack surrounding gate MOSFET (JLGSSRG) for application as biosensor. *Journal of Computational Electronics* 16 (3): 556–567.

52 Amano, H., Akasaki, I., Kozawa, T. et al. (1988). Electron beam effects on blue luminescence of zinc-doped GaN. *Journal of Luminescence* 40: 121.

53 Amano, H., Sawaki, N., Akasaki, I., and Toyoda, Y. (1986). Metalorganic vapor phase epitaxial growth of a high quality GaN film using an AlN buffer layer. *Applied Physics Letters* 48: 353.

54 Henini, M. and Razeghi, M. (2005). *Optoelectronic Devices: III-Nitrides*, 1e. Elsevier Science.

55 Sze, S.M. and Ng, K.K. (2006). *Physics of Semiconductor Devices*. Wiley-Interscience.

56 Pankove, J. I., Miller, E. A. and Berkeyheiser, J. E. 1971, GaN electroluminescent diodes, *1971 International Electron Devices Meeting*, Washington, DC, USA, pp. 78–78, https://doi.org/10.1109/IEDM.1971.188400.

57 Amano, H., Kito, M., Hiramatsu, K., and Akasaki, I. (1989). p-type conduction in Mg-doped GaN treated with low-energy electron beam irradiation (LEEBI). *Japanese Journal of Applied Physics* 28: L2112–L2114.

58 Nakamura, S., Senoh, M., and Mukai, T. (1993). p-GaN/n-InGaN/n-GaN double-heterostructure blue-light-emitting diodes. *Japanese Journal of Applied Physics* 32: L8–L11.

59 Biswas, K., Sarkar, A., and Sarkar, C.K. (2015). Impact of barrier thickness on analog, RF & linearity performance of nanoscale DG heterostructure MOSFET. *Superlattices and Microstructures* 86: 95–104.

60 Bari, S., De, D., and Sarkar, A. (2015). Effect of gate engineering in JLSRG MOSFET to suppress SCEs: an analytical study. *Physica E: Low-dimensional Systems and Nanostructures* 67: 143–151.

61 Nakamura, S. and Krames, M.R. (2013). History of gallium–nitride-based light-emitting diodes for illumination. *Proceedings of the IEEE* 101 (10): 2211–2220.

62 Razeghi, M., Bayram, C.; Vashaei, Z.; Cicek, E., and McClintock, R. (2010). III-nitride optoelectronic devices: from ultraviolet detectors and visible emitters towards terahertz intersubband devices. *2010 23rd Annual Meeting of the IEEE Photonics Society*, Denver, CO, USA (7–11 November 2010). *IEEE Photonics Society. IEEE*. pp. 351–352.

63 Fujita, S., Sakamoto, A., and Tanabe, S. (2008, 2008). Luminescence characteristics of YAG glass-ceramic phosphor for white LED. *IEEE Journal of Selected Topics in Quantum Electronics* 14: 1387–1391.

64 Schmid, G. (2011). *Nanoparticles: From Theory to Application*. Wiley.

65 Achermann, M., Petruska, M.A., Kos, S. et al. (2004). Energy-transfer pumping of semiconductor nanocrystals using an epitaxial quantum well. *Nature* 429: 642.

66 Rissin, D.M., Kan, C.W., Campbell, T.G. et al. (2010). Single-molecule enzyme-linked immunosorbent assay detects serum proteins at sub femto molar concentrations. *Nature Biotechnology* 28 (6): 595–599. https://doi.org/10.1038/nbt.1641.

67 Placzek, E.A., Plebanek, M.P., Lipchik, A.M. et al. (2010). A peptide biosensor for detecting intracellular Abl kinase activity using matrix-assisted laser desorption/ionization time-off light mass spectrometry. *Analytical Biochemistry* 397 (1): 73–78. https://doi.org/10.1016/j.ab.2009.09.048.

68 Thévenot, D.R., Toth, K., Durst, R.A., and Wilson, G.S. (2001). Electrochemical biosensors: recommended definitions and classification. *Biosensors and Bioelectronics* 16 (1–2): 121–131. https://doi.org/10.1081/AL-100103209.

69 Kong, R.M., Ding, L., Wang, Z. et al. (2015). A novel aptamer functionalized MoS_2 nano sheet fluorescent biosensor for sensitive detection of prostate specific antigen. *Analytical and Bioanalytical Chemistry* 407 (2): 369–377. https://doi.org/10.1007/s00216-014-8267-9.

70 Im, M., Ahn, J.-H., Han, J.-W. et al. (2011). Development of a point-of-care testing platform with a nano gap embedded separated double-gate field effect transistor array and its readout system for detection of avian influenza. *IEEE Sensors Journal* 11 (2): 351–360. https://doi.org/10.1109/JSEN.2010.2062502.

71 Komarova, N.V., Andrianova, M.S., Gubanova, O.V. et al. (2015). Development of a novel enzymatic biosensor based on an ion-selective field effect transistor for the detection of explosives. *Sensors and Actuators B: Chemical* 221: 1017–1026. https://doi.org/10.1016/j.snb.2015.07.015.

72 Dorvel, B.R., Reddy, B. Jr.,, Go, J. et al. (2012). Silicon nano wires with high-k hafnium oxide dielectrics for sensitive detection of small nucleic acid oligomers. *ACS Nano* 6 (7): 6150–6164. https://doi.org/10.1021/nn301495k.

73 Fu, W., Nef, C., Tarasov, A. et al. (2013). High mobility graphene ion-sensitive field-effect transistors by non-covalent functionalization. *Nanoscale* 5: 12104–12110. https://doi.org/10.1039/C3NR03940D.

74 So, H.M., Won, K., Kim, Y.H. et al. (2005). Single-walled carbon nano tube biosensors using aptamers as molecular recognition elements. *Journal of the American Chemical Society* 127 (34): 11906–11907. https://doi.org/10.1021/ja053094r.

75 Chu, C.H., Sarangadharan, I., Regmi, A. et al. (2017). Beyond the Debye length in high ionic strength solution: direct protein detection with field-effect transistors (FETs) in human serum. *Scientific Reports* 7: 5256-1–5256-15. https://doi.org/10.1038/s41598-017-05426-6DO.

76 Kirste, R., Rohrbaugh, N., Bryan, I. et al. (2015). Electronic biosensors based on III-nitride semiconductors. *Annual Review of Analytical Chemistry* 8: 149–169. https://doi.org/10.1146/annurev-anchem-071114-040247.

77 Poonia, R., Bhat, A.M., Periasamy, C., and Sahu, C. (2022). Performance analysis of MOS-HEMT as a biosensor: a dielectric modulation approach. *Silicon* 4 (15): 10023–10036. https://doi.org/10.1007/s12633-022-01742-3.

10

First Principles Theoretical Design on Graphene-Based Field-Effect Transistors

Yoshitaka Fujimoto

Graduate School of Engineering, Kyushu University, Fukuoka, Japan

10.1 Introduction

Ever since the successful exfoliation of single-layer graphene from multilayer graphite, graphene has received much interest from the viewpoints of both fundamental nanoscience and relevant applications because it shows various unique properties such as high mechanical strength and optical transparency [1–3]. Specifically, due to its a linear energy dispersion near the Fermi energy and an extremely high carrier mobility, graphene is a candidate for a potential device material for next-generation nanoelectronics [4–7].

One of the most effective methods to modify the electronic properties of carbon-based materials should be the doping of foreign atoms into its hexagonal carbon network [8–12]. Specifically, boron and nitrogen are known to be one of the most efficient dopants for carbon-based materials. Actually, it is reported that the doping with boron atom should induce the acceptor states near the valence band edge, whereas the substitution of the nitrogen atom should give rise to the donor states near the conduction band edge [13–19]. Therefore, boron-doped graphene would behave as the p-type semiconductors, and nitrogen-doped graphene would show the n-type ones [8]. On the other hand, the substitutional doping would reduce the conductivity of the graphene-based materials, resulting in the degradation of the performance of the graphene-based electronic devices [20, 21]. Actually, the B-atom doping significantly reduces the quantum conductance of graphene layers around the Fermi energy [10–12, 22]. In the case of carbon nanotubes, it has been reported that the B-atom doping affects the quantum conductance in the valence bands, while the N-atom doping affects that in the conduction bands [23, 24]. Therefore, in order to develop high-performance graphene-based electronic devices, the doping effect on electronic structures and transport properties has to be studied.

Advanced Nanoscale MOSFET Architectures: Current Trends and Future Perspectives,
First Edition. Edited by Kalyan Biswas and Angsuman Sarkar.

Like graphene, hexagonal boron-nitride (*h*-BN) has a honeycomb lattice structure consisting of alternating boron and nitrogen atoms, and is often called white graphite due to its similar mechanical properties and the color of *h*-BN [25–28]. Unlike metallic graphene, *h*-BN is known to have a wide-gap semiconducting property (~6 eV) [18, 29–32]. It is reported that *h*-BN is often used as a substrate for graphene since it shows good transport properties than conventional SiO_2 [33–36]. This is due to the absence of dangling bonds, an atomically flat sheet, and a small lattice mismatch (about 1.8%) [37, 38]. Nonetheless, the recent scanning tunneling microscopy (STM) study suggested that there exist intrinsic defects in underlying *h*-BN substrates of graphene/*h*-BN heterostructures and that those defects in *h*-BN should affect transport properties of graphene [39]. Therefore, the relationship between these defects in graphene/*h*-BN and electronic and transport properties of graphene should be clarified both theoretically and experimentally because defects in semiconductors often affect the electronic transport properties of the systems.

This chapter reviews our first principles density-functional study that clarifies the electronic properties of graphene and graphene/*h*-BN heterostructure for designing the high-performance graphene-based devices. The chapter mainly consists of two parts. In the first part, the electronic structures and STM images of doped graphene are reported, and then the transport properties of graphene are revealed. In the next part, the energetics and the electronic structures of graphene/*h*-BN heterostructures are studied. Moreover, STM images of graphene/*h*-BN heterostructures are demonstrated, and it is discussed that an impurity dopant in the underlying *h*-BN affects the electronic properties of graphene.

10.2 Graphene

This section shows how electronic properties of graphene are changed by doping with boron and nitrogen atoms. It also reveals how the dopant atom affects the transport properties of graphene.

10.2.1 Electronic Structure

Figure 10.1a shows the energy-band structure of the pristine graphene. There is linear dispersion around the Fermi energy and they touch at the Dirac point. This means that the electrons and holes behave as massless Dirac fermions, and they can move at a speed of 1/300 times that of light. Graphene possesses no energy gap [40]. Figures 10.1b,c show the energy-band structures of substitutionally B-doped and N-doped graphenes, respectively. The number of electrons in B atom is one less than that of C atom, while that in the N atom is one more. The B atom

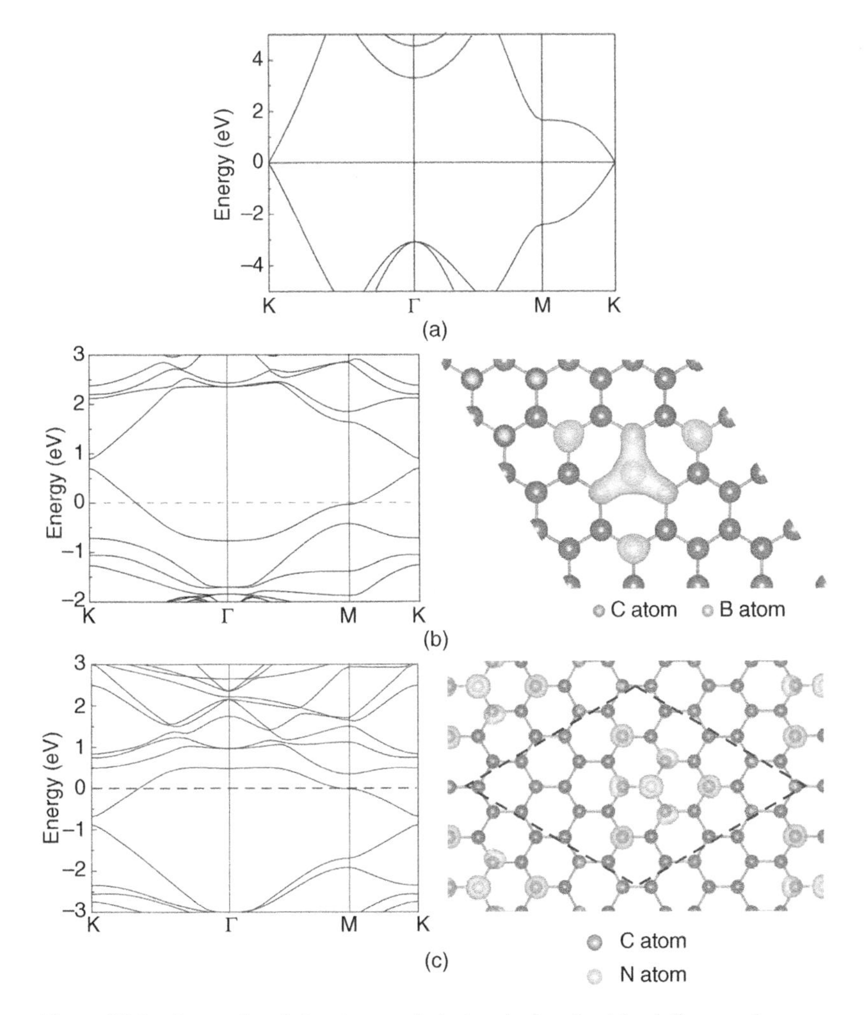

Figure 10.1 Energy band structure and electronic density: (a) pristine graphene, (b) B-doped graphene and (c) N-doped graphene. The Fermi energy is set to zero. Source: (b) Reproduced with permission from Fujimoto and Saito [10], copyright 2019 the Institute of Physics. (c) Reproduced with permission from Fujimoto and Saito [41], copyright 2011 the American Physical Society.

and the N atom can induce the acceptor-like state and the donor-like state near the Fermi energy, respectively. Therefore, the B-doped graphene would become p-type semiconductors, and the N-doped graphene would be n-type semiconductors. In Figure 10.1b,c, the isosurfaces of electron densities of B-doped and N-doped graphenes at Γ point, respectively, are shown. The electron densities of B and N impurities are delocalized around the B atom and the N atom, respectively.

There are differences between B and N defects in the spatial distributions of electron densities: the electron density of the B defect is extended to three C atoms around the B atom and consists of a triangle-like shape, whereas that of the N defect is distributed above N atom and three C atoms and appear to be individual four spots [41]. Interestingly, this delocalization in the electron density is also observed in other defects with three-fold symmetry, such as impurity-doped and impurity-adsorbed graphenes [42].

10.2.2 Scanning Tunneling Microscopy

STM is a powerful tool for obtaining the local electronic structures around atomic vacancies and impurities at atomic levels. The STM images of B-doped and N-doped graphenes are demonstrated. The STM images of defective graphene are obtained using the Tersoff–Hamann (TH) approach [43]. It is assumed that the tunneling current is proportional to the local density of states (LDOS) of the surface at the tip position [44, 45]. The STM images can be generated from the isosurface of the spatial distribution integrated by the LDOS $\rho(\boldsymbol{r}, E)$ at spatial points $\boldsymbol{r} = (x, y, z)$ and energy E over the energy range from the Fermi energy E_{F} to $E_{\mathrm{F}} + eV$ with applied voltage V:

$$I \sim \int_{E_{\mathrm{F}}}^{E_{\mathrm{F}}+eV} \rho(\boldsymbol{r}, E)\mathrm{d}E \tag{10.1}$$

The negative and positive bias voltages reflect the occupied and unoccupied electronic states, respectively. We refer to this "energy-integrated local density of states" as EI-LDOS [46–48].

The STM images of B-doped bilayer and N-doped graphenes obtained by Eq. (10.1) are shown in Figure 10.2. The STM image of B-doped bilayer graphene has a triangle-shaped bright area, which shows similar images without depending on the stacking patterns (Figure 10.2b) [10]. The triangle-shaped bright area in the simulated STM image is similar to that in the B-doped monolayer graphene observed experimentally [38]. The STM image of the N-doped graphene shows sharp contrast to that of a the B-doped one. We now examine the STM image of N-impurity atom in a monolayer graphene. Three bright protrusions can be seen at the three C atoms around the N atom. The dark area in the STM image is located above the N atom as if the N atom were absent, although the spatial distribution of the impurity-induced state extends above the N atom as well (Figure 10.1c). This discrepancy comes from how the distributions of the LDOSs protrude above the N atom and the surrounding three C atoms; the LDOS above the three C atoms are more delocalized than that above the N atom. Consequently, the STM image at the three C atoms near the N atom shows three bright spots, whereas the STM image at the N atom appears to be dark [41].

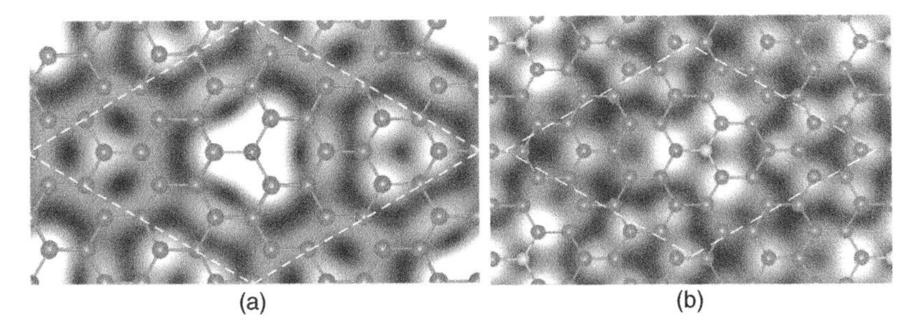

Figure 10.2 Scanning tunneling microscopy of (a) B-doped bilayer graphene and (b) N-doped graphene. The STM images are generated at the bias voltage of +0.5 eV. Source: (a) Fujimoto and Saito [9]. Reproduced from [9], with permission of Elsevier/CC BY 3.0. (b) Fujimoto and Saito [41]. Reproduced from [41], with permission of IOP Publishing Ltd/CC BY 3.0.

10.2.3 Electronic Transport

We now examine how the adsorption of molecules as well as the impurity dopant change the electron transport properties of graphene. To evaluate the transport properties of various graphene systems, a periodic boundary condition is imposed along the zigzag direction of a graphene with periodic length L_x. For the calculations of the electrical conductances, the scattering wavefunctions of the systems are calculated by the overbridging boundary-matching (OBM) method, where the systems are composed of B-doped graphene and two semi-infinite graphenes attached to it [49–53], using the real-space finite-difference approach [54–56]. The conductance $G(E)$ is associated with the transmission coefficient $T(E, k_x)$ by the Landauer–Büttiker formula [57]:

$$G(E) = \frac{2e^2}{h} \int_{-\pi/L_x}^{\pi/L_x} \frac{dk_x}{2\pi/L_x} T(E, k_x) \tag{10.2}$$

where e and h represent the electron charge and Planck's constant, respectively.

The conductances of pristine graphene, and B-doped graphene with and without NO and NO_2 molecules are exhibited in Figure 10.3a by using Eq. (10.2) [10]. The conductance of the pristine graphene exhibits a linear dispersion and "V"-shaped structure. These agree with experimental results [1, 2]. The B atom doped in graphene scatters electrons and therefore reduces the conductance of graphene in the range between $E = -0.5$ eV and $+0.5$ eV. In the cases of $E = +0.5$ eV and $E = -0.5$ eV, the conductances of the B-doped graphene are lower by ~30% compared with those of the pristine one. The reduction rate of the conductance for the B-doped graphene shows almost the same value. When a NO_2 molecule is adsorbed to the B-atom impurity in graphene, the conductance reduces by 40% for the energy $E = +0.5$ eV. When a NO molecule is adsorbed, the conductance considerably reduces by about 50%. In the case of the energy

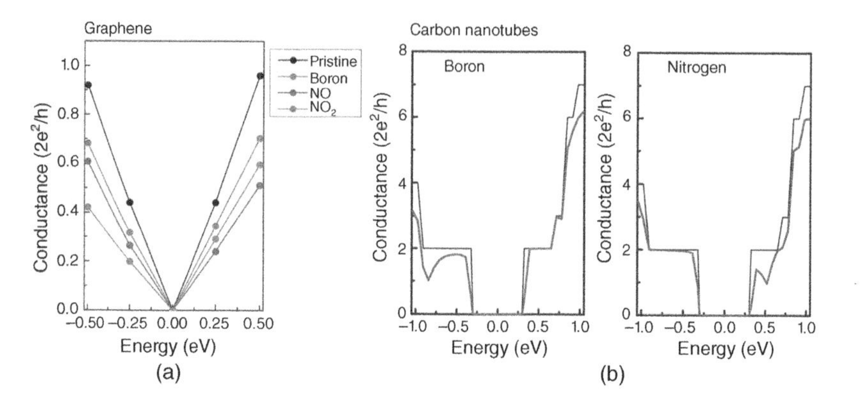

Figure 10.3 Conductance as a function of energy: (a) graphene and (b) (10,0) carbon nanotube. The Fermi energy is set to zero. Source: (a) Reproduced with permission from Fujimoto and Saito [10], copyright 2019 the Institute of Physics. (b) Reproduced with permission from Fujimoto and Saito [23], copyright 2022 the Electrochemical Society.

$E = -0.5$ eV, the conductance decreases by more than 50% for the adsorption of a NO_2 molecule. The adsorption of a NO molecule gives rise to 35% reduction in the conductance. Thus, the conductance of graphene largely changes by the adsorption of molecules and the introduction of the impurity dopant. Furthermore, the variation of the conductance would depend on the type of adsorbate and the dopant atoms. In Figure 10.3b, the conductance of (10,0) carbon nanotube (CNT) is exhibited. Being different from graphene, the conductance of CNT shows step-like structure with an energy gap. When a B atom is doped to the CNT, electrons are scattered by the B atom impurity in the CNTs. The introduction of the B atom into the CNTs gives rise to the acceptor states, and therefore a dip related to the B impurity appears around $E = -0.84$ eV in the conductance curve within the valence bands, whereas there exist no dips within the conduction bands. When N atom is doped the CNTs, donor states appear below the conduction bands. Therefore, a dip in the conductance curve emerges around $E = +0.52$ eV in the conduction bands, and no dip appears in the valence bands [23]. Accordingly, the conductances of graphene and CNT sizably decrease by the introduction of B and N atoms and the adsorption of NO and NO_2 molecules, and therefore the existence of the impurities and the adsorbates would lead to the degradation of the performance of electronic devices such as field-effect transistors.

10.3 Graphene/*h*-BN Hybrid Structure

In this section, the energetics and electronic properties of graphene on doped *h*-BN heterostructure are exhibited. In the two-layer system of graphene/*h*-BN heterostructure, effects of the stacking patterns and dopant sites are examined. In the four-layer system, the effect of the doped *h*-BN layer number will be discussed [58].

10.3.1 Atomic Structure

In the present work, four kinds of two-layer systems and a four-layer system are treated. The two-layer system consists of the h-BN monolayer and the graphene monolayer, as shown in Figure 10.4a–d, and a four-layer system consists of h-BN trilayer and the graphene monolayer, as shown in Figure 10.4e. In the case of the two-layer systems, four different stacking patterns are considered: Aa, Ab, Ab', and rotated stackings. The commensurate case is assumed here, and the graphene layer and the h-BN layer are expanded slightly and compressed after the optimization,

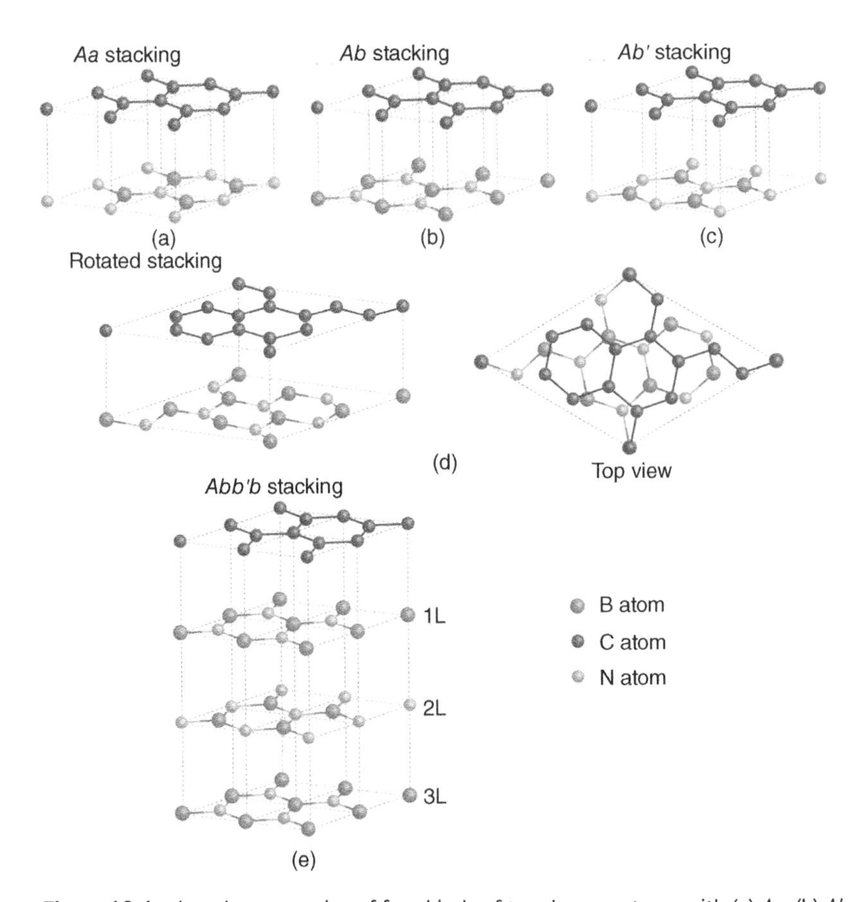

Figure 10.4 Atomic geometries of four kinds of two-layer systems with (a) *Aa*, (b) *Ab*, (c) *Ab'*, and (d) rotated stackings, and (e) that of a four-layer system with *Abb'b* stacking. In the *b' h*-BN layer, B atoms are placed at the N atom sites of the *b* layer, and vice versa. In the rotated two-layer system, the rotated graphene with an angle of 21.8° is placed on the h-BN monolayer. In the four-layer system, graphene monolayer is stacked on the h-BN trilayer consisting of 1L, 2L, and 3L (see text). Dotted lines are guides to the eye to indicate the unit cell. Source: Reproduced with permission from Haga et al. [58], copyright 2019 the American Physical Society.

respectively. In the case of the *Aa* stacking, C atoms in the graphene layer are just positioned above B and N atoms in the *h*-BN layer (Figure 10.4a). In the case of the *Ab* stacking, one of the two C atoms in the graphene layer is located on top of the B atom and the other C atom located above the center of a hexagon in the *h*-BN layer (Figure 10.4b). In the case of the *Ab* stacking, one of the two C atoms is placed above the N atom, while the other is located above the center of a hexagon of the *h*-BN layer (Figure 10.4c). Thus, the capital letter (A) denotes the relative position of the graphene layer, and the small letters (a, b, and b') denote the relative positions and the relative directions of the *h*-BN layers. In the case of the rotated system (Figure 10.4d), the 21.8° rotated graphene is laid on the *h*-BN monolayer, and there are 14 C atoms in the graphene layer, seven B atoms and seven N atoms in the *h*-BN monolayer. In order to simulate the interactions between the graphene and *h*-BN layers with incommensurate stacking, this rotated system is considered. The *aa′* stacking pattern is known to be often observed in the *h*-BN bulk [59, 60]. In the case of the four-layer system (Figure 10.4e), the graphene is laid on the *h*-BN trilayer with the *aa′* stacking pattern corresponding to the typical stacking in the bulk *h*-BN [61].

Next, graphene/C-doped *h*-BN hybrid structures are treated. The *h*-BN layer has two kinds of substitution sites for C-atom doping (B and N sites), and we name the C atom substitutionally doped in these sites as C_B and C_N, respectively. For the *Aa*-stacked two-layer system, the C atom in the graphene layer is positioned above the C atom doped at the B site or N site in the *h*-BN layer. For the *Ab* stacked two-layer system, the C atom in the graphene layer is located just above the C atom doped at the B site in the *h*-BN layer, and the C atom doped at the N site is below the center of the hexagon of the graphene layer. The rotated two-layer system has seven different substitution sites for C atom doping, both at the B site and at the N site (Figure 10.4d). In this work, we have examined the C doping at two kinds of B sites and also two kinds of N sites among seven sites in each case. However, the doping-site dependence of the band structure as well as that of the STM image is found to be very small, and we will show the results for the C doping at B and N sites in the *h*-BN layer, which are nearly below the center of the hexagon of the graphene layer as shown in Figure 10.5.

For the four-layer system doped with C atom, only the C atom dopings at the N site are treated. The C_N atom is doped below the center of the hexagon of the graphene layer for the top layer (1L) or the bottom layer (3L) in the *h*-BN trilayer, while the C_N atom is doped below the C atom of the graphene layer for the middle layer (2L).

10.3.2 Structure and Energetics

We start by introducing the optimized atomic structures and the energetics for the undoped graphene/*h*-BN two-layer and four-layer structures. To discuss the

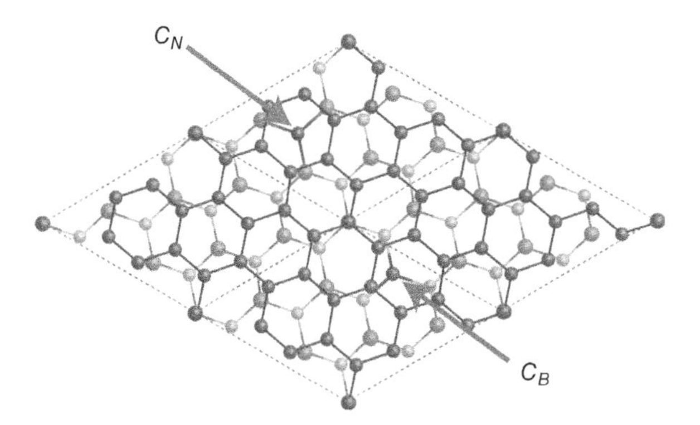

Figure 10.5 Top view of atomic structure of C-doped rotated two-layer system with 2×2 supercell having 112 atoms. The C substitution sites of B atom and N atom nearly under the center of hexagon of the graphene layer are indicated by arrows with C_B and C_N, respectively. The dotted line denotes the unit cell of the pristine phase. Source: Reproduced with permission from Haga et al. [58], copyright 2019 the American Physical Society.

energetics, the formation energy E_f is calculated by

$$E_f = (E_{tot} - E_{gra} - mE_{h\text{-BN}})/N \tag{10.3}$$

where E_{tot}, E_{gra}, and $E_{h\text{-BN}}$ denote total energies of the graphene/*h*-BN systems, the graphene monolayer, and the *h*-BN monolayer, respectively, m is the number of *h*-BN layers of the graphene/*h*-BN structures, and N is the number of atoms in the supercell.

Table 10.1 shows total energies and interlayer distances between graphene and *h*-BN layers. It is found that the optimized in-plane lattice constants are all 2.468 Å for the two-layer systems with three stacking patterns, which are the average value of optimized lattice constants of the graphene monolayer (2.446 Å) and the *h*-BN monolayer (2.490 Å). In the case of the rotated two-layer system, the optimized in-plane lattice constant is found to be 6.530 Å, and is just the average value of B–N (C–C) bond length of 1.425 Å, which takes the value of the other three two-layer systems. In the case of the four-layer system, the optimized in-plane lattice constant is found to be 2.479 Å. One can see that the *Ab* stacking is the most stable and it has the shortest interlayer distance (3.23 Å) among the four types of two-layer systems (see Table 10.1). The rotated two-layer system is found to be the second stable structure, and the energy difference is only 0.88 meV/atom. On the other hand, the interlayer distance is sizably long, compared with that of the *Ab*-stacking. The *Ab′* stacking is the third stable structure with total energy difference of 4.62 meV/atom. Any interlayer interactions would give rise to the energy difference between the *Ab* and the *Ab′* stackings: The interaction between the N atom in the *h*-BN layer and the C atom in graphene layer is less preferable energetically compared with that between the B atom and the C atom. Such

Table 10.1 Interlayer distances and formation energies of two-layer and four-layer systems for graphene/*h*-BN heterostructures.

System	Stacking pattern	Interlayer distance (Å)	Formation energy (eV)
Two-layer	*Aa*	3.52	−0.21
	Ab	3.23	−5.65
	Ab′	3.46	−1.03
	Rotated	3.39	−4.77
Four-layer	*Abb′b*	3.22	−9.26

For the four-layer system, the interlayer distance between graphene and the top layer of the *h*-BN (1L) is listed.
Source: Reproduced with permission from Haga et al. [58], copyright 2019 the American Physical Society.

behavior has already been suggested in theoretical calculations [9]. On the other hand, there are both the B–C type and the B–N type interlayer interactions for the rotated two-layer system, which is shown in Figure 10.4. Hence, the rotated structure becomes slightly unfavorable in energy than the *Ab* stacking structure. The optimized interlayer distance for the four-layer structure between graphene and the top layer of *h*-BN is found to be 3.22 Å, and that between the neighboring layers of the *h*-BN trilayer are found to be both 3.24 Å. The formation energy of the four-layer structure is much lower than that of the *Ab*-stacked two-layer systems. This is because two more sets of interlayer interactions among the trilayer of the *h*-BN produce the sizable energy gain.

Let us consider the atomic configurations of the two-layer systems composed of graphene and C-doped *h*-BN at the B site. The optimized in-plane bond lengths between the C atom and the neighboring N atoms for the C-doped *Aa*, *Ab*, and *Ab′* stacked two-layer systems are almost the same as the C-doped rotated two-layer system (1.36 Å). The B–N bond length of the undoped graphene/*h*-BN systems is much longer than the C–N ones. The C-doped *h*-BN layer at the B site of the heterostructure is almost flat as in the case of the C-doped *h*-BN monolayer at the B site. We also consider the atomic configurations of the two-layer systems, where C atom in *h*-BN is doped at the N site. The in-plane bond lengths between the C atom of the *h*-BN layer and the neighboring B atoms of the C-doped two-layer systems are also almost the same as those of the rotated two-layer system (1.47 Å). The C_N atom does not reside on the *h*-BN planar sheet. For two-layer systems with the *Aa* and *Ab* stackings, the C_N atom moves away from the *h*-BN sheet to the graphene sheet by 0.62 and 0.59 Å, respectively, while, for the *Ab*-stacked two-layer system, the C_N atom also approaches to the graphene sheet by only 0.05 Å. These structural deformations may mean that the attractive interaction between the impurity C atom in the *h*-BN layer and the C atom in the neighboring graphene layer is favorable as discussed later. In the case of the *Aa* and *Ab* stacking patterns, the C atom in graphene is positioned on top of the impurity C atom doped to the *h*-BN sheet. In the case of the *Ab* stacking pattern, the impurity C atom doped in the *h*-BN sheet resides below the center of the hexagon of graphene. For the rotated two-layer system, the C atom doped in the *h*-BN sheet approaches the graphene sheet by 0.08 Å. This is reasonable since, in this case, the C_N is located not directly below the hexagon center site of the graphene layer but rather close to there. For four-layer system, we consider the C-doped *h*-BN layer at N site, and it is found that the in-plane B–C bond lengths after structural optimizations are 1.48 Å. Irrespective of the depth of the doped *h*-BN sheet from the top graphene sheet, the impurity C atom is present in the *h*-BN plane. It has been reported that the substitution of a C atom to the *h*-BN monolayer at the N site is unfavorable in energy compared to that at the B site.

10.3.3 Electronic Structure

Here, the electronic bands of the *Ab*-stacked and the rotated two-layer systems, and the four-layer system are examined. In Figure 10.6a–c, electronic bands of the pristine and C-doped *h*-BN at B site and at N site of the *Ab*-stacked two-layer systems are exhibited, respectively. For the pristine case, there is a finite band gap of 0.06 eV near the apex of the Dirac cone at the Γ point (Figure 10.6a). This is because the interlayer interactions with the *h*-BN layer breaks the sublattice symmetry of the C atoms of graphene. The C atoms doped in *h*-BN at the B site and the N site should induce the donor-like and the acceptor-like levels near the Fermi energy, as depicted in Figure 10.6b,c, respectively, which agrees well with previous theoretical reports [19, 28]. For the C-doped case at the B site, one can see that there is an almost flat band around 0.5 eV above Fermi energy (Figure 10.6b). Thereby, the electrons occupied at this donor-like level completely move to the graphene. On the other hand, the almost flat band is also induced near the Fermi energy. Therefore, the hole is transferred from the acceptor-like level to the graphene for C-doped case at the N site (Figure 10.6c).

Figure 10.6d–f shows the electronic bands of the rotated two-layer systems. In the case of the pristine rotated two-layer system, there is a nearly zero band gap (Figure 10.6d). The C atoms doped in the *h*-BN sheet of the rotated system at the B site and the N site can induce the donor-like and the acceptor-like levels, respectively, near the Fermi energy (Figure 10.6e,f). Like the *Ab*-stacked two-layer system, the donor-like level also appears above the Fermi energy, and the electrons move from the donor-like level to graphene. Interestingly, the carrier

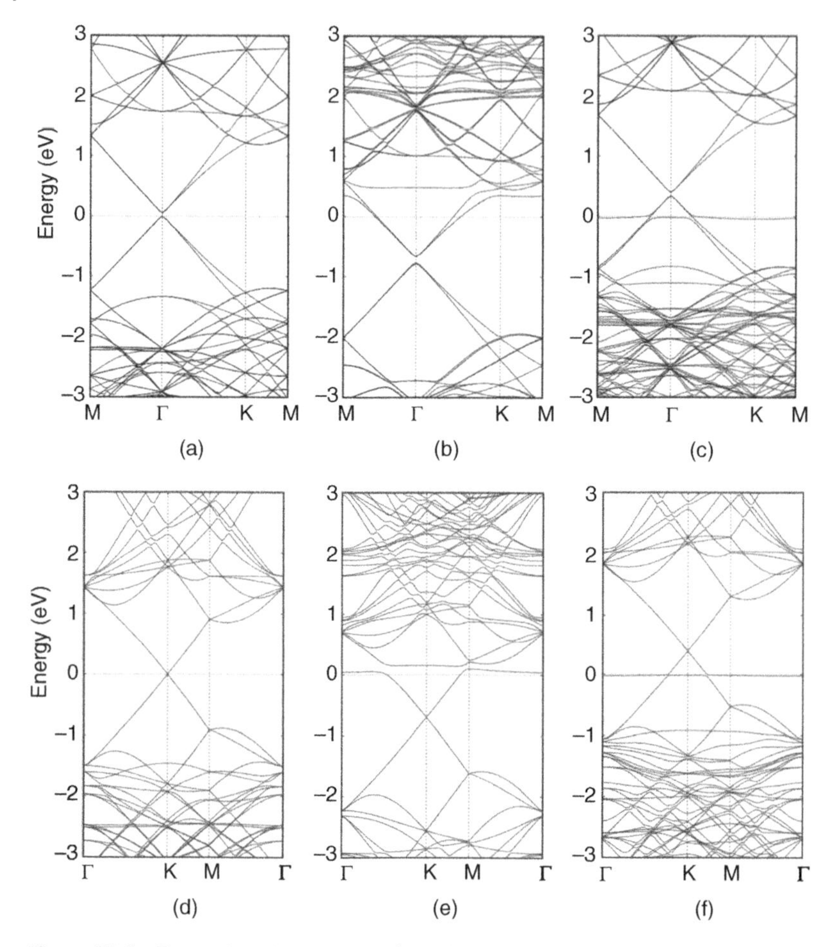

Figure 10.6 Energy band structures of the *Ab*-stacked graphene/*h*-BN two-layer systems: (a) pristine *h*-BN case, (b) C-doped *h*-BN at the B site case, and (c) that at the N site case. Energy band structures of the rotated graphene/*h*-BN two-layer system: (d) pristine *h*-BN case, (e) C-doped *h*-BN at the B site case, and (f) that at the N site case. The Fermi level is set to zero. In the rotated systems, the 2×2 supercell is used not only for the band structure of doped case but also for that of the pristine case. Source: Reproduced with permission from Haga et al. [58], copyright 2019 the American Physical Society.

concentrations on graphene layers show asymmetric behaviors, which depends on the C-doped *h*-BN sheets at the B or the N site.

The electronic bands of the C-doped four-layer heterostructures are also examined. The electronic bands near the Fermi energy of C-doped four-layer heterostructures at the N site in (a) top layer (1L), (b) middle layer (2L), and (c) bottom layer (3L) of *h*-BN trilayer are exhibited in Figure 10.7a–c. The

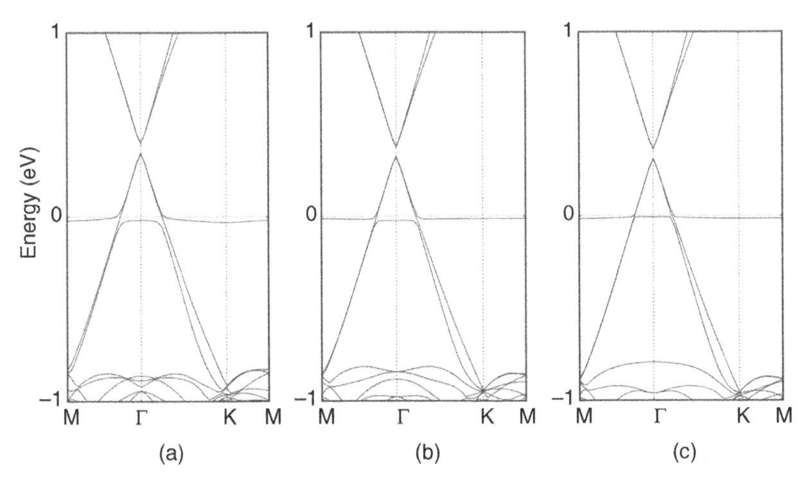

Figure 10.7 Energy band structures near the Fermi level of four-layer systems doped with C atoms at the N site in (a) top layer (1L), (b) middle layer (2L), and (c) bottom layer (3L) of *h*-BN trilayer. The Fermi level is set to zero. Source: Reproduced with permission from Haga et al. [58], copyright 2019 the American Physical Society.

acceptor-like levels appear around the Fermi energy, irrespective of the depth of the Fermi energy measured from the apex of the Dirac cone and the C-doped *h*-BN layer number. Like the *Ab*-stacked two-layer heterostructure, the electrical hole is not occupied completely, and the apex of the Dirac cone relatively moves toward the Fermi energy. Thereby, the hole carriers on the graphene would have the same carrier density, irrespective of the C-doped *h*-BN layer number. On the other hand, it can be seen that the acceptor-like level varies from a dispersive band to a flat one with increasing distance between C-doped *h*-BN and the graphene sheets. This is because the interlayer interaction between the acceptor-like level of the C-doped *h*-BN sheet and π-orbital level of the graphene becomes weaker with increasing the distance between the graphene layer and the doped *h*-BN sheet. Therefore, the spatial distribution of the STM images would change, which depends on the doped *h*-BN layer number as shown later.

10.3.4 Scanning Tunneling Microscopy

Simulated STM images for the *Aa*, *Ab'*, and rotated stacked graphene/C-doped *h*-BN two-layer heterostructures are examined. In Figure 10.8a,b, we exhibit the constant height STM images of the graphene layer deposited on the C-doped *h*-BN layer at the B and the N sites for the *Ab*-stacked two-layer heterostructures, respectively. Here, the STM images are generated at the bias voltages of +0.5 and −0.5 eV for C-doped cases at the B site and the N site, respectively. For the graphene sheet

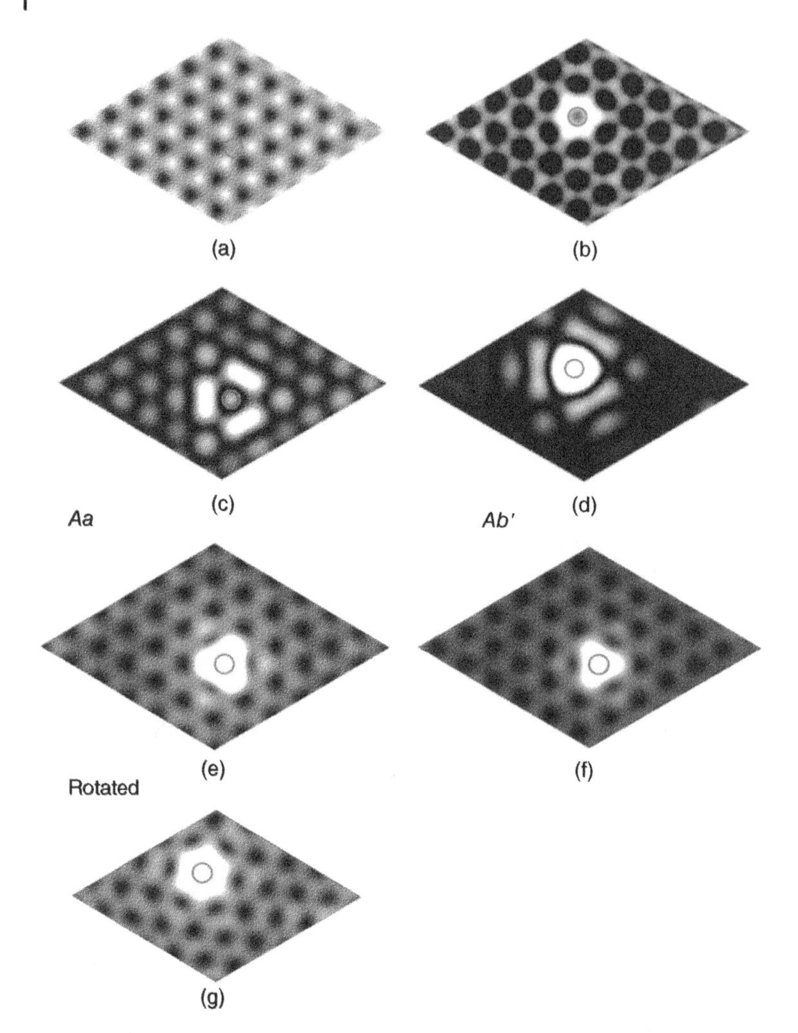

Figure 10.8 Simulated STM images of the *Ab*-stacked two-layer system to be taken over the graphene layer with (a) C_B and (b) C_N, and these to be taken under the *h*-BN layer with (c) C_B and (d) C_N. The STM images are constructed applied bias voltages of +0.5 eV for (a) and (c), and −0.5 eV for (b) and (d). Simulated STM images of graphene layer on C-doped *h*-BN at N site for the (e) *Aa*-stacked, (f) *Ab′*-stacked, and (g) rotated two-layer system. The STM images are constructed at applied bias voltage of −0.5 eV for all three cases. The circle in each figure denotes the position of the C atom doped in the *h*-BN layer. Source: Reproduce from [58], with permission of American Physical Society.

on the C-doped *h*-BN sheet at the B site (C_B), three bright spots appear in the STM images as if the C atom doped in the underlying *h*-BN sheet were not present, and those spots are just on top of C atoms in graphene sheet (Figure 10.8a). The other-type C atom is located above the B atom of the *h*-BN sheet and it appears as weakly bright spots in the STM image. It is noted that the STM image appears to be the dark area at the center of the hexagon of the graphene sheet. The STM image mainly reflects π-orbital states of graphene while it does not reflect the C_B atom. On the other hand, the STM image of the graphene deposited on the C-doped *h*-BN at the N site (C_N) is considerably different from that at the B site (C_B). A hexagonally shaped, bright region shows up above the impurity C atom doped in the underlying *h*-BN sheet. Thus, it is found that the C_N atom in the underlying *h*-BN sheet can be observed clearly by the STM method, while C_B is not observable.

We also examine the STM image of the *h*-BN sheet with C_B impurity of the *Ab*-stacked two-layer heterostructure. The STM images of the C-doped *h*-BN sheet at the B site and the N site for the *Ab*-stacking patterns are exhibited in Figure 10.8c,d, respectively. In the case of the C_B atom, a small bright spot appears at the C-atom impurity inside three bright heavy lines between two B atoms. On the other hand, in the case of the C_N atom, a large triangle-shaped bright region appears inside three bright fine lines between two N atoms. It will be shown later that this STM image mainly reflects the C-atom impurity state, consisting of the triangular-shaped spatial distributions of the LDOS. These are in good agreement with our results reported previously on C-doped *h*-BN monolayers [28].

We further examine the STM images of three different stacking patterns: *Aa*, *Ab′*, and the rotated two-layer systems. The STM images of the graphene sheet deposited on C-doped *h*-BN at the N site are exhibited for the *Aa*-stacked, *Ab*-stacked, and the rotated two-layer heterostructures, respectively, in Figure 10.8e–g. There is a triangular bright region in the STM image for the *Aa* and *Ab′* stacking patterns, while there exists a hexagonal bright region for the rotated stacking pattern, as in the case of the *Ab* stacking pattern. For the two-layer systems with *Aa*, *Ab*, *Ab′* and rotated stacking patterns, C_N impurities in the *h*-BN are detectable by using the STM measurements, even if it is observed through the graphene layer.

We here study the relationship between the C-atom impurity-induced states and the STM images of the graphene sheet deposited on the C-doped *h*-BN sheet. In Figure 10.9a,b, the cross sections of contour plots of the EI-LDOS of the *Ab*-stacked two-layer systems doped at the B and N sites are shown, respectively. One finds that the spatial distributions of the EI-LDOS near the C-atom impurity for the N doping site have largely changed since the impurity state induced by the C atom is extended considerably, which affects the EI-LDOSs above and below the *h*-BN sheet (Figure 10.9b). On the other hand, for the B-site doping case, one also finds that the spatial distribution of EI-LDOS above the graphene layer is almost

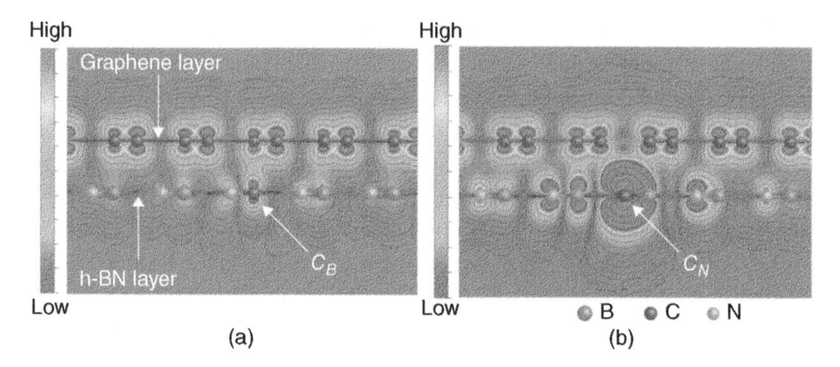

Figure 10.9 Contour maps of EI-LDOS of graphene/C-doped *h*-BN at (a) B site and (b) N site for *Ab*-stacked two-layer systems. The LDOS are integrated from E_F to $E_F + 0.5$ eV in (a) and $E_F - 0.5$ eV to E_F in (b). Contour lines are drawn in logarithmic scales. Source: Reproduced with permission from Haga et al. [58], copyright 2019 the American Physical Society.

unaffected by the C_B-atom impurity since the C-atom impurity-induced state is localized (Figure 10.9a). Thus, the extended C_N-atom impurity state modifies the shape of the topmost contour line above the graphene sheet (Figure 10.9b), producing the bright region in the STM image. On the other hand, the localized C_B impurity state does not almost affect the contour line above the graphene sheet (Figure 10.9a), which does not modify the STM image. Their difference would be attributed to the sizes of B and N atoms.

Let us discuss the relationship between the STM images of the graphene sheet deposited on the C-doped *h*-BN trilayer and the C-doped layer number of the *h*-BN trilayer. Figure 10.10a–c exhibit the STM images and the contour plots of the EI-LDOS of the C-doped four-layer system, where C atoms are doped in the top layer, the middle layer, and the bottom layer of *h*-BN trilayers at the N site, respectively. For these three cases, it is interesting that the STM images change in brightness as well as the shape of the bright region. As in the case of the *Ab*-stacked two-layer heterostructure of the graphene/C-doped *h*-BN at the N site (Figure 10.8b), for doped case in the top *h*-BN layer, there is a hexagonally arranged bright region above the C impurity in the underlying *h*-BN (see Figure 10.10a). For the C-doped case at the middle *h*-BN sheet, the STM image has a bright region with the three-folded rotational symmetry near the C-atom impurity. In the case of C_N atom doped in the top and the middle *h*-BN sheets, one can find that the spatial distributions of the EI-LDOS above the C atom doped in the *h*-BN sheets are largely modified in the vacuum region above the graphene layer. For the doped case in the bottom *h*-BN sheet, there also exists a hexagonally shaped bright region around the C_N impurity in the *h*-BN sheet in the STM image. Thus, the STM images vary with depending on the C-doped *h*-BN sheet number,

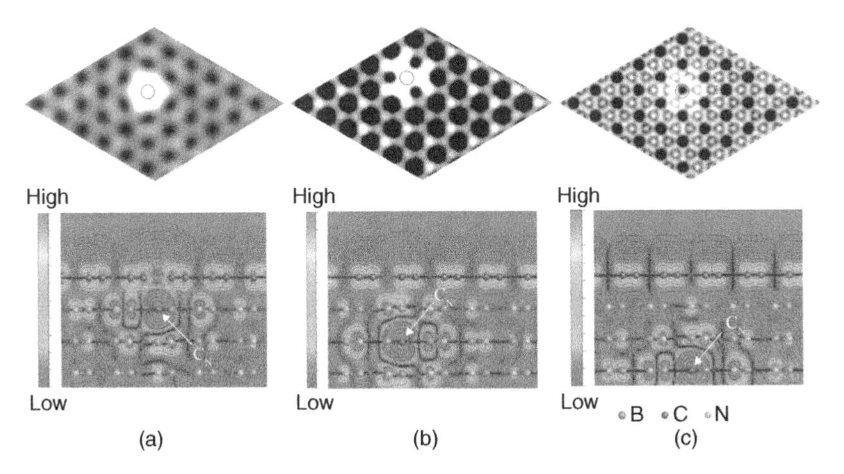

Figure 10.10 Simulated STM images (upper panels) and contour maps (lower panels) of *Abb′b* stacked four-layer systems over the graphene layer stacked on C-doped *h*-BN at the N site in (a) top layer (1L), (b) middle layer (2L), and (c) bottom layer (3L) of *h*-BN trilayers. The STM images are constructed at an applied bias voltage of −0.2 eV. The brightness of each image is tuned independently, so the contrast around the C_N region becomes clear. On the other hand, contour maps are drawn with the same logarithmic scale for all three maps, indicating clearly that (b) is much brighter than (c), and (a) is also much brighter than (b). The circles drawn in the bright areas of the STM image denote the position of the C atom doped in the *h*-BN layers. Source: Reproduced from [58], with permission of American Physical Society.

which could detect the C-doped *h*-BN sheet number. The bright spot in the STM image relating to the C_N-atom-induced acceptor state in the *h*-BN sheet might correspond to the bright dots, as already discussed in the experimental observation [39]. Furthermore, it has also been reported that there are variations in the intensity of bright dots in the STM experiment [39]. The C_N-induced acceptor states in the deep inside *h*-BN sheet below the graphene could modify the STM images above graphene sheet, and therefore this could mean that the brightness of the dots depends on the depth as well as the shape of the C_N-doped *h*-BN sheet.

10.4 Conclusions

This chapter has reviewed electronic and transport properties of the graphene layers and the heterostructures consisting of graphene and C-doped *h*-BN layers on the basis of our first principles electronic structure and transport studies. We have shown that doping of the B and N atoms affect the electronic structures, STM images, as well as the transport properties. It is shown that the B atom and N atom induce the acceptor-like state and the donor-like state near the Fermi energy, respectively. In the STM images, the B-doped graphene and the N-doped graphene have triangular area around the B atom and three individual spots near the N atom, respectively. It is also shown that the conductance of graphene significantly decreases by introduction of B-atom impurities.

The chapter has further reviewed the electronic properties and the STM images of heterostructures consisting of the graphene and C-doped *h*-BN layers. The C atom doped at the B site or N site in underlying *h*-BN layers induces asymmetric p-type and n-type doping behaviors of the graphene/C-doped *h*-BN heterostructures. Since the *h*-BN atomic layers are the 2D semiconductor materials, this asymmetric behavior induced by C-atom doping is of great importance for designing the future 2D-layered nanoelectronic devices. When the C atom is doped at the N site in the underlying *h*-BN, the C-atom impurity state is observed as a bright region in the STM images. On the other hand, it does not appear when the C atom is doped at the B site. Moreover, the C-atom impurity states appear as distinctive bright shapes in the STM images, which depends on the C-doped layer number in multilayer *h*-BN substrates. Thus, the C-atom impurities buried in *h*-BN layers doped at the N site are observable by using STM methods. The chapter has revealed the usefulness of the STM technique for detecting the C_N dopant not only at the topmost layer but also at even deeper layers when the graphene is placed on top of the *h*-BN layers.

Acknowledgments

This work was partly supported by JSPS KAKENHI Grant Nos. JP17K05053 and JP21K04876. Computations were partly done at Institute for Solid State Physics, the University of Tokyo.

References

1 Novoselov, K.S., Geim, A.K., Morozov, S.V. et al. (2004). *Science* 306: 666.
2 Morozov, S.V., Novoselov, K.S., Katsnelson, M.I. et al. (2008). *Physical Review Letters* 100: 016602.
3 Geim, A.K. and Novoselov, K.S. (2007). *Nature Materials* 6: 183.
4 Zhang, Y., Tan, Y.W., Stormer, H.L., and Kim, P. (2005). *Nature* 438: 201.
5 Berger, C., Song, Z., Li, X. et al. (2006). *Science* 312: 1191.
6 Castro, E.V., Novoselov, K.S., Morozov, S.V. et al. (2007). *Physical Review Letters* 99: 216802.
7 Zhang, Y., Tang, T.-T., Girit, C. et al. (2009). *Nature* 459: 820.
8 Fujimoto, Y. and Saito, S. (2016). *Chemical Physics* 478: 55.

9 Fujimoto, Y. and Saito, S. (2015). *Surface Science* 634: 57.
10 Fujimoto, Y. and Saito, S. (2019). *Japanese Journal of Applied Physics* 58: 015005.
11 Fujimoto, Y. and Saito, S. (2020). *Applied Surface Science Advances* 1: 100028.
12 Onodera, M., Isayama, M., Taniguchi, T. et al. (2020). *Carbon* 167: 785.
13 Fujimoto, Y. and Saito, S. (2011). *Physica E* 43: 677.
14 Fujimoto, Y. and Saito, S. (2016). *Physical Review B* 94: 245427.
15 Fujimoto, Y., Koretsune, T., and Saito, S. (2014). *Journal of the Ceramic Society of Japan* 122: 346.
16 Ayala, P., Arenal, R., Loiseau, A. et al. (2010). *Reviews of Modern Physics* 82: 1843.
17 Fujimoto, Y. and Saito, S. (2015). *Journal of the Ceramic Society of Japan* 123: 576.
18 Fujimoto, Y. and Saito, S. (2016). *Journal of the Ceramic Society of Japan* 124: 584.
19 Haga, T., Fujimoto, Y., and Saito, S. (2021). *Physical Review Materials* 5: 094003.
20 Weidner, M., Fuchs, A., Bayer, T.J.M. et al. (2019). *Advanced Functional Materials* 29: 1807906.
21 Gilbert, S.M., Pham, T., Dogan, M. et al. (2019). *2D Materials* 6: 021006.
22 Fujimoto, Y. (2021). *Modern Physics Letters B* 35: 2130001.
23 Fujimoto, Y. and Saito, S. (2022). *Journal of the Electrochemical Society* 169: 037512.
24 Fujimoto, Y. (2022). *Journal of Electrochemical Science and Engineering* 12: 431.
25 Sichel, E.K., Miller, R.E., Abrahams, M.S., and Buiocchi, C.J. (1976). *Physical Review B* 13: 4607.
26 Blase, X., Rubio, A., Louie, S.G., and Cohen, M.L. (1994). *EPL* 28: 335.
27 Yasuda, K., Wang, X., Watanabe, K. et al. (2021). *Science* 372: 1458.
28 Fujimoto, Y. and Saito, S. (2016). *Physical Review B* 93: 045402.
29 Watanabe, K., Taniguchi, T., and Kanda, H. (2004). *Nature Materials* 3: 404.
30 Watanabe, K., Taniguchi, T., Niiyama, T. et al. (2009). *Nature Photonics* 3: 591.
31 Arnaud, B., Lebegue, S., Rabiller, P., and Alouani, M. (2006). *Physical Review Letters* 96: 026402.
32 Cassabois, G., Valvin, P., and Gil, B. (2016). *Nature Photonics* 10: 262.
33 Chen, J.-H., Jang, C., Xiao, S. et al. (2008). *Nature Nanotechnology* 3: 206.
34 Dean, C.R., Young, A.F., Meric, I. et al. (2010). *Nature Nanotechnology* 5: 722.
35 Xue, J., Sanchez-Yamagishi, J., Bulmash, D. et al. (2011). *Nature Materials* 10: 282.
36 Zomer, P.J., Dash, S.P., Tombros, N., and van Wees, B.J. (2011). *Applied Physics Letters* 99: 232104.
37 Liu, L., Feng, Y.P., and Shen, Z.X. (2003). *Physical Review B* 68: 104102.

38 Zhao, L., Levendorf, M., Goncher, S. et al. (2013). *Nano Letters* 13: 4659.
39 Wong, D., Velasco, J. Jr., Ju, L. et al. (2015). *Nature Nanotechnology* 10: 949.
40 Neto, A.H.C., Guinea, F., Peres, N.M.R. et al. (2009). *Reviews of Modern Physics* 81: 109.
41 Fujimoto, Y. and Saito, S. (2011). *Physical Review B* 84: 245446.
42 Wehling, T.O., Balatsky, A.V., Katsnelson, M.I. et al. (2007). *Physical Review B* 75: 125425.
43 Tersoff, J. and Hamann, D.R. (1983). *Physical Review Letters* 50: 1998.
44 Tersoff, J. and Hamann, D.R. (1985). *Physical Review B* 31: 805.
45 Fujimoto, Y., Okada, H., Inagaki, K. et al. (2003). *Japanese Journal of Applied Physics* 42: 5267.
46 Okada, H., Fujimoto, Y., Endo, K. et al. (2001). *Physical Review B* 63: 195324.
47 Fujimoto, Y., Okada, H., Endo, K. et al. (2001). *Materials Transactions* 42: 2247.
48 Fujimoto, Y. and Oshiyama, A. (2013). *Physical Review B* 87: 075323.
49 Fujimoto, Y. and Hirose, K. (2003). *Physical Review B* 67: 195315.
50 Fujimoto, Y. and Hirose, K. (2003). *Nanotechnology* 14: 147.
51 Fujimoto, Y., Asari, Y., Kondo, H. et al. (2005). *Physical Review B* 72: 113407.
52 Fujimoto, Y., Hirose, K., and Ohno, T. (2005). *Surface Science* 586: 74.
53 Ono, T., Tsukamoto, S., Egami, Y., and Fujimoto, Y. (2011). *Journal of Physics: Condensed Matter* 23: 394203.
54 Chelikowsky, J.R., Troullier, N., and Saad, Y. (1994). *Physical Review Letters* 72: 1240.
55 Hirose, K., Ono, T., Fujimoto, Y., and Tsukamoto, S. (2005). *First-Principles Calculations in Real-Space Formalism, Electronic Configurations and Transport Properties of Nanostructures*. London: Imperial College Press.
56 Ono, T., Fujimoto, Y., and Tsukamoto, S. (2012). *Quvantum Matter* 1: 4.
57 Büttiker, M., Imry, Y., Landauer, R., and Pinhas, S. (1985). *Physical Review B* 31: 6207.
58 Haga, T., Fujimoto, Y., and Saito, S. (2019). *Physical Review B* 100: 125403.
59 Pease, R.S. (1950). *Nature (London)* 165: 722.
60 Constantinescu, G., Kuc, A., and Heine, T. (2013). *Physical Review Letters* 111: 036104.
61 Jin, C., Lin, F., Suenaga, K., and Iijima, S. (2009). *Physical Review Letters* 102: 195505.

11

Performance Analysis of Nanosheet Transistors for Analog ICs

Yogendra P. Pundir[1,2], Arvind Bisht[2], and Pankaj K. Pal[2]

[1] *Department of Electronics and Communication Engineering, HNB Garhwal (A Central) University, Srinagar Garhwal, Uttarakhand, India*
[2] *Department of Electronics Engineering, National Institute of Technology Uttarakhand, Srinagar Garhwal, Uttarakhand, India*

11.1 Introduction

The consistent improvements in functionality and performance of integrated circuits (ICs) have fueled the digital revolution [1–3]. Alongside the impressive growth of data usage, the data traffic and the required data transfer speeds have also been on the rise [4]. These remarkable technological developments are possible solely on the foundations provided by each new generation of metal oxide field-effect transistors. The transistor dimensions are shrunk to achieve performance and cost advantage while ensuring enough electrostatic gate control over the channel to mitigate the short-channel effects (SCEs) [2]. Despite the physical limitations posed by available photolithography technologies, every new technology node provides power, performance, area, and cost (PPAC) advantages in comparison to the previous node [5]. Device-technology co-optimization (DTCO) and countless innovations in technology make the PPAC improvements possible [5, 6].

The tri-gate architecture of FinFETs provided relief from the increased SCE issues faced by the planar metal–oxide–semiconductor field-effect transistors (MOSFETs) as the gate length shrunk below 25 nm. The FinFETs have been the workhorse of IC technology since the 22 nm technology node [7]. At scaled nodes, the FinFETs suffer from poor device electrostatics, large device parasitic, large leakage power, and sub-optimal low-voltage operation [8, 9]. Recently, semiconductor manufacturers have proposed nanosheet transistor (NST) technology as a successor to the FinFETs for continuing the IC scaling at sub 7 nm nodes [10, 11]. NSTs use vertically stacked multiple nanosheet channels with a gate-all-around

Advanced Nanoscale MOSFET Architectures: Current Trends and Future Perspectives, First Edition. Edited by Kalyan Biswas and Angsuman Sarkar.

(GAA) structure and, therefore, offer several advantages over the FinFETs [12–14]. These advantages include a more significant drive current for the same area footprint, better gate electrostatics at downscaled gate lengths, and continuously variable channel widths. While FinFETs require more fins for larger current, nanosheets can be vertically stacked for the purpose. The technology transition from FinFETs to NSTs is affordable as NSTs use processing tools and manufacturing methods similar to FinFETs [11]. Also, the circuit designers can simply replace FinFETs with NSTs in layouts without modifying the footprints [15].

11.2 Evolution of Nanosheet Transistors

In 1965, Gordon Moore postulated Moore's law, observing that the number of transistors in an IC die doubles every year and later updated it to double every two years [16, 17]. Using clever circuit architecture, feature-down scaling, and bigger die with high yield, the Silicon industry has lived up to Moore's prediction [4, 10, 18–20]. Limitations of lithography technology have slowed down Moore's Law and transistors' downsizing, but the quest for an ideal transistor is here to stay [4, 21]. Multiple technology challenges requiring the focus of the semiconductor community for the attainment of ideal 3D transistors are shown in Figure 11.1 [4, 21].

Scaling down the transistor dimension brings multiple benefits to the circuits using them [22]. A shorter gate length makes the transistor faster. Also, the shorter

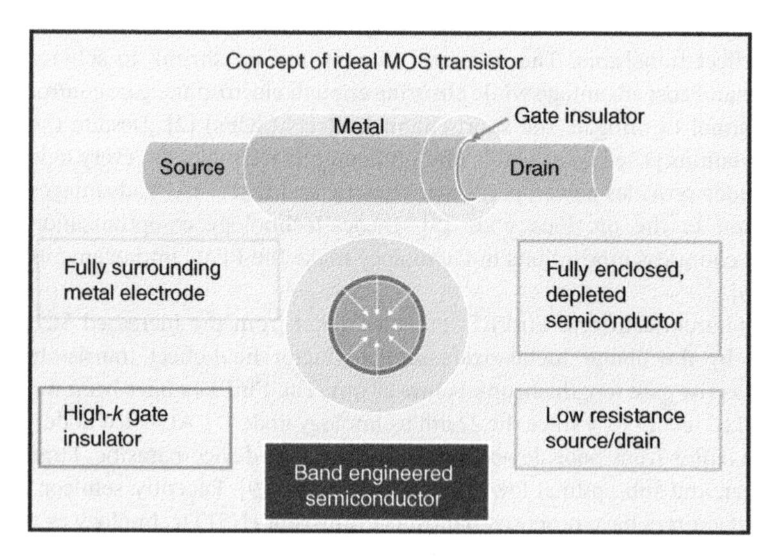

Figure 11.1 Concept of the completely refurbished ideal 3D transistor. Source: Adapted from [4, 21].

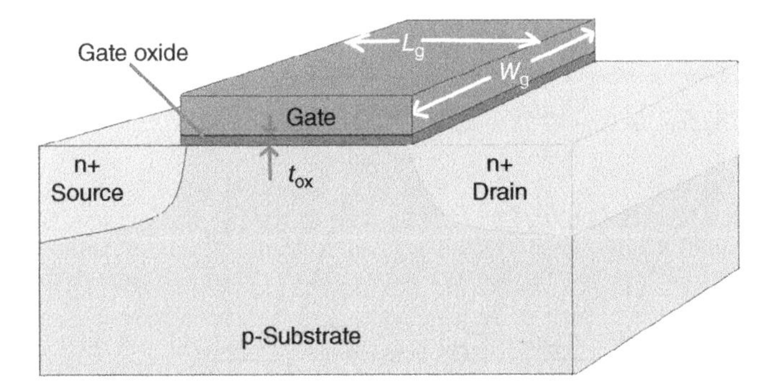

Figure 11.2 Structure of planar N-channel MOSFET.

transistor dimensions mean that the wires connecting them are also shorter. Scaling thus results in smaller wire parasitic capacitance and helps to faster circuit performance. If transistor gate length $L_{g,}$ as defined in the planar MOSFET structure shown in Figure 11.2, is scaled down by factor S, then generally, the transistor speed and chip area improve by factor S and S^2, respectively. With length scaling as the chip area becomes smaller, more dies are possible from the same wafer after fabrication. The resulting smaller dies also have fewer defects, and thus, the yield is improved [23]. Therefore, transistor scaling can achieve the same functionality at lower production costs.

11.2.1 Short-Channel Effects and Their Mitigation

For long-channel bulk MOSFETs, the roles of the vertical and horizontal electric fields are distinct and independent. The vertical field is applied via the gate terminal, and the minimum gate voltage (V_{GS}) required to strongly invert the channel is called threshold voltage (V_{TH}). The vertical field controls the amount of channel charge, while the horizontal field drives the current [24, 25]. Both fields do not interfere with each other, and thus the threshold voltage of MOSFET is independent of drain voltage (V_{DS}) for long channel MOSFETs. However, as the transistor gate length is shrunk, the drain's electric field begins to aid the field by the gate, and the roles of the drain and gate in MOSFET are no longer independent. The application of drain voltage for short channel lengths reduces the barrier at the source end, as shown in Figure 11.3. This effect is called drain-induced-barrier-lowering (DIBL).

The electrostatic integrity of the MOSFET device is represented by the subthreshold swing (SS) and DIBL. Subthreshold Swing is the amount of gate voltage change required to cause an order of magnitude change in drain current (I_{DS}). DIBL expresses the change in the MOSFET's threshold voltage by applying increased drain voltage. The switching ability of MOSFET can be represented as

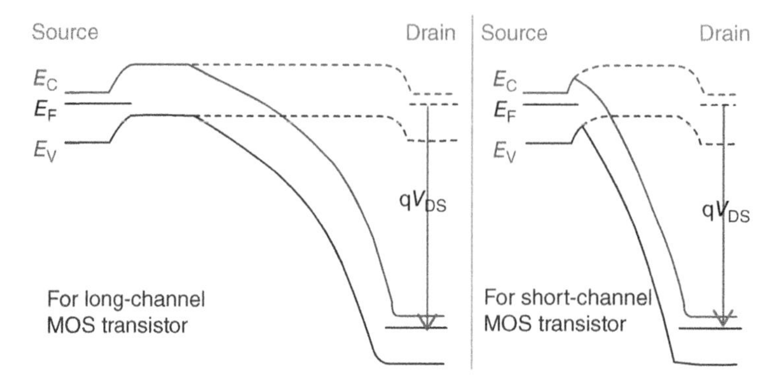

Figure 11.3 Energy band diagram beneath the gate oxide.

the ratio of its ON current to OFF current (I_{ON}/I_{OFF}). These parameters are defined in Figure 11.4. For shorter channel lengths, the influence of the drain field is more substantial. Therefore, the gate's control on the channel barrier is not strong enough to turn the device off in an effective manner. This results in sizeable static power and poor switching ability, I_{ON}/I_{OFF}. For attaining adequate $I_{ON}/I_{OFF,}$ of the order 10^5 or better, with small supply voltages, we need a MOS device with a small SS value. SS metric shows how fast the device can be switched ON and OFF. Due to their thermodynamic career transport, the conventional MOSFETs have a theoretical low limit on SS of 60 mV/decade at room temperature [26]. This prevailing lower limit on the SS restricts how lower the supply voltage can be scaled.

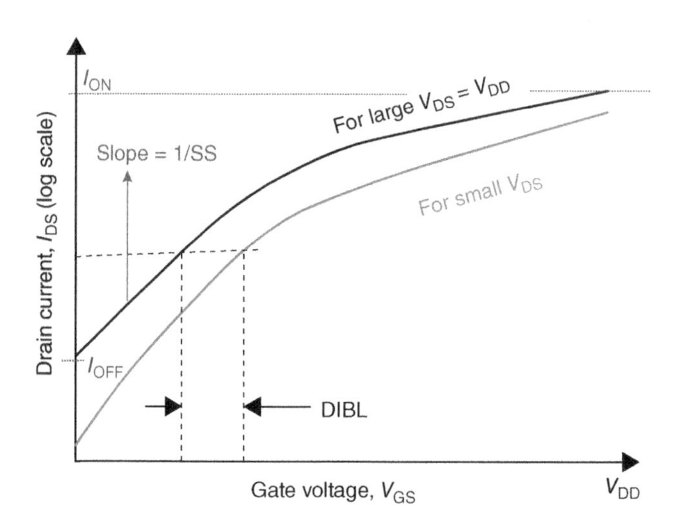

Figure 11.4 Description of SS and DIBL in transfer characteristics of MOSFET.

The techniques to mitigate the SCEs in the planar devices are summarized in [27] as follows: (i) Creation of retrograde (delta) profile under the channel; (ii) S/D extensions (tip); (iii) Adding halo/pocket implants under the S/D extensions; and (iv) Engineering the S/D extension for the smallest depth (x_j) while ensuring low resistance [27–30]. The use of thinner gate oxides can restore the control of the gate over the channel. However, the gate-oxide thickness (t_{ox}) cannot be decreased arbitrarily. Because thin oxide also results in quantum-mechanical tunneling and, consequently, a large gate-leakage current. High-k dielectrics, such as hafnia oxide (HfO_2) with relative permittivity $k = 22$, have been adopted in place of SiO_2 ($k = 3.9$) as gate-oxide [31–33]. High-k oxides allow the use of sufficiently thick gate-oxide to mitigate quantum-mechanical tunneling while ensuring adequate control of the gate over the channel. Source/drain junctions are kept shallow, especially near the gate interface, so that the effects of the drain field on the channel charge carriers and the source-side barrier height are minimized. Channel doping can be engineered using complex vertical and horizontal profiles to restrain electric field lines emanating from the source to the drain [34].

However, realizing complex doping profiles is process-wise costly and also results in smaller ON currents along with larger band-to-band tunneling (BTBT) and gate-induced-drain-leakage (GIDL) currents [35]. Further, larger doping in the channel results in higher V_{TH} and lower mobility due to increased scattering of charge carriers, resulting in a smaller drain current [24]. Also, a larger depletion capacitance increases the SS of the device [36]. Large doping near-source/drain extension aggravates the BTBT leakage problem [37]. Further, the drain current (I_{DS}) of MOSFETs depends not only on the channel resistance but also on the source and drain resistances. If source/drain junction depths are decreased to mitigate SCEs, then doping in source/drain regions must be increased to keep the resistance low [38]. However, the maximum allowed doping is limited by the solid-solubility of dopants ($\sim 10^{20}$ cm^{-3}) in semiconductors [39]. In addition, the realization of shallow junctions is challenging due to the annealing steps needed for dopant activation during fabrication [39]. Finally, as the gate length becomes very short, the channel volume also becomes very small. The limited number of charge carriers in a small volume of a channel results in large stochastic variations called random-dopant-fluctuation (RDF), among the similarly drawn devices after their fabrication [40].

11.2.2 The FinFET Technology

The electrostatic control of the gate over the channel can be increased by either bringing the gate closer to the substrate or by increasing the number of gates. The gate can be brought closer by reducing the body thickness. Using a thin body ensures that the bulk is not far away from the gate, resulting in better control of the gate on leakage current. Ultra-thin-body silicon-on-insulator (UTB-SOI) MOSFETs use a thin buried oxide layer to control the short SCEs. UTBSOI MOSFETs also offer low channel doping and V_{TH} tuning by body bias. These devices require manufacturing technology similar to planar devices. However, their small drive current results in poor layout efficiency and makes them less useful [41]. Multi-gate transistor structures allow more than one gate to control the channel, suppress the SCEs, and improve the electrostatic integrity and scalability [42]. Using multiple gates allows the gate lengths to scale below 25 nm and helps extend Moore's law [7]. Figure 11.5 shows the evolution of multi-gate MOSFET structures ultimately resulting in NST.

While a double-gate transistor structure was explored to keep the CMOS scaling going, the tri-gate structure of FinFETs could deliver sufficient performance gains [37, 43, 44] to justify the economic investments required for new technology [45, 46]. Thus, the industry underwent a sea change in the last decade from planar MOSFET to non-planar FinFETs. FinFETs use a very thin body and have the gate controlling the channel charge from three sides. The tri-gate structure of FinFETs results in excellent channel electrostatics and, therefore, smaller off currents even for low threshold voltage transistors [27, 47]. Thus, FinFETs provided a high I_{ON}/I_{OFF} current ratio along with small I_{OFF}, and have been used since the 22 nm technology node [46, 48, 49].

FinFETs can provide larger effective widths by the use of taller fins. Further, a better fin process control and gate-stack engineering for multi-V_{TH} tuning, instead of implant-based V_{TH} control, is possible. Thus, the FinFETs have lesser V_{TH} variations and operate with lower power supply voltage. FinFETs also have a lower body effect than planar devices. Recent technology innovations such as pitch-splitting (litho-etch-litho-etch), cut masks, spacer-based pattering: self-aligned double patterning, and self-aligned quadruple patterning [50], along with the fin-depopulation, have kept the momentum for FinFETs to continue [31, 46, 51, 52]. The 7 nm node FinFETs are the first ones to use EUV (13.5 nm) technology, along with prevalent deep UV (193 nm) immersion lithography, for volume production [53–55]. Recently, FinFETs at sub-7 nm technology nodes have also been demonstrated [19, 56–58]. However, the cost of production has not reduced remarkably with the decreased size [59].

Taller fins increase effective width and compensate for the increase in contact resistances due to smaller contact areas [60]. But the height of the fins is limited by the difficulty in realizing the metal interconnect layers above the device and the imperfections of patterning techniques. Further, at a very small fin thickness, the carrier's mobility deteriorates due to interface scattering and quantum confinement [14]. Despite the challenges, every new technology node FinFETs uses narrower and taller fins along with better junction optimizations to improve device electrostatics [14]. The FinFETs have progressed from 22 to 5 nm nodes to achieve

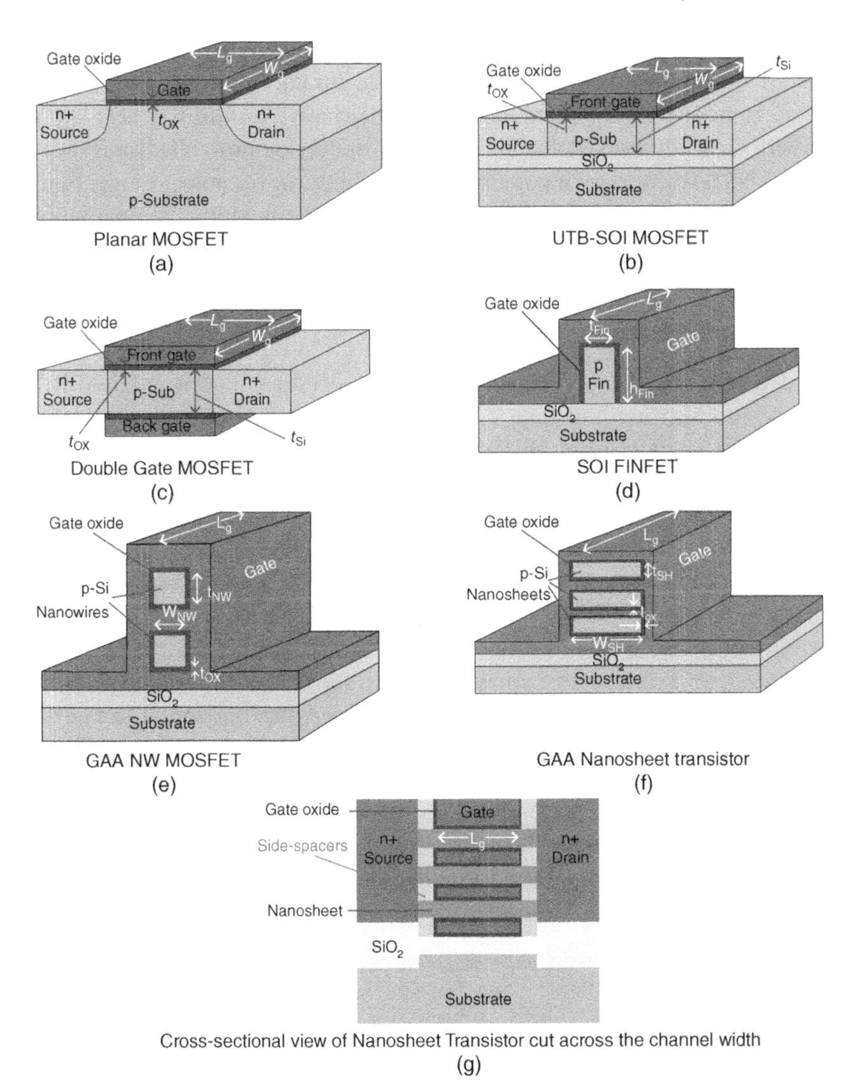

Figure 11.5 Evolution of MOSFET structures. (a) Planar MOSFET. (b) UTB-SOI MOSFET. (c) Double gate MOSFET. (d) SOI FINFET. (e) GAA NW MOSFET. (f) GAA nanosheet transistor. (g) Cross-sectional view of nanosheet transistor cut across the channel width.

PPAC scaling objectives. The realization of thin and tall fins very close to each other to achieve the target cell-area scaling has also become increasingly more challenging [54].

Therefore, despite numerous innovations, the FinFETs seem to have reached their ultimate electrostatic and physical limits [61]. The FinFETs are believed to be operating at the smallest possible supply voltages and delivering their best possible performance per energy dissipation. Therefore, beginning at the sub-7 nm nodes, GAA devices have been explored as a viable alternative to FinFETs [14] GAA devices with vertically stacked thin and wide nanosheet channels called nanosheet transistors (NSTs) have shown potential for scalability, lower power dissipation, smaller operating voltage, and larger ON current while retaining compatibility with the existing circuit and manufacturing technology [12, 62, 63].

11.2.3 Advent of Nanosheet Transistors

Toshiba demonstrated the first GAAFET in 1988, a vertical nanowire GAAFET, and termed it a Surrounding Gate Transistor (SGT) [64]. GAA device has the gate controlling channel from all four sides [65]. This allows GAA devices to offer excellent gate electrostatics [66]. GAA MOSFETs can use either nanowires or nanosheets, as shown in Figure 11.5, and be called nanowire FET (NWFET) or Nanosheet FET (NSFET or NST), respectively. The structure of NST is also shown in Figure 11.6. While nanowires have circular or close to a square cross-section, nanosheets have widths much larger than thickness. The NST is a version of the NWFET in which the gate surrounds the transistor's channel and has layered silicon nanosheets rather than nanowires. If the FinFET is considered to be a vertical structure, the nanosheet FET can be regarded as a horizontal or lateral structure [37]. The channel orientation in GAA MOSFET can be either vertical or horizontal. Hu and Li [67] shows that nanosheet devices offer better effective drive currents than nanowire-type devices while the nanowires provide the best gate electrostatics [68, 69]. Lateral GAA devices, in comparison to their vertical counterparts, have process technology similar to the current technology

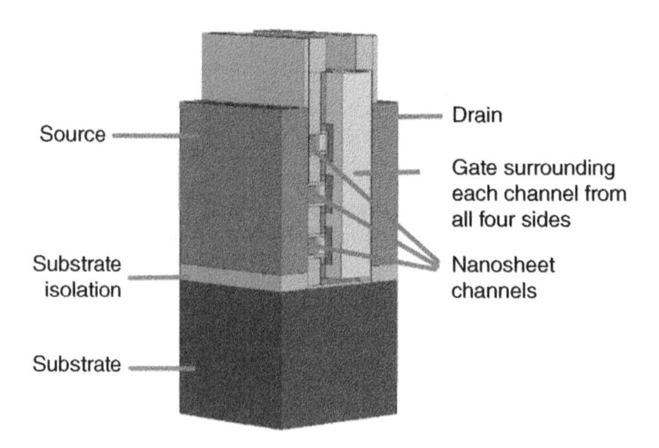

Figure 11.6 Structure of GAA nanosheet transistor.

for FinFETs and thus save on new investment requirements [14, 70]. Also, the improved gate electrostatics of lateral NST over the FinFETs result in smaller SS, better overdrive voltage, lower supply voltage, and increased scalability [49, 62, 71, 72]. P-channel NSTs show significant current improvements over P-channel FinFETs due to compressive strain, and thus NSTs have better-matched performance for N- and P-channel devices [13] GAA technology allows multiple threshold voltage devices using different work-function metals [73]. Many studies have shown the performance advantages of lateral GAA NSTs over FinFETs [13, 74–76]. Therefore, the IRDS-2021 reports predict the lateral GAA NSTs to be at the forefront along with FinFETs soon [4, 78].

NSTs are called multi bridge channel field-effect transistors (MBC-FETs) by Samsung [74]. Samsung has worked on MBC-FET since the 2000s and has already announced to use MNC-FETs from its 3 nm technology node [15]. Samsung reports that the NSTs at the 3 nm node will lead to the next generation of SRAM and allow for smaller voltage read operation [77] While FinFETs require more fins for larger current, NSTs can have multiple vertically stacked nanosheets. The technology transition from FinFETs to NSTs is affordable as NSTs use processing tools and manufacturing methods similar to FinFETs. Circuit designers can simply replace FinFETs with NSTs in layouts without modifying footprints [15]. For NSTs, the W_{eff} can be increased without compromising on SCE, and the DC performance can be improved by increasing the W_{SH}. Samsung also reports that NSTs have comparable time-dependent dielectric-breakdown (TDDB) values to the FinFETs and similar device temperatures as FinFETs [78]. Intel calls its NSTs nano-ribbon transistors and expects to manufacture these from the Intel20A node, marking the start of Angstrom-Era [79, 80]. IBM has also showcased chips fabricated at 2 nm nodes using the NSTs. IBM's 2 nm process is also the first to apply EUV lithography on the chip's front-end-of-line (FEOL) fabrication steps, where transistors and other structures are made (nanosheets and gate). Typically, the EUV patterning is used in the middle-of-line (MEOL) and back-end-of-line (BEOL) stages. By 2023, TSMC plans to introduce a 2 nm GAAFET node [59].

The NSTs have been generally accepted as a potential successor to FinFETs due to the number of benefits offered. Thanks to their GAA structure, nanosheet FETs with better electrostatics provide a better current density and a smaller area. Also, the nanosheets can be stacked vertically to achieve a larger ON current without increasing the area footprint. Thus, it offers a larger effective width for the same area footprint. The similarity between the process flows of NSTs and FinFETs allows the NSTs to share the existing fabrication infrastructure for FinFETs, thus saving a significant investment on the technology cost. NST allows for continuously variable width and restores the flexibility lost earlier in the case of FinFETs, for circuit designers [81].

11.3 TCAD Modeling of Nanosheet Transistor

TCAD tools are helpful in the semiconductor device industry for a variety of tasks. These tasks include understanding the semiconductor device operation, fabrication process optimization, debugging the process-related faults, device optimization to avoid expensive wafer reruns in the foundry, analyzing the impact of various process and statistical variability sources, and the performance analysis of novel devices [82]. Synopsys Sentaurus TCAD is one such complete graphical operating environment that, with the process, device, and circuit simulation tools, helps in the predictive evaluation of the impact of new materials and structures on the electrical characteristics of semiconductor devices. Device modeling can be performed using multiple approaches, such as drift-diffusion (DD), Non-equilibrium Green functions (NEGF), and Monte-Carlo (MC) methods. More accurate models require more computing resources; therefore, the choice of model is a trade-off between the accuracy and computational complexity [83].

DD model is used to model near-equilibrium device behavior and assumes carriers to be in thermal equilibrium. DD model uses Poisson's equation, carrier transport equation in DD regime with the current continuity equation. DD model is suitable for device dimensions sufficiently large to discard any quantum effects [83]. DD model is particularly useful for technology nodes above 100 nm and is simple and computationally efficient. Mobility is one of the key parameters determining the transistor current and thus is critical for calibration. As mobility is determined by different scattering mechanisms at play, the TCAD model has to take these scattering mechanisms, e.g., surface roughness, and lattice/phonon scattering, into account [25, 83, 84].

DD formalism in TCAD includes low-field mobility, perpendicular-field mobility effects, and lateral-field mobility effects. Low-field mobility combined with the effects of perpendicular-field (E_{norm}) affects the transfer characteristics at small drain voltage (V_{DS}). The application of high drain voltage further limits mobility to lower values; this is known as the velocity saturation effect occurring due to a high lateral electric field [84]. At scaled dimensions, the DD model fails to correctly model I_{ON} due to its inability to capture the velocity overshoots [85]. Higher-order Boltzmann's transport equation (BTE) and hydrodynamic set of models can be used to consider velocity overshoots but at significant computational overhead. Another approach is to use the MC method to get a direct solution for BTE to know energy and momentum distribution along with nonequilibrium transport effects [86, 87]. Further, at nanoscale dimensions, quantum-mechanical effects are to be considered for accurately modeling the mobile charge carriers and the gate capacitance. As the dimensions of modern transistors are now at a quantum-mechanical length scale, the wave nature of carriers leads to a change in threshold voltage and reduction of gate capacitance [83]. Schrödinger-equation-based models, e.g., density gradient (DG) model, and modified local-density approximation (MLDA) model, can be included to incorporate the corrections required for quantum mechanical effects [83]. The channel charge can be more accurately estimated using NEGF or Poisson–Schrödinger solver, but at an acute computational cost [88]. Nevertheless, a DD model with calibrated ballistic mobility and high-field velocity saturation models can also be used to describe FinFETs with channel length values as small as 5 nm [86] and saves computation time.

NSTs of 5 nm technology-node with N- and P-channel [71] having three vertically stacked nano-sheet channels and a GAA structure are realized using the SDE tool in Sentaurus TCAD [89]. The three-dimensional (3D) view of NST is shown in Figure 11.7a. A single stack of three sheets is an optimal configuration [90] and is the one experimentally demonstrated by [71]. Silicon is still the optimal choice of material used for nanosheet channels [91]. Source, Drain, and Channel regions have silicon with (100) transport direction for N-channel NST. For the P-channel NST, $Si_{0.5}Ge_{0.5}$ with (110) transport direction can be used for the source, drain, and nanosheet channels [92]. Silicon dioxide (SiO_2) is used as an interfacial layer between the high-*k* gate dielectric, i.e. hafnium oxide (HfO_2), and the silicon (100). The side spacers use low-*k* dielectric with relative permittivity of 5 [93]. Dielectric layers of SiO_2 under both source and drain regions are used to compress the leakage through the substrate [71, 94]. A dielectric underneath only the source and drain regions is enough to provide benefits comparable to complete oxide isolation with the substrate [71]. Figure 11.7b shows a two-dimensional (2D) schematic view of a nanosheet channel that indicates important physical dimensions used for the calibrated NST. The values of essential device dimensions and physical parameters [71, 95] that are used for calibrating the TCAD model for NST are listed in Table 11.1. Doping in source and drain extension regions is modeled as Gaussian doping profiles with a 1 nm/decade slope [96].

DD transport model with Poisson equation and current continuity equation are solved together self-consistently using the SDEVICE tool of Sentaurus TCAD. *The density-gradient* model is included to account for the quantum confinement effects [89]. Bandgap-narrowing effects for silicon are considered by using the *Old-Slot-Boom* model. Mobility models in the DD module include doping-dependent *Low-field ballistic mobility*, *Inversion and accumulation layer mobility* (IALMob) with auto-orientation, *Perpendicular field mobility* with *Remote-phonon-scattering*, and *High-field saturation velocity* model. The *Thin-layer mobility* model is used to capture the structural confinement of charge carriers in 5 nm thick nanosheets. The *Inversion and accumulation layer mobility* model is required to incorporate the Coulomb scattering, phonon scattering, and surface roughness scattering. The *low-field ballistic field* mobility model includes the quasi-ballistic effects on the charge carrier mobility. The degradation of mobility due to the trap-charges is incorporated by

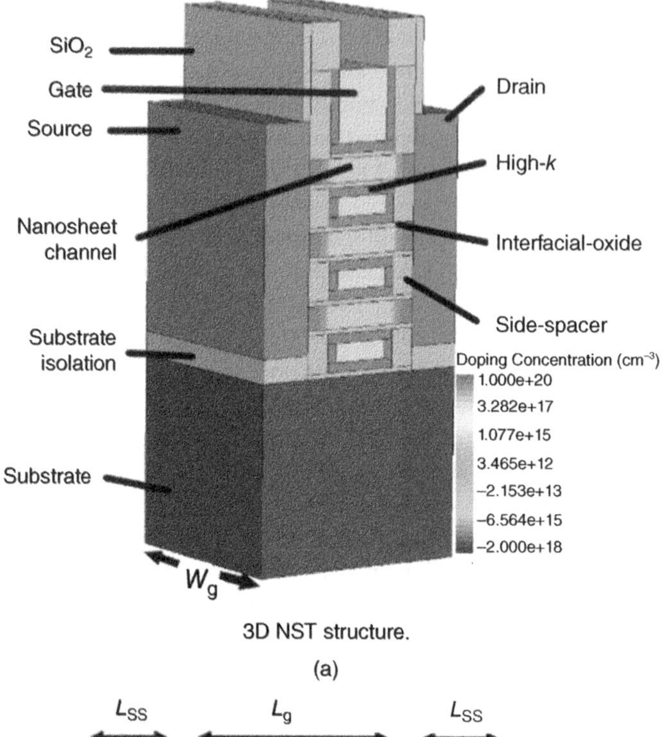

3D NST structure.
(a)

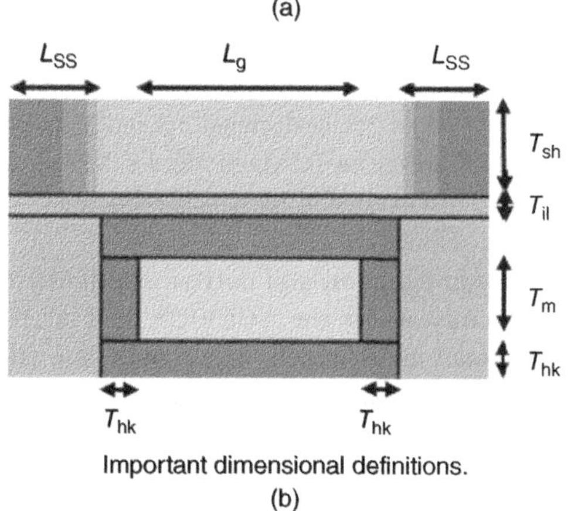

Important dimensional definitions.
(b)

Figure 11.7 TCAD model of nanosheet transistor with three vertically stacked channels. (a) 3D NST structure. (b) Important dimensional definitions.

Table 11.1 NST Device Dimensions and Physical Parameters for TCAD Calibration.

Device Parameters	**Values**
Gate's length (L_g)	12 nm
Spacer's length (L_{ss})	5 nm
Nanosheet's width (W_{sh})	45 nm
Nanosheet's thickness (T_{sh})	5 nm
High-*k* layer's thickness (T_{hk})	2 nm
Interfacial layer's thickness (T_{il})	1 nm
Metal layer's thickness (T_m)	4 nm
Sheet pitch	10 nm
Relative permittivity of high-*k*	22
Relative permittivity of low-*k*	5
Relative Permittivity of Interfacial Layer	3.9
Source/drain doping concentration (phosphorous) for N-NST	1×10^{20} cm^{-3}
Channel doping concentration (boron) for N-NST	1×10^{15} cm^{-3}
Metal-gate work-function for N-NST	4.34 eV
Source/drain doping concentration (boron) for P-NST	5×10^{20} cm^{-3}
Channel doping concentration (phosphorous) for P-NST	1×10^{15} cm^{-3}
Metal-gate work-function for P-NST	4.4 eV
Resistivity of source/drain contact	10^{-9} Ω cm^2
Power supply voltage (V_{DD})	0.7 V

activating the calculations of transverse-electric-field (E_{normal}) along with the *remote-phonon-scattering*, *thin-layer mobility*, and *inversion & accumulation layer mobility* models. The models used for recombination are the *Hurkx band-to-band* model, *Shockley–Read–Hall* model with doping dependence, and *Auger* recombination model. The stress-induced mobility enhancements are incorporated using the *Piezo* model [13, 83, 95].

The physical parameters of the 5 nm node NST are optimized/benchmarked using the approach explained in the cited literature [13, 97] so that the results from the TCAD setup match the experimentally measured data [71]. One approach for the TCAD calibration is as follows. The low-field mobility and the metal work-function values are calibrated to match the I_{OFF} (i.e. I_{DS} at $V_{GS} = 0$ V and $|V_{DS}| = V_{DD}$). Low-field mobility impacts the subthreshold current of NSTs, especially when these are operating in the linear region ($|V_{DS}| = 50$ mV). The extension region's Gaussian doping profiles are calibrated to match the DIBL

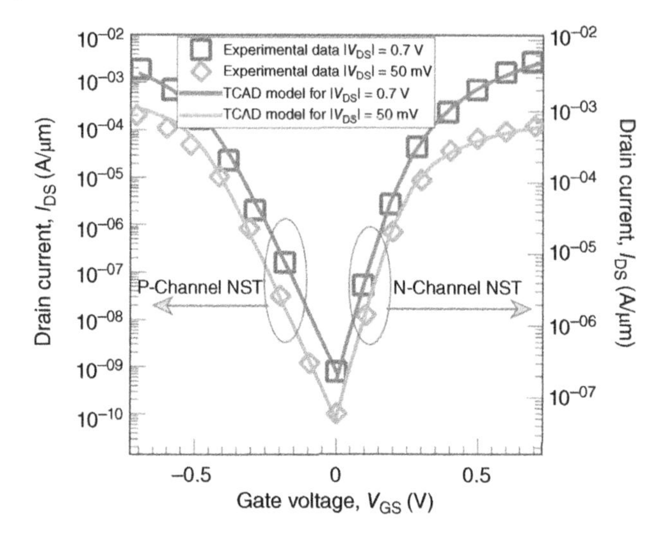

Figure 11.8 Results of TCAD simulation setup calibration with experimental data. Source: Adapted from Barbé et al. [11].

and subthreshold swing values with the experimental observations [71]. The ON current value (i.e. I_{DS} at $|V_{GS}| = |V_{DS}| = V_{DD}$, is matched by calibrating the charge carrier saturation velocity and the resistance of the drain/source contacts. The transfer characteristics ($I_{DS} - V_{GS}$) for the simulated and reference experimental N- and P-channel NSTs [71] are shown in Figure 11.8. The drain current values, shown in Figure 11.8, are drain current per unit micron width of NSTs and expressed in A/μm unit. The values of drain voltage used for linear and saturation curves are 50 mV and 0.7 V, respectively. It can be observed that the TCAD simulated results for NST satisfactorily match the measured results for the actual NST. Such a calibrated TCAD model can be used for characterizing the NST performance. Different performance parameters of NST for analog applications are explained in Section 11.4 and can be extracted using the SDEVICE and INSPECT tools of Sentaurus TCAD.

11.4 Transistor's Analog Performance Parameters

The DC performance of a transistor represents the relationship between terminal voltages and currents. It is typically measured from the transfer (I_{DS} vs. V_{GS}) and output (I_{DS} vs. V_{DS}) characteristics of a transistor. Terminal voltages and current definitions are shown in Figure 11.9. For comparison of different device structures, the current levels and the related DC performance parameters are sometimes normalized by the channel width or perimeter in the case of multi-gate MOSFETs. Also, for comparison among different devices, the off-state current, I_{OFF} (I_{DS} at $V_{GS} = 0$ V and $V_{DS} = V_{DD}$) is fixed. The value of I_{OFF} is dependent on the target applications, e.g., standard performance (SP), high performance (HP), and low power (LP) [13, 95]. In practice, the V_{TH} is adjusted to meet either of these performance needs. The V_{TH} can be measured using methods explained in [98, 99].

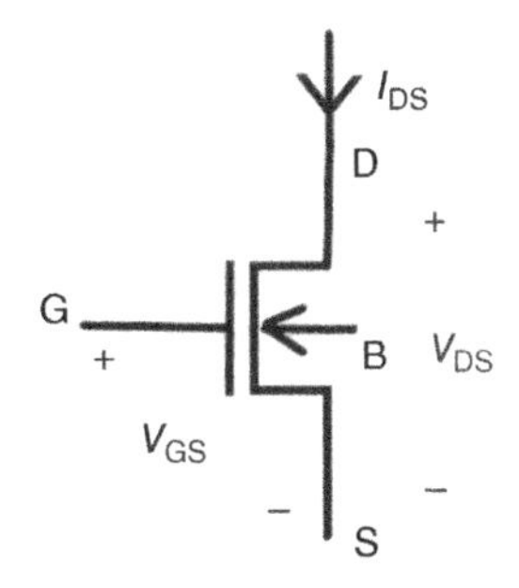

Figure 11.9 The symbol for N-channel MOSFET.

A small-signal model of MOSFET is shown in Figure 11.10, which includes small-signal parameters such as g_m and g_{ds} that are defined next.

11.4.1 Transconductance

The ability of the transistor to modulate its drain current with the input voltage is characterized in terms of performance parameter transconductance (g_m). Mathematically, the transconductance is defined as:

$$g_m = \frac{dI_{DS}}{dV_{GS}} \tag{11.1}$$

and shows the rate of change of I_{DS} with change in V_{GS}.

Assuming a simple Shockley's square law model of MOSFET, the formulas of g_m can be obtained [100] as follows:

$$g_m = \sqrt{2\mu_n C_{OX} \frac{W}{L} I_{DS}} \tag{11.2}$$

$$g_m = \frac{2I_{DS}}{V_{GS} - V_{TH}} \tag{11.3}$$

$$g_m = \mu_n C_{OX} \frac{W}{L} (V_{GS} - V_{TH}) \tag{11.4}$$

If the bias current I_{DS} is already fixed from the power dissipation constraints, then as per equation (11.2) the g_m is proportional to the square root of the W/L

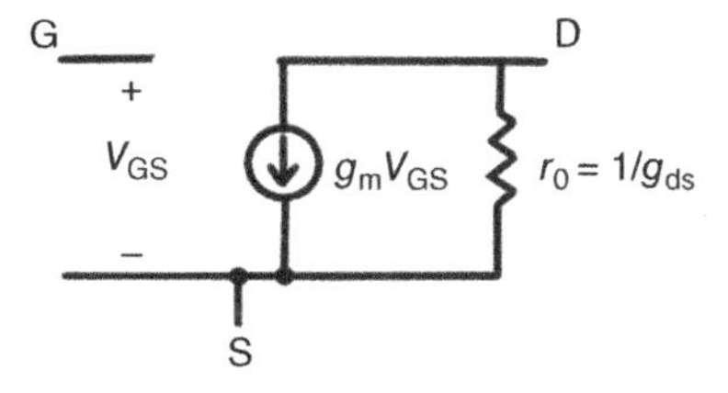

Figure 11.10 Low-frequency small-signal model of MOSFET.

ratio of a transistor. Equation (11.3) shows that for a given W/L and $I_{DS,}$ the g_m is inversely proportional to the gate overdrive voltage ($V_{OV} = V_{GS} - V_{TH}$). If the I_{DS} is not fixed, then the g_m has linear relations with the W/L ratio and V_{OV} [100].

11.4.2 Output Conductance

Performance parameter output conductance (g_{ds}) shows how well the transistor acts as a constant current source. A smaller value of g_{ds} translates to a more significant output impedance ($r_o = 1/g_{ds}$) and is a desirable property for a transistor to act as a current source. It shows the extent of change in I_{DS} with changing $V_{DS.}$ The g_{ds} is defined as a derivative of I_{DS} to V_{DS} as shown in Eq. (11.5) below.

$$g_{ds} = \frac{dI_{DS}}{dV_{DS}} \tag{11.5}$$

For MOSFETs, the value of g_{ds} is primarily governed by the phenomena of channel length modulation and DIBL [101]. For nanoscale gate lengths, the DIBL becomes the predominant physical cause of the increase in I_{DS} with increasing V_{DS} and determines the value of g_{ds}. A parameter known as Early Voltage (V_A) is typically used to compare transistors and is defined as follows.

$$V_A = \frac{I_{DS}}{g_{ds}} \tag{11.6}$$

Early voltage for MOSFETs is the voltage at which the tangent to the I_{DS} vs. V_{DS} crosses the voltage axis [102, 103]. A larger value of early voltage in technology shows its superiority in terms of gate control over channels and means better output conductance. The value of V_A is dependent on channel length and bias current. DIBL in very short transistors causes V_{TH} to decrease with an increase in drain voltage. Thus, it results in a larger drain current and, consequently, a larger g_{ds}. The resulting increase in g_{ds} values due to DIBL is worse in the case of devices in weak inversion which have larger g_m/I_D increase. In circuits, such degradations of g_{ds} may be countered by keeping the V_{DS} value well above the V_{DSAT} ($V_{GS} - V_{TH}$).

11.4.3 Intrinsic Gain

Intrinsic gain for a MOS transistor is defined as the drain voltage to gate voltage ratio when the source is grounded, and an ideal current source load is tied to the drain terminal. The ratio of output voltage (V_{DS}) and input voltage (V_{GS}) under no-load condition ($R_L = \infty$) is called intrinsic voltage gain (A_{v0}) and can be formulated as Eq. (11.7).

$$A_{v0} = \frac{g_m}{g_{ds}} \tag{11.7}$$

It is of critical importance to ensure the stability of the transistor bias points in circuits [104]. The use of feedback is a prominent circuit-level technique to improve the bias stability but at the cost of overall amplifier gain [100]. Thus, the transistors need to possess a sizeable intrinsic gain such that a sufficient open-loop gain can provide for the required closed-loop gain. The intrinsic gain can also be defined as a product of the transconductance and output impedance ($A_v = g_m r_o$, where r_o is the reciprocal of g_{ds}). And, it is desirable to have a larger value of g_m and a smaller value of g_{ds} so that a large value of A_v is achieved [100].

11.4.4 Transconductance Efficiency

The ratio of g_m and I_{DS} is defined as transconductance efficiency (g_m/I_D). The value of bias current I_{DS} determines power dissipation ($P = V_{DD}I_{DS}$), and thus, g_m/I_D shows the ability of the transistor to convert V_{GS} variations into I_{DS} variations at a particular bias current. There are methods using g_m/I_D for the design of amplifier circuits to achieve optimum gain and power dissipation [105–107]. Another helpful performance metric, gain-efficiency, showing the intrinsic gain ability of transistors at a given power dissipation, can be defined as A_v/I_{DS}. For short-channel transistors, the transconductance efficiency becomes poor as the velocity saturation effect kicks in. Due to velocity saturation, the I_{DS} varies with V_{GS} but in a more linear manner compared to the squared fashion in the case of long-channel transistors. Due to weaker g_m/I_D, a larger I_{DS} is required to attain any desired g_m value.

11.4.5 Discharge Time

The discharge time (t_d) is defined as the time taken by a transistor, at $V_{GS} = V_{DD}$, to discharge a load capacitor connected at the drain to $1/e$ times its initial voltage [108]. A circuit is shown in Figure 11.11a, which is used to measure t_d as described

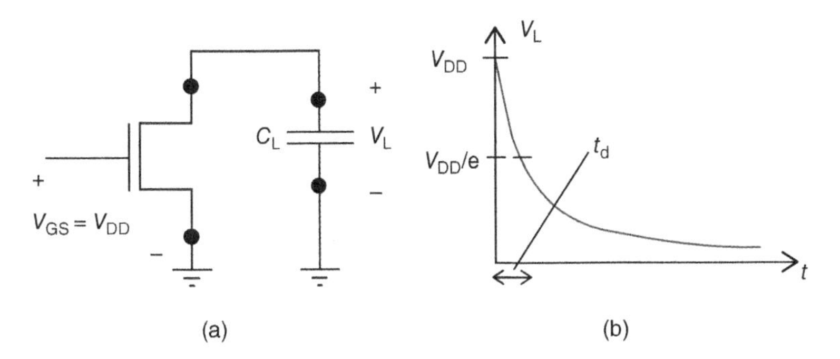

Figure 11.11 Measurement of discharge time (t_d). (a) Circuit for t_d measurement. (b) Description of t_d.

in Figure 11.11b. With the decrease in value of capacitor voltage, the V_{DS} of MOSFET also falls, and accordingly, the drain current will decrease. In effect, the t_d value is proportional to C_L/I_{eff}, where the I_{eff} is the average current drive strength of MOSFET. A lower value of t_d indicates a larger average current drive of the transistor.

11.4.6 Small Signal Capacitances and AC Model

The physical structure of MOSFET results in intrinsic capacitances. The effect of such capacitances appears when the device is used for high-frequency operation. A small-signal AC model, as shown in Figure 11.12, includes different inter-terminal capacitive components [100, 109]. These capacitances include C_{gd} between gate and drain, C_{gs} between gate and source, C_{gb} between gate and body, C_{sb} between source and body, and C_{db} between drain and body.

11.4.7 Transit Frequency

The transit frequency (f_T) is an important transistor parameter for high-bandwidth applications. The capacitances slow down the high-frequency behavior of the device, and the current gain reduces with an increase in frequency. The frequency at which the current gain from gate to drain terminal, with both drain and source at ground potential, falls to unity values is called a transit, or cut-off frequency denoted as f_T. The f_T is sometimes referred to as unit gain frequency. The value of f_T can be approximated [109–111] to be as follows:

$$f_T \approx \frac{g_m}{2\pi(C_{gd} + C_{gs})} = \frac{g_m}{2\pi C_{gg}} \tag{11.8}$$

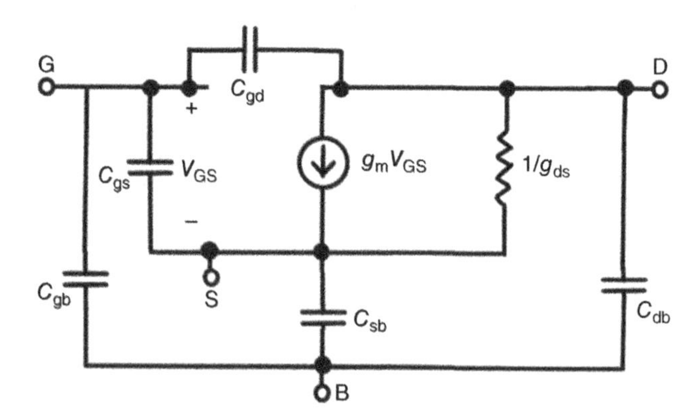

Figure 11.12 High-frequency model of MOSFET.

f_T can be increased by increasing the ratio of the intrinsic component of C_{gg}, i.e., $C_{ox,}$ which contributes to drain current, to the extrinsic parasitic components. Another possibility is to increase the saturation velocity of charge carriers. Different performance metrics can be derived using the A_v, f_T, and g_m/I_D parameters to represent the combined DC and AC performance. For example, the gain-frequency product (*GFP*), transconductance-frequency product (*TFP*), and gain-transconductance-frequency product (*GTFP*) [112, 113] are defined as per the following equations:

$$\mathrm{GFP} = A_v \times f_T \tag{11.9}$$

$$\mathrm{TFP} = \frac{g_m}{I_{DS}} \times f_T \tag{11.10}$$

$$\mathrm{GTFP} = A_v \times \frac{g_m}{I_{DS}} \times f_T \tag{11.11}$$

In practice, the actual high-frequency performance available from the device while operating in a circuit is much below the f_T. This occurs due to the neglect of extrinsic capacitive elements. Nevertheless, the f_T and related parameters present helpful trends in the performance of the transistor [114]. Gain and transit frequency have a dependence on gate length as well. In the absence of velocity saturation, with an increase in $L_{g,}$ the A_v increases linearly with L_g, while f_T decreases in proportion to $L_g{}^2$ [104]. However, in the presence of velocity saturation, the fall in A_v values with increasing L_g becomes more rapid, while f_T decreases in proportion to L_g.

11.5 Challenges and Perspectives of Modern Analog Design

The system-on-chip (SoC) used in mobile phones and the memory chips are the primary economic driver of today's semiconductor industry [16]. And, each IC technology node is optimized for superior logic and static-random-access-memory (SRAM) performance to ensure the area and cost advantages for digital applications. Consequently, the performance of analog and mixed-signal (AMS) circuits such as PLLs, datalink I/Os, voltage regulators, data converters, and band-reference circuit blocks does not necessarily get primary attention from the newer nodes [17, 18]. Further, for the advanced communication technologies in popular demand such as 5G, a larger density of components is required to meet the application requirements. Also, new radio frequency (RF) application domains such as wireless sensor networks (WSNs) and internet of things (IoTs) have peculiar constraints to meet. The great challenges are related to the enormous amount of data, network connectivity, and time-bound decisions and responses [28]. Furthermore, with the benefits of low fabrication cost, low-power consumption, and high-level integration, the CMOS technology has become a competitive choice for RF circuits also [19, 20]. Moreover, the AMS blocks of SoCs are also realized alongside their digital counterparts using the same fabrication process. Thus, the analog designers have to live with the technology that is optimized for memory/digital applications. The reported NSTs-based technology/structures, therefore, need investigation from an AMS perspective. Therefore, the evaluation of NSTs for analog circuit applications becomes a crucial requirement for facilitating modern analog circuit design.

Analog circuit designers have gate length (L_g), gate width (W_g), circuit topology, and transistor bias current or overdrive voltage as performance knobs to design analog circuit blocks. These design variables are decided by the desired circuit performance and power dissipation limits. While the choice of maximum bias current is constrained by the power dissipation, the gate width is decided by the available overdrive voltage and the minimum bias current required to meet the desired performance from a circuit. While extensive SPICE-based simulations and other design tools are used to finalize the IC designs, the IC performance after fabrication depends on process, voltage, and temperature effects.

The process-induced variations in the transistor performance are the manifestations of various imperfections in the fabrication process, such as work-function variations due to metal thickness variation and metal gate granularity, line-edge-roughness, gate-edge-roughness, and sheet-thickness-variation. Although, the NSTs are considered less prone to variability issues because of epitaxially controlled nanosheet thickness [10, 75, 115]. But wafer-to-wafer, die-to-die, and device-to-devi variations do exist in reality and result in speed and performance variability among similarly drawn NSTs.

The power supply is required for an integrated circuit to operate. On-chip power supply regulators are used along with filters to attain the best supply voltage stability. However, the ac (noise) component of supply voltage, i.e., ripple voltage, is unfortunately always present and manifests as variations in the analog circuit's output [116]. Although the designers use local decoupling near circuit blocks to solve the issue of supply noise it is often not enough. One approach is to use separate voltage regulators. For the digital and analog circuit blocks in an AMS ICs, to limit the switching noise of digital blocks from reaching the analog blocks. Further, the variations in temperature along with the supply voltage and process-related variations are the pitfalls for almost all ICs and must be considered to ensure a working IC after fabrication.

Further, the advent of low-power communication, sensor interfacing circuits, and the need for portability have resulted in crucial limitations on power dissipation and circuit area [117]. Analog circuits are, therefore, required to operate at minimum power, i.e., with the smallest possible current consumption [118] For a given bias current, the analog performance parameters depend on channel length and the extent of inversion, i.e., the current density in the channel. The extent of inversion is represented by an inversion coefficient (I_C) in modern MOS models [119, 120]. Some circuit designers use the I_C in place of V_{OV} to design analog circuits with the optimum performance to current consumption ratio [103, 118, 121]. When I_C is used instead of overdrive voltage, the performance parameters such as g_m, g_m/I_D, and f_T appear in straight lines on log–log scales and present an easy insight for the circuit designers [122]. Further, the supply voltage has also scaled with CMOS technology, although at a sluggish rate. Small supply voltage ($V_{DD} < 1$ V) has pushed the MOSFETs to operate in moderate inversion region ($0.1 < I_C < 10$), and sometimes in weak inversion ($I_C < 0.1$) too. Therefore, the transistor operation is no longer guaranteed to be in strong inversion ($I_C > 10$).

Several investigations [118, 123–127] on the performance of transistors operating in moderate to weak inversion and the reported possible improvements are summarized as follows. The g_m value is proportional to the I_C and the square root of I_C in weak and strong inversion, respectively. However, the g_m values saturate to a constant value of $W_g C_{ox} v_{sat}$ as the device enters a velocity-saturated regime. Circuit designs like voltage-controlled oscillators and low-noise amplifiers that are dominated by g_m should, therefore, be designed accordingly. Similarly, the g_m/I_D value increases as the device reaches weaker inversion and is largest at $I_C = 1$. And, biasing the transistor for $I_C < 1$ does not improve the g_m/I_D. The f_T values increase as I_C increases and finally saturate to a maximum value determined by v_{sat}, L_g, and parasitic resistances and capacitances.

Long devices in moderate inversion result in good A_v, minimum mismatch, and minimum flicker noise. The best DC performance is achieved if the length is reduced to moderate levels and the device is in moderate inversion. The resulting benefits include moderately low V_{OV}, moderately high g_m, and g_m/I_D, moderate f_T, and minimum gate-referred thermal noise. The largest V_A and r_o are available for long devices operating in strong inversion. However, as the L_g is reduced while the device is still in saturation, a better g_m stability and a minimum drain flicker noise current density are achieved. The best AC performance with the maximum possible f_T is achievable for short devices operating in very strong inversion. AC performance is superior for short devices as the device capacitances are minimal. Also, a minimum layout area is achievable for short devices in strong inversion.

In modern low-voltage processes-based ICs, transistors are increasingly being biased in moderate inversion to achieve high transconductance efficiency, low drain saturation voltage (V_{DSAT}), moderate bandwidth, and sufficient immunity to velocity saturation [103, 128]. The g_m increases with an increase in I_{DS} and W/L, so the same g_m can be achieved with larger W/L at lower I_{DS}, thus lowering the power dissipation. If the compromise on the area is acceptable, the transistor can be biased in moderate inversion with lower I_{DS}. The same technique can be used to achieve a given gain-bandwidth product (*a.k.a.* unit gain frequency, $\omega_u = g_m/C_L$, where C_L is load capacitance) for an amplifier [124]. However, the designer needs to be wary of the increased parasitic capacitance with larger *W/L* and the need for a tightly controlled V_{TH} for the devices working in the moderate inversion [129] Further, aiming at power efficiency as bias current levels are brought down, the device loses speed (bandwidth) performance. It is shown in [124] that the device biased at the higher side of moderate inversion offers the best *TFP* value implying optimum gain, power, and speed values. Optimization of *TFP* is of critical importance for low-power radio frequency applications such as low noise amplifiers (LNAs) [130]. Additionally, at moderate inversion, the noise figures are also minimal [126, 131]. Therefore, moderate inversion can be used to optimize A_v, input-referred thermal noise (S_{VG}), ω_u, phase margin, and settling time in OP AMPs [128].

Other than proper biasing to achieve optimum performance, many other dedicated circuit techniques exist [132]. These include capacitance cancellation that provides better speed at the same power consumption. Bandwidth for circuits like optical receivers can be increased by a factor of up to 3 without any severe mismatching issues [133]. Resistance cancellation is a ubiquitously used technique in voltage-controlled oscillators (VCOs) to compensate for resistance loss in inductors. It can also be used to enhance the gain-bandwidth product of voltage amplifiers [134]. Pole zero cancellation in three-stage amplifiers results in large power savings. Noise cancellation is another technique that uses two signal paths with opposite phases from the noise source to the output [135, 136] Lastly, the distortion cancellation technique has been shown to reduce distortion in differential amplifiers, current mirrors, and OP AMP circuits [137, 138]. Further, there are digital-assisted analog circuits like VCO-based PLLs that use time-to-digital conversion for measuring the required correction voltage to VCO [139]. While large-size devices can be used to solve the issue of a mismatch, it costs the power and speed as parasitic also increase with the size. So digital correction methods can be used to mitigate the mismatch effects [140] This improves the overall mismatch and noise performance of the Analog-to-Digital Converter (ADC) [132] Flicker noise can be reduced by using a large gate area but at the premium of large capacitances that require larger currents to drive them if speed needs are to be met. Alternately, [141, 142] demonstrates the use of autozeroing, chopper stabilization, and correlated double sampling methods to reduce the Flicker noise.

The operating region of the transistor determines g_m/I_D ratio. Power dissipation of the device from an analog perspective is not dependent on minimum gate length. Therefore, the scaling of the transistor does not necessarily result in superior analog performance. Further, as the supply voltage is scaled to lower values with scaling, lower voltage swings lead to poorer signal-to-noise ratio (SNR), and the task of attaining low distortion and high linearity has become more challenging [143]. Also, at small dimensions, the mismatch-related issues are more severe. The g_m/I_D ratio also does not change with the minimum gate length scaling, but the f_T value improves. This aspect of scaling helps with the g_m/I_D vs. f_T trade-off, i.e., with lower length same bandwidth can be achieved at smaller $I_{DS,}$ and consequently, the power dissipation can be reduced. As the minimum gate lengths are downscaled, the relation of g_m/I_D with I_C does not change, but the possible f_T value at any I_C increases [143] In many analog circuits, especially that use saturated output devices, transistors are required to be provided with sufficient headroom to ensure their operation in saturation. With a smaller supply voltage, the available voltage swing is limited and cannot be improved even if power consumption is compromised [144].

References

1 Zhang, Z., Wang, X., and Yan, Y. (2021). A review of the state-of-the-art in electronic cooling. *e-Prime* 1: 100009.

2 Bohr, M.T. and Young, I.A. (2017). CMOS scaling trends and beyond. *IEEE Micro* 37 (6): 20–29.

3 Jung, E. (2018). 4th industrial revolution and foundry : challenges and opportunities. In: *2018 IEEE International Electron Devices Meeting (IEDM)*, 1–10. IEEE.

4 IRDS (2021). International roadmap for devices and systems 2021 edition executive summary. https://irds.ieee.org/editions/2021/executive-summary.

5 Han, G. and Hao, Y. (2021). Design technology co-optimization towards sub-3 nm technology nodes. *Journal of Semiconductors* 42 (2): 020301–020301.

6 Loke, A.L.S., Lee, C.K., and Leary, B.M. (2019). Nanoscale CMOS implications on analog/mixed-signal design. In: *Proceedings of the Custom Integrated Circuits Conference*, 1–57.

7 Ferain, I., Colinge, C.A., and Colinge, J.P. (2011). Multigate transistors as the future of classical metal-oxide-semiconductor field-effect transistors. *Nature* 479 (7373): 310–316.

8 Salahuddin, S., Ni, K., and Datta, S. (2018). The era of hyper-scaling in electronics. *Nature Electronics* 1 (8): 442–450.

9 Wu, H., Gluschenkov, O., Tsutsui, G., et al. (2019). Parasitic resistance reduction strategies for advanced CMOS FinFETs beyond 7 nm. *Technical Digest - International Electron Devices Meeting, IEDM*, December 2018, pp. 35.4.1–35.4.4.

10 Loubet, N., Hook, T., Montanini, P., et al. (2017). Stacked nanosheet gate-all-around transistor to enable scaling beyond FinFET. *Digest of Technical Papers – Symposium on VLSI Technology*, pp. T230–T231.

11 Barbé, J., Barraud, S., Rozeau, O., et al. (2019). Stacked Nanowires/ Nanosheets GAA MOSFET From Technology to Design Enablement To cite this version : from Technology to Design Enablement.

12 Yakimets, D., Garcia Bardon, M., Jang, D., et al. (2018). Power aware FinFET and lateral nanosheet FET targeting for 3 nm CMOS technology. *Technical Digest - International Electron Devices Meeting, IEDM*, pp. 20.4.1–20.4.4.

13 Yoon, J.-S., Jeong, J., Lee, S., and Baek, R.-H. (2018). Systematic DC/AC performance benchmarking of sub-7-nm node FinFETs and nanosheet FETs. *IEEE Journal of the Electron Devices Society* 6: 942–947.

14 Jang, D., Yakimets, D., Eneman, G. et al. (2017). Device exploration of nanosheet transistors for sub-7-nm technology node. *IEEE Transactions on Electron Devices* 64 (6): 2707–2713.

15 Srivastava, S. and Acharya, A. (2024). Challenges and future scope of gate-all-around (GAA) transistors: physical insights of device-circuit interactions. In: *Device Circuit Co-Design Issues in FETs*, 231–258. CRC Press.

16 Moore, G.E. (1965). Cramming more components onto integrated circuits. *Electronics (Basel)* 38 (8): 114–119.

17 Guarnieri, M. (2016). The unreasonable accuracy of Moore's law [historical]. *IEEE Industrial Electronics Magazine* 10 (1): 40–43.

18 Barraud, S., Lapras, V., Previtali, B., et al. (2018). Performance and design considerations for gate-all-around stacked-NanoWires FETs. *Technical Digest – International Electron Devices Meeting, IEDM*, (001), pp. 29.2.1–29.2.4.

19 Yeap, G., Chen, X., Yang, B.R., et al. (2019). 5 nm CMOS production technology platform featuring full-fledged EUV, and high mobility channel FinFETs with densest 0.021 μm^2 SRAM cells for mobile SoC and high performance computing applications. *Technical Digest – International Electron Devices Meeting, IEDM*, 2019-Decem, pp. 879–882.

20 Chandel, R. (2001). VLSI microfabrication technologies and MEMS. *IETE Journal of Education* 42 (1–4): 33–41.

21 Gargini, P. (2017). *Roadmap Evolution: from NTRS to ITRS, From ITRS 2.0 to IRDS*. The Making of an Industry.

22 Guar, A. and Mahmoodi, H. (2015). Impact of technology scaling on performance of domino logic in nano-scale CMOS. *IEEE/IFIP International Conference on VLSI and System-on-Chip, VLSI-SoC*, 07–10 October, pp. 295–298.

23 Guldi, R., Winter, T., Sridhar, N., et al. (1999). Systematic and random defect reduction during the evolution of integrated circuit technology. *10th Annual IEEE/SEMI. Advanced Semiconductor Manufacturing Conference and Workshop. ASMC 99 Proceedings (Cat. No. 99CH36295)*, pp. 2–7.

24 Taur, Y. and Ning, T.H. (2009). *Fundamentals of Modern VLSI Devices*. Cambridge University Press.

25 Sze, S.M. and Ng, K.K. (2006). *Physics of Semiconductor Devices*, 3e. Wiley.

26 Salahuddin, S. and Datta, S. (2008). Use of negative capacitance to provide voltage amplification for low power nanoscale devices. *Nano Letters* 8 (2): 405–410.

27 Kuhn, K.J. (2012). Considerations for ultimate CMOS scaling. *IEEE Transactions on Electron Devices* 59 (7): 1813–1828.

28 Yu, B., Wann, C.H.J., Nowak, E.D. et al. (1997). Short-channel effect improved by lateral channel-engineering in deep-submicronmeter MOSFET's. *IEEE Transactions on Electron Devices* 44 (4): 627–634.

29 Nishiyama, A., Matsuzawa, K., and Takagi, S.I. (2001). SiGe source/drain structure for the suppression of the short-channel effect of sub-0.1-μm p-channel MOSFETs. *IEEE Transactions on Electron Devices* 48 (6): 1114–1119.

30 Kim, S.D., Park, C.M., and Woo, J.C.S. (2002). Advanced source/drain engineering for box-shaped ultrashallow junction formation using laser annealing and pre-amorphization implantation in sub-100-nm SOI CMOS. *IEEE Transactions on Electron Devices* 49 (10): 1748–1754.

31 Mistry, K., Allen, C., Auth, C. et al. (2007). A 45 nm logic technology with high-k+ metal gate transistors, strained silicon, 9 Cu interconnect layers, 193nm dry patterning, and 100% Pb-free packaging. In: *2007 IEEE International Electron Devices Meeting*, 247–250. IEEE.

32 Först, C.J., Ashman, C.R., Schwarz, K., and Blöchl, P.E. (2004). The interface between silicon and a high-*k* oxide. *Nature* 427 (6969): 53–56.

33 Devrani, V. and Srivastava, V.M. (2012). Advancement of MOSFET with the application of hafnium. *2012 International Conference on Computer Communication and Informatics, ICCCI 2012*, pp. 10–13.

34 Jacobs, J.B. and Antoniadis, D. (1995). Channel profile engineering for MOSFET's with 100 nm channel lengths. *IEEE Transactions on Electron Devices* 42 (5): 870–875.

35 Agarwal, A., Kim, C.H., Mukhopadhyay, S., and Roy, K. (2004). Leakage in nano-scale technologies: mechanisms, impact and design considerations. *Proceedings of the Design Automation Conference*, pp. 6–11.

36 Salahuddin, S. and Datta, S. (2007). Use of negative capacitance to provide a sub-threshold slope lower than 60 mV/decade. arXiv:0707.2073v1.

37 Angelov, G.V., Nikolov, D.N., and Hristov, M.H. (2019). Technology and Modeling of Nonclassical Transistor Devices. *Journal of Electrical and Computer Engineering* 2019.

38 Saraswat, K.C., Kim, D., Krishnamohan, T., and Pethe, A. (2007). Performance limitations of Si bulk CMOS and alternatives for future ULSI. *Journal of the Indian Institute of Science* 87 (3): 387–399.

39 Plummer, J.D. and Griffin, P.B. (2001). Material and process limits in silicon VLSI technology. *Proceedings of the IEEE* 89 (3): 240–258.

40 Chiang, M.H., Lin, J.N., Kim, K., and Te Chuang, C. (2007). Random dopant fluctuation in limited-width FinFET technologies. *IEEE Transactions on Electron Devices* 54 (8): 2055–2060.

41 Kuhn, K.J. (2009). CMOS scaling beyond 32 nm: challenges and opportunities. *Proceedings of the* Design *Automation Conference*, pp. 310–313.

42 Sood, H., Srivastava, V.M., and Singh, G. (2018). Advanced MOSFET technologies for next generation communication systems - perspective and challenges: a review. *Journal of Engineering Science and Technology Review* 11 (3): 180–195.

43 Suzuki, K., Tanaka, T., Tosaka, Y. et al. (1993). Scaling theory for double-gate SOI MOSFET's. *IEEE Transactions on Electron Devices* 40 (12): 2326–2329.

44 Colinge, J.P. and Chandrakasan, A. (2008). FinFETs and other multi-gate transistors.

45 Bhattacharya, D. and Jha, N.K. (2015). FinFETs: from devices to architectures. *Digitally-Assisted Analog and Analog-Assisted Digital IC Design*. 2014: 21–55.

46 Auth, C., Allen, C., Blattner, A., et al. (2012). A 22 nm high performance and low-power CMOS technology featuring fully-depleted tri-gate transistors, self-aligned contacts and high density MIM capacitors. *Digest of Technical Papers - Symposium on VLSI Technology*, pp. 131–132.

47 Farkhani, H., Peiravi, A., Kargaard, J.M., and Moradi, F. (2014). Comparative study of FinFETs versus 22nm bulk CMOS technologies: SRAM design perspective. *International System on Chip Conference*, pp. 449–454.

48 Auth, C. (2012). 22-nm Fully-depleted tri-gate CMOS transistors. *Proceedings of the Custom Integrated Circuits Conference.*

49 Spessot, A., Parvais, B., Rawat, A., et al. (2020). Device scaling roadmap and its implications for logic and analog platform. *2020 IEEE BiCMOS and Compound Semiconductor Integrated Circuits and Technology Symposium, BCICTS 2020*, (Fig 1).

50 Choi, Y.K., King, T.J., and Hu, C. (2002). A spacer patterning technology for nanoscale CMOS. *IEEE Transactions on Electron Devices* 49 (3): 436–441.

51 Auth, C., Aliyarukunju, A., Asoro, M., et al. (2018). A 10 nm high performance and low-power CMOS technology featuring 3rd generation FinFET transistors, self-aligned quad patterning, contact over active gate and cobalt local interconnects. *Technical Digest – International Electron Devices Meeting, IEDM*, pp. 29.1.1–29.1.4.

52 Datta, S. (2018). Ten nanometre CMOS logic technology. *Nature Electronics* 1 (9): 500–501.

53 Xie, Q., Lin, X., Wang, Y. et al. (2015). Performance comparisons between 7-nm FinFET and conventional bulk CMOS standard cell libraries. *IEEE Transactions on Circuits and Systems II: Express Briefs* 62 (8): 761–765.

54 Xie, R., Montanini, P., Akarvardar, K., et al. (2016). A 7 nm FinFET technology featuring EUV patterning and dual strained high mobility channels. *2016 IEEE Electron Device Meeting (IEDM)*, 12 (c), pp. 2.7.1–2.7.4.

55 Ha, D., Yang, C., Lee, J., et al. (2017) Highly manufacturable 7 nm FinFET technology featuring EUV lithography for low power and high performance applications. *Digest of Technical Papers – Symposium on VLSI Technology*, pp. T68–T69.

56 Ryckaert, J., Baert, R., Verkest, D., et al. (2019). Enabling sub-5 nm CMOS technology scaling thinner and taller! *Technical Digest - International Electron Devices Meeting, IEDM*, December 2019, pp. 685–688.

57 TSMC (2020). 5 nm Technology. pp. 1–4.

58 TSMC (2022). 3 nm Technology. pp. 3–6.

59 Singh, K. (2021). Gate-All-Around (GAA) FET – Going Beyond The 3 Nanometer Mark.

60 Bardon, M.G., Sherazi, Y., Schuddinck, P., et al. (2017). Extreme scaling enabled by 5 tracks cells: holistic design-device co-optimization for FinFETs and lateral nanowires. *Technical Digest – International Electron Devices Meeting, IEDM*, pp. 28.2.1–28.2.4.

61 Saini, G. and Rana, A.K. (2011). Physical scaling limits of FinFET structure: a simulation study. *International Journal of VLSI Design & Communication Systems* 2 (1): 26–35.

62 Lee, Y.M., Na, M.H., Chu, A., et al. (2018). Accurate performance evaluation for the horizontal nanosheet standard-cell design space beyond 7 nm technology. *Technical Digest – International Electron Devices Meeting, IEDM*, pp. 29.3.1–29.3.4.

63 Eyben, P., MacHillot, J., Kim, M., et al. (2019). 3D-carrier profiling and parasitic resistance analysis in vertically stacked gate-all-around Si nanowire CMOS transistors. *Technical Digest – International Electron Devices Meeting, IEDM*, December 2019.

64 Takato, H., Sunouchi, K., Okabe, N. et al. (1988). High performance CMOS surrounding gate transistor (SGT) for ultra high density LSIs. In: *Technical Digest., International Electron Devices Meeting*, 222–225. IEEE.

65 Colinge, J.-P., Gao, M.H., Romano, A., et al. (1990) Silicon-on-insulator "Gate-All-Around" MOS device. *1990 IEEE SOS/SOI Technology Conference. Proceedings*, pp. 137–138.

66 Sharma, D. and Vishvakarma, S.K. (2015). Analyses of DC and analog/RF performances for short channel quadruple-gate gate-all-around MOSFET. *Microelectronics Journal* 46 (8): 731–739.

67 Hu, W. and Li, F. (2021). Scaling beyond 7 nm node: an overview of gate-all-around FETs. *2021 9th International Symposium on Next Generation Electronics, ISNE 2021*.

68 Mertens, H., Ritzenthaler, R., Pena, V., et al. (2018). Vertically stacked gate-all-around Si nanowire transistors: key Process Optimizations and Ring Oscillator Demonstration. *Technical Digest - International Electron Devices Meeting, IEDM*, pp. 37.4.1–37.4.4.

69 Nagy, D., Espineira, G., Indalecio, G. et al. (2020). Benchmarking of FinFET, nanosheet, and nanowire FET architectures for future technology nodes. *IEEE Access* 8: 53196–53202.

70 Ritzenthaler, R., Mertens, H., Pena, V., et al. (2019). Vertically stacked gate-all-around Si nanowire CMOS transistors with reduced vertical nanowires separation, new work function metal gate solutions, and DC/AC performance optimization. *Technical Digest - International Electron Devices Meeting, IEDM*, 2018-December, pp. 21.5.1–21.5.4.

71 Loubet, N., Hook, T., Montanini, P. et al. (2017). T17-5 (Late News) stacked nanosheet gate-all-around transistor to enable scaling beyond FinFET T230 T231. *VLSI Technology* 5 (1): 14–15.

72 Liebmann, L., Zeng, J., Zhu, X., et al. (2016). Overcoming scaling barriers through design technology cooptimization. *Digest of Technical Papers – Symposium on VLSI Technology*, September 2016, pp. 3–4.

73 Barraud, S., Previtali, B., Vizioz, C., et al. (2020). 7-Levels-stacked nanosheet GAA transistors for high performance computing. *Digest of Technical Papers - Symposium on VLSI Technology*, June 2020, pp. 9–10.

74 Lee, S.-Y., Kim, S.-M., Yoon, E.-J. et al. (2003). A novel multibridge-channel MOSFET (MBCFET): fabrication technologies and characteristics. *IEEE Transactions on Nanotechnology* 2 (4): 253–257.

75 Song, S.C., Colombeau, B., Bauer, M., et al. (2019). 2 nm node: benchmarking FinFET vs nano-slab transistor architectures for artificial intelligence and next gen smart mobile devices. *Digest of Technical Papers – Symposium on VLSI Technology*, June 2019, pp. T206–T207.

76 Veloso, A., Eneman, G., de Keersgieter, A., et al. (2021). Nanosheet FETs and their potential for enabling continued Moore's law scaling. *2021 5th IEEE Electron Devices Technology and Manufacturing Conference, EDTM 2021*, pp. 2–4.

77 Moore, S.K. (2021). Samsung's 3-nm tech shows nanosheet transistor advantage nanosheet devices allow tuning of memory cell design in a way FinFETs can't.

78 Bae, G., Bae, D., Kang, M., et al. (2018). 3 nm GAA technology featuring multi-bridge-channel FET for low power and high performance applications. *2018 IEEE International Electron Devices Meeting (IEDM)*, pp. 28.7.1–28.7.4.

79 Moore, S.K. (2021). Intel: back on Top by 2025 ? Silicon giant bets on nanosheet devices and new on-chip power distribution tech.

80 Navaraj, W.T., Yogeswaran, N., Vinciguerra, V., and Dahiya, R. (2017). Simulation study of junctionless silicon nanoribbon FET for high-performance printable electronics. *2017 European Conference on Circuit Theory and Design, ECCTD 2017*, pp. 12–15.

81 Hook, T.B. (2018). Power and technology scaling into the 5 nm node with stacked nanosheets. *Joule* 2 (1): 1–4.

82 Dutta, T., Medina-Bailon, C., Rezaei, A., et al. (2021). TCAD Simulation of Novel Semiconductor Devices, pp. 1–4.

83 Synopsys (2019). Sentaurus Device User Guide, Version Q-2019.12.

84 Dimitrijev, S. (2013). *Principle of Semiconductor Devices, 2*e. Oxford University Press.

85 Talib, A.A. (2018). *Modelling and Simulation Study of NMOS Si Nanowire Transistors*. School of Engineering, University of Glasgow.

86 Choi, M., Moroz, V., Smith, L., and Huang, J. (2015). Extending drift-diffusion paradigm into the era of FinFETs and nanowires. *International Conference on Simulation of Semiconductor Processes and Devices, SISPAD*, October 2015, pp. 242–245.

87 Martinez, A. and Barker, J.R. (2020). Quantum transport in a silicon nanowire FET transistor: hot electrons and local power dissipation. *Materials* 13 (15): 3326.

88 Park, H., Kim, J., Choi, W., et al. (2019). NEGF simulations of stacked silicon nanosheet FETs for performance optimization. *2019 International Conference on Simulation of Semiconductor Processes and Devices (SISPAD)*, pp. 1–3.

89 Pundir, Y.P., Saha, R., and Pal, P.K. (2020). Effect of gate length on performance of 5nm node N-channel nano-sheet transistors for analog circuits. *Semiconductor Science and Technology* 36 (1): 015010.

90 Feng, P., Song, S.C., Nallapati, G. et al. (2017). Comparative analysis of semiconductor device architectures for 5-nm node and beyond. *IEEE Electron Device Letters* 38 (12): 1657–1660.

91 Yao, J., Li, J., Luo, K. et al. (2018). Physical insights on quantum confinement and carrier mobility in Si, $Si_{0.45}Ge_{0.55}$, Ge gate-all-around NSFET for 5 nm technology node. *IEEE Journal of the Electron Devices Society* 6: 841–848.

92 Moroz, V., Huang, J., and Choi, M. (2017). FinFET/nanowire design for 5 nm/3 nm technology nodes: channel cladding and introducing a "bottleneck" shape to remove performance bottleneck. *2017 IEEE Electron Devices Technology and Manufacturing Conference, EDTM 2017 - Proceedings*, **3**, pp. 67–69.

93 Pundir, Y.P., Bisht, A., Saha, R., and Pal, P.K. (2021). Air-spacers as analog-performance booster for 5 nm-node N-channel nanosheet transistor. *Semiconductor Science and Technology* 36 (9): 095037.

94 Jegadheesan, V., Sivasankaran, K., and Konar, A. (2019). Impact of geometrical parameters and substrate on analog/RF performance of stacked nanosheet field effect transistor. *Materials Science in Semiconductor Processing* 93: 188–195.

95 Yoon, J.S., Jeong, J., Lee, S., and Baek, R.H. (2018). Multi-V_{th} strategies of 7-nm node Nanosheet FETs with Limited Nanosheet Spacing. *IEEE Journal of the Electron Devices Society* 6: 861–865.

96 Pundir, Y.P., Bisht, A., Saha, R. et al. (2022). Effect of temperature on performance of 5-nm node nanosheet transistors for analog applications. *Silicon* 14 (16): 10581–10589.

97 Yoon, J.S., Jeong, J., Lee, S., and Baek, R.H. (2019). Punch-through-stopper free nanosheet FETs with crescent inner-spacer and isolated source/drain. *IEEE Access* 7: 38593–38596.

98 Ortiz-Conde, A., García-Sánchez, F.J., Muci, J. et al. (2013). Revisiting MOSFET threshold voltage extraction methods. *Microelectronics Reliability* 53 (1): 90–104.

99 Loke, A.L.S., Wu, Z., Moallemi, R., et al. (2010). Constant-current threshold voltage extraction in HSPICE for nanoscale CMOS analog design. *Synopsys Users Group (SNUG)*, pp. 1–19.

100 Razavi, B. (2017). *Design of Analog CMOS Integrated Circuits*, 2e. McGraw Hill Education.

101 Cheng, Y. and Hu, C. (1999). *MOSFET Modeling and BSIM3 User ' s Guide*, vol. 3, 6221. Springer Science & Business Media.

102 Passi, V. and Raskin, J.P. (2017). Review on analog/radio frequency performance of advanced silicon MOSFETs. *Semiconductor Science and Technology* 32 (12): 123004.

103 Binkley, D.M. (2007). Tradeoffs and optimization in analog CMOS design. *2007 14th International Conference on Mixed Design of Integrated Circuits and Systems*, pp. 47–60.

104 Gray, P.R., Hurst, P.J., Lewis, S.H., and Meyer, R.G. (2009). *Analysis and Design of Analog Integrated Circuits*, 5e. Wiley.

105 Dammak, H.D., Bensalem, S., Zouari, S. et al. (2008). Design of folded cascode OTA in different regions of operation through g_m/I_D methodology. *International Journal of Electronics and Computer Science Engineering* 2 (9): 28–33.

106 Sabry, M.N., Omran, H., and Dessouky, M. (2018). Systematic design and optimization of operational transconductance amplifier using g_m/I_D design methodology. *Microelectronics J* 75: 87–96.

107 Giustolisi, G. and Palumbo, G. (2021). A g_m/I_D-based design strategy for IoT and ultra-low-power OTAs with fast-settling and large capacitive loads. *Journal of Low Power Electronics and Applications* 11 (2): 21.

108 Lu, Y.-C., and Hu, V.P.-H. (2019). Evaluation of analog circuit performance for ferroelectric SOI MOSFETs considering interface trap charges and gate length variations. *2019 Silicon Nanoelectronics Workshop (SNW)* 8, pp. 1–2.

109 Yoon, J.S. and Baek, R.H. (2020). Device design guideline of 5-nm-node FinFETs and nanosheet FETs for analog/RF applications. *IEEE Access* 8: 189395–189403.

110 Sarkar, A., Kumar Das, A., De, S., and Kumar Sarkar, C. (2012). Effect of gate engineering in double-gate MOSFETs for analog/RF applications. *Microelectronics Journal* 43 (11): 873–882.

111 Sirohi, A., Sahu, C., and Singh, J. (2019). Analog/RF performance investigation of dopingless fet for ultra-low power applications. *IEEE Access* 7: 141810–141816.

112 Tayal, S., Nandi, A., and Nandi, A. (2017). Analog/RF performance analysis of inner gate engineered junctionless Si nanotube. In: *Superlattices and Microstructures*, vol. 111, 862–871. Elsevier.

113 Verma, Y.K., Mishra, V., and Gupta, S.K. (2021). Analog/RF and linearity distortion analysis of MgZnO/CdZnO quadruple-gate field effect transistor (QG-FET). *Silicon* 13: 91–107.

114 Binkley, D.M. (2008). *Tradeoffs and Optimization in Analog CMOS Design*. Wiley.

115 Yoon, J.S., Lee, S., Lee, J. et al. (2020). Reduction of process variations for sub-5-nm node Fin and nanosheet FETs using novel process scheme. *IEEE Transactions on Electron Devices* 67 (7): 2732–2737.

116 Behzad, R. (2017). *Desig of Analog CMOS Integrated Circuits*. McGraw Hill.

117 Raj, N., Sharma, R.K., Jasuja, A., and Garg, R. (2010). A low power OTA for biomedical applications. *Multidisciplinary Journals in Science and Technology, Journal of Selected Areas in Bioengineering (JSAB)* 1–5.

118 Sansen, W. (2015). Minimum power in analog amplifying blocks: presenting a design procedure. *IEEE Solid-State Circuits Magazine* 7 (4): 83–89.

119 Enz, C., Chicco, F., and Pezzotta, A. (2017). Nanoscale MOSFET modeling: part 1: the simplified EKV model for the design of low-power analog circuits. *IEEE Solid-State Circuits Magazine* 9 (3): 26–35.

120 Chauhan, Y.S., Venugopalan, S., Chalkiadaki, M.A. et al. (2014). BSIM6: analog and RF compact model for bulk MOSFET. *IEEE Transactions on Electron Devices* 61 (2): 234–244.

121 Sansen, W. (2013). Analog design procedures for channel lengths down to 20 nm. *Proceedings of the IEEE International Conference on Electronics, Circuits, and Systems*, (2), pp. 337–340.

122 Sansen, W. (2018). Biasing for zero distortion: using the EKV/BSIM6 expressions. *IEEE Solid-State Circuits Magazine* 10 (3): 48–53.

123 Mangla, A., Chalkiadaki, M.A., Fadhuile, F. et al. (2013). Design methodology for ultra low-power analog circuits using next generation BSIM6 MOSFET compact model. *Microelectronics Journal* 44 (7): 570–575.

124 Enz, C., and Chicco, F. (2017). Nanoscale MOSFET Modeling part 2. *IEEE Solid-State Circuits Magazine*, Summer (November), pp. 73–81.

125 Binkley, D.M., Hopper, C.E., Tucker, S.D. et al. (2003). A CAD methodology for optimizing transistor current and sizing in analog CMOS design. *IEEE Transactions on Computer-Aided Design of Integrated Circuits and Systems* 22 (2): 225–237.

126 Enz, C., Chalkiadaki, M.A., and Mangla, A. (2015). Low-power analog/RF circuit design based on the inversion coefficient. *European Solid-State Circuits Conference*, October 2015, pp. 202–208.

127 Afacan, E. (2019). Inversion coefficient optimization based Analog/RF circuit design automation. *Microelectronics Journal* 83: 86–93.

128 Yang, Y., Binkley, D.M., and Li, C. (2012). Using moderate inversion to optimize voltage gain, thermal noise, and settling time in two-stage CMOS amplifiers. *ISCAS 2012 – 2012 IEEE International Symposium on Circuits and Systems*, (4), pp. 432–435.

129 Filanovsky, I.M. and Oliveira, L.B. (2019). On sensitivity of bias operation point in transistors with moderate inversion. *Midwest Symposium on Circuits and Systems*, August 2019, pp. 168–171.

130 Taris, T., Begueret, J.B., and Deval, Y. (2011) A 60 μW LNA for 2.4 GHz wireless sensors network applications. *Digest of Papers – IEEE Radio Frequency Integrated Circuits Symposium*, pp. 4–7.

131 Chalkiadaki, M.A. and Enz, C.C. (2015). RF small-signal and noise modeling including parameter extraction of nanoscale MOSFET from weak to strong inversion. *IEEE Transactions on Microwave Theory and Techniques* 63 (7): 2173–2184.

132 Sansen, W. (2015). Analog CMOS from 5 micrometer to 5 nanometer. *IEEE International Solid-State Circuits Conference Digest* of *Technical Papers*, **58**, pp. 22–27.

133 Hermans, C., Tavernier, F., and Steyaert, M. (2006). A gigabit optical receiver with monolithically integrated photodiode in 0.18 μm CMOS. *ESSCIRC 2006 – Proceedings of the 32nd European Solid-State Circuits Conference*, pp. 476–479.

134 Ohri, K.B. and Callahan, M.J. (1979). Integrated PCM Codec. *IEEE Journal of Solid-State Circuits* 14 (1): 38–46.

135 Bruccoleri, F., Klumperink, E.A.M., and Nauta, B. (2004). Wide-band CMOS low-noise amplifier exploiting thermal noise canceling. *IEEE J Solid-State Circuits* 39 (2): 275–282.

136 Murphy, D., Darabi, H., Abidi, A. et al. (2012). A blocker-tolerant, noise-cancelling receiver suitable for wideband wireless applications. *IEEE Journal of Solid-State Circuits* 47 (12): 2943–2963.

137 Zhang, H., Fan, X., and Sinencio, E.S. (2009). A low-power, linearized, ultra-wideband LNA design technique. *IEEE Journal of Solid-State Circuits* 44 (2): 320–330.

138 Mahrof, D.H., Klumperink, E.A.M., Ru, Z. et al. (2014). Cancellation of OpAmp virtual ground imperfections by a negative conductance applied to improve RF receiver linearity. *IEEE Journal of Solid-State Circuits* 49 (5): 1112–1124.

139 Murmann, B. (2008). A/D converter trends: power dissipation, scaling and digitally assisted architectures. *Proceedings of the Custom Integrated Circuits Conference* (CICC), pp. 105–112.

140 Kinget, P.R. (2005). Device mismatch and tradeoffs in the design of analog circuits. *IEEE Journal of Solid-State Circuits* 40 (6): 1212–1224.

141 Temes, G.C. (1996). Autozeroing and correlated double sampling techniques. *Analog Circuit Design* 45–64.

142 Enz, C.C. and Temes, G.C. (1996). Circuit techniques for reducing the effects of Op-Amp imperfections: autozeroing, correlated double sampling, and chopper stabilization. *Proceedings of the IEEE* 84 (11): 1584–1614.

143 Kinget, P.R. (2015). Scaling analog circuits into deep nanoscale CMOS: obstacles and ways to overcome them. *Proceedings of the Custom Integrated Circuits Conference*, November 2015.

144 Stopjakova, V., Rakus, M., Kovac, M. et al. (2018). Ultra-low voltage analog IC design: challenges, methods and examples. *Radioengineering* 27 (1): 171–185.

12

Low-Power Analog Amplifier Design using MOS Transistor in the Weak Inversion Mode

Soumya Pandit and Koyel Mukherjee

Centre of Advanced Study, Institute of Radio Physics and Electronics, University of Calcutta, Kolkata, India

12.1 Introduction

With the widespread use of advanced technologies like Internet of Things (IoT), or application domains like bio-medical implantable instruments, and also the ever-increasing demand for more and more portable/battery-operated electronic devices, the implementation of integrated circuits within strict power budget has become the primary design goal for all analog and mixed-signal circuit designers. The design of low power analog amplifier circuit has an inherent trade-offs with the area-budget, noise, and speed/frequency requirements of the circuit [1]. The optimization of power consumption in an analog circuit is not a stand-alone phenomenon. The power dissipation in a circuit is closely related with the signal-to-noise ratio of the circuit as is theoretically derived below [2, 3].

Let us consider a conceptual model of a common source amplifier circuit with the load represented by a conceptual resistor R. The drain current noise of a MOS transistor in terms of the conductance g_m is expressed as

$$i_{dn}^2 = 4k_B T \cdot \gamma g_m \cdot B_n \tag{12.1}$$

where γ is a constant whose value is equal to $2/3$ for long-channel MOS transistor, k_B is the Boltzmann constant, T is the absolute temperature, and B_n is the system noise bandwidth. The output noise voltage is written as

$$v_{dn}^2 = i_{dn}^2 R^2 \tag{12.2}$$

If the output full-scale voltage (peak-to-peak) is denoted by V_{FS}, the r.m.s voltage at the output will be

$$v_s = \frac{V_{\text{FS}}}{2.\sqrt{2}} \tag{12.3}$$

Advanced Nanoscale MOSFET Architectures: Current Trends and Future Perspectives,
First Edition. Edited by Kalyan Biswas and Angsuman Sarkar.

The signal-to-noise ratio is therefore, written as

$$\mathrm{SNR} = \frac{V_{\mathrm{FS}}^2 \cdot g_m}{32k_B T\gamma A_v^2 \cdot B_n} \tag{12.4}$$

Here A_v is the gain of the amplifier circuit. The transconductance g_m of a MOS transistor is related to the overdrive voltage as $g_m \propto v_{od}$ where v_{od} is the equivalent effective overdrive voltage. For a planar MOS transistor working in the strong inversion mode, $v_{od} \approx (v_{\mathrm{GS}} - V_T)/2$. In the weak inversion mode, it becomes 2 to 3 times $k_B T/q$.

The power consumption of an analog circuit is written as

$$P = I_D \cdot V_{\mathrm{FS}} = 32k_B T\gamma A_v^2 \cdot B_n \cdot \mathrm{SNR}\,\frac{v_{od}}{V_{\mathrm{FS}}} \tag{12.5}$$

The two important observations from this conceptual model are as follows: Firstly, there is a linear relation between the power dissipation in a circuit and the targeted signal-to-noise ratio. Therefore, lowering the power dissipation beyond a certain limit without compromising with the signal-to-noise ratio is not possible. Secondly, scaling down the supply voltage, which is often the major route to minimizing the power dissipation leads to increased power dissipation if the overdrive voltage is not being reduced. It is important to give insight regarding the choice of operating mode of a MOS transistor, i.e. whether to operate in the weak inversion mode or in the strong inversion mode. From the fundamental definitions, it follows that the overdrive voltage can be reduced considerably by working the MOS transistor in weak inversion mode. This theoretical conjecture drives analog designers to work in the weak inversion mode of operation for MOS transistors.

12.2 Review of the Theory of Weak Inversion Mode Operation of MOS Transistor

A bulk planar MOS transistor is considered to be working in the weak inversion mode when the overdrive voltage v_{od} is below the threshold voltage by at least 72 mV, i.e. $v_{od} < -72\,\mathrm{mV}$ [4].

12.2.1 Drain Current Model in the Weak Inversion Mode

The drain current model of a bulk planar MOS transistor operating in the weak inversion saturation mode is written as [2]

$$I_D = \mu C_{ox}\,(\eta - 1)\,U_T^2 \left(\frac{W}{L}\right) \exp\left(\frac{V_{\mathrm{GS}} - V_T}{\eta U_T}\right) \tag{12.6}$$

Here W and L are effective channel width and length respectively, μ is the mobility of the current carriers inside the channel of the transistor, C_{ox} is the oxide capacitance per unit area, $U_T = k_B T/q$ is the thermal voltage, V_{GS} is the applied gate-to-source voltage, V_T is the threshold voltage, and η is the sub-threshold swing factor. This expression of the drain current assumes that the applied drain-to-source voltage has exceeded its saturation limit, which for a transistor operating in the weak inversion mode is approximately equal to $4U_T \approx 100 - 104\,\mathrm{mV}$. During the weak inversion mode of operation by a MOS transistor, a depletion region is formed in the substrate underneath the interface. The MOS capacitor can thus be considered to be the equivalent capacitance of two series connected capacitors: the oxide capacitor C_{ox} and the depletion layer capacitor C_{dm}. The sub-threshold swing factor is defined as [2]

$$\eta = 1 + \frac{C_{dm}}{C_{ox}} \tag{12.7}$$

neglecting the interface state charges related capacitive effect. The parameter which quantifies how sharply a transistor is turned by the applied gate voltage is referred to as the sub-threshold swing S, expressed as follows [2]

$$S = \left(\frac{d\left(\log_{10} I_D\right)}{dV_{\mathrm{GS}}} \right)^{-1} = 2.3 \frac{\eta k_B T}{q} \approx \eta \times 60\,\mathrm{mV} \tag{12.8}$$

The drain current expression is written in more compact form as follows

$$I_D = I_o \left(\frac{W}{L} \right) \exp\left(\frac{V_{\mathrm{GS}} - V_T}{\eta U_T} \right) \tag{12.9}$$

where I_o is referred to as the technology current constant. The transconductance of a MOS transistor operating in the weak inversion mode is written as

$$g_m = \frac{I_D}{\eta U_T} \tag{12.10}$$

The primary advantages that may be acquired by operating a MOS transistor in the weak inversion mode are summarized as follows:

- Since the effective overdrive voltage is small, the supply voltage can be scaled down considerably, keeping in consideration the trade-off between the signal-to-noise ratio and power dissipation, as discussed in the theoretical conjecture presented earlier.
- The transport of current carriers follows a diffusion mechanism rather than drift mechanism, and therefore, the drain current is very less, even in the range of a maximum of a few hundred of nanoampere (nA) only. In a standard 0.18 μm digital CMOS technology and with unity aspect ratio, just at the onset of the inversion, the maximum drain current flowing through the channel is only 400 nA

for an n-channel MOSFET and 165 nA for a p-channel MOSFET. This results in a bias current requirement that is much smaller than that of a conventional mode of strong inversion. Thus, the overall power consumption is minimized.
- Transconductance efficiency (g_m/I_D) is maximum in the deep weak inversion mode. This helps to achieve the maximum gain with minimum power consumption.

However, these advantages come at the cost of certain design challenges, which are now summarized below.

- The drain current of a MOS transistor working in the weak inversion mode is strongly dependent on temperature-dependent factors like technology current constant I_o and thermal voltage U_T. This puts extra overhead to the designers in implementing additional temperature compensation/correction circuitry.
- A transistor operating in the weak inversion mode is capable of handling low-frequency input signals only. The diffusion mechanism of the charge carriers restricts the application of high-frequency input signals.
- At low frequencies, the flicker noise component becomes dominant in the noise characteristics of a MOS transistor. Therefore, designers look for larger geometry of the transistor device to minimize the effect [5]. This in turn brings a trade-off between low-power and silicon area of the device.

12.2.2 Concept of Inversion Coefficient

Inversion coefficient (IC) is a mathematical parameter that numerically quantifies the level of channel inversion in a MOS transistor. Larger value of inversion coefficient means higher degree of channel inversion. This parameter is mathematically defined as [6]

$$\mathrm{IC} = \frac{I_D}{\left[I_o\,(W/L)\right]} \tag{12.11}$$

where, I_D is the drain current, IC is the inversion coefficient, and W and L are the effective width and channel length, respectively, of a MOSFET.

IC value less than 0.1 signifies weak inversion, that is between 0.1 and 10 signify moderate inversion, and values above 10 signify to strong inversion [6]. In the inversion coefficient based design methodology, drain current, inversion coefficient and channel length are considered as the three independent degrees of design freedom to the circuit designers.

Figures 12.1 and 12.2 show the relationship between I_D, IC, and L of an n-channel and p-channel MOS transistor, respectively. It is observed that I_D varies linearly with IC, but does not show much dependence on the channel length before strong inversion takes place, i.e. IC < 10.

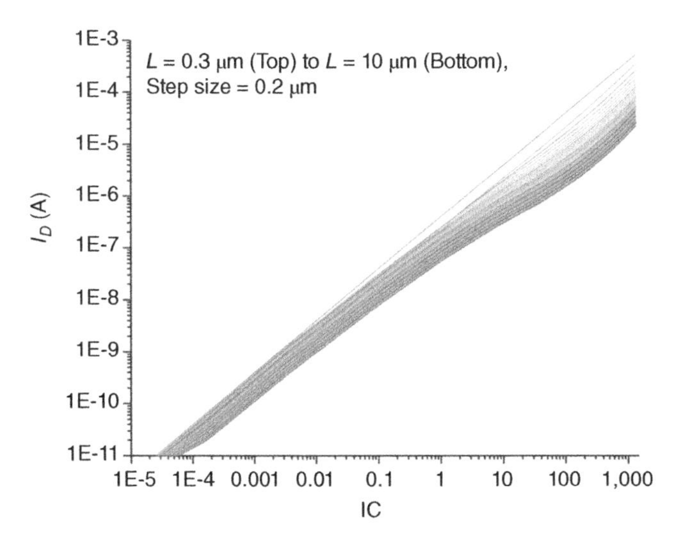

Figure 12.1 Drain current vs. inversion coefficient plot with channel length as parameter of an n-channel MOS transistor, 180 nm digital CMOS technology of SCL.

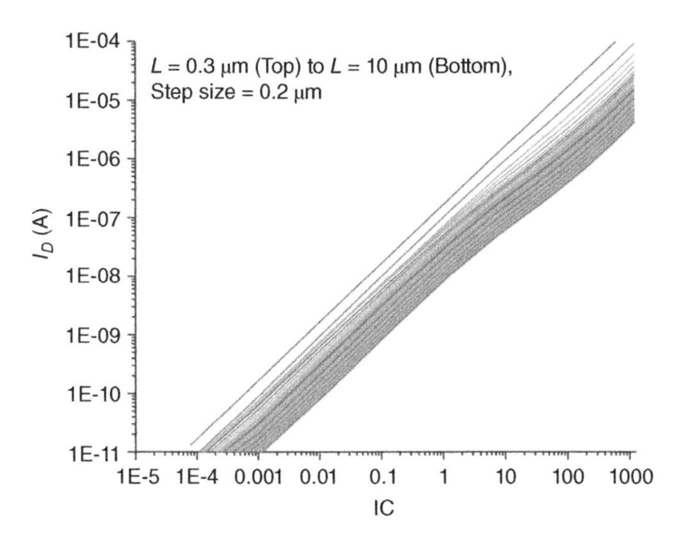

Figure 12.2 Drain current vs. inversion coefficient plot with channel length as parameter of a p-channel MOS transistor, 180 nm digital CMOS technology of SCL.

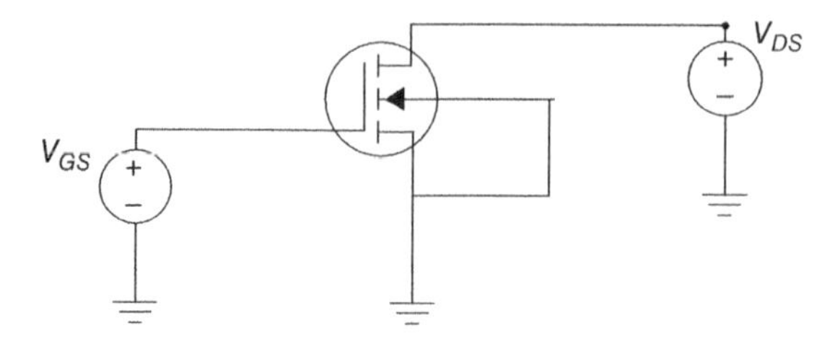

Figure 12.3 A typical test setup for determining the technology current I_o.

12.2.3 Parameter Extraction

12.2.3.1 Technology Current Constant I_o

From (12.11), the technology current constant I_o may be considered as the normalized drain current when the value of the inversion coefficient is 1.

The simulation set-up is shown in Figure 12.3. The drain-to-source voltage, V_{DS}, is kept sufficiently higher compared to the gate-to-source voltage in order to ensure operation in saturation region. The gate-to-source voltage V_{GS} is swept for the entire allowable voltage range while keeping the aspect ratio at unity. From the simulation results, the g_m/I_D plot is obtained. In the chosen 180 nm digital CMOS technology of SCL, the aspect ratio $W = L = 5\ \mu m$ is considered during the experiment. V_{DS} is kept fixed at 1.8 V and V_{GS} is varied from 0 to 1.8 V.

In the g_m/I_D vs. I_D plot, the extrapolated tangents through the weak inversion portion and that, through the strong inversion portion, intersect each other at a point, which defines the technology current constant, I_o, on the I_D axis [6, 7]. In Figure 12.4, for an n-channel MOS transistor, the I_D corresponding to this intersection point is approximately $I_{on} = 400 \times 10^{-9}$ A. On the other hand for a

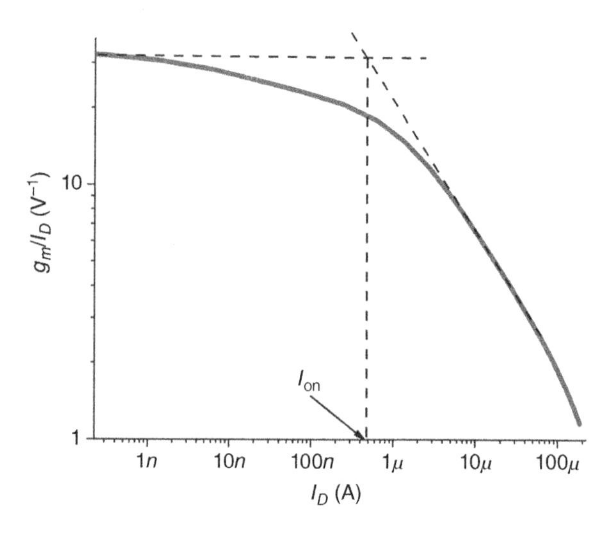

Figure 12.4 g_m/I_D vs. I_D plot of an n-channel MOSFET with $W = L = 5\ \mu m$ and $V_{DS} = 1.8$ V.

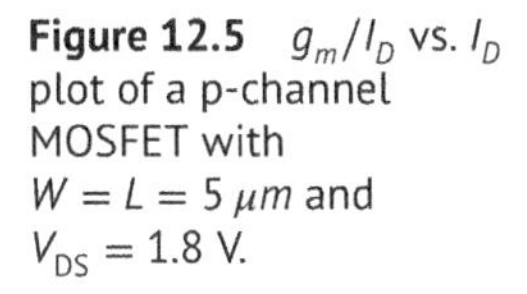
Figure 12.5 g_m/I_D vs. I_D plot of a p-channel MOSFET with $W = L = 5\ \mu m$ and $V_{DS} = 1.8$ V.

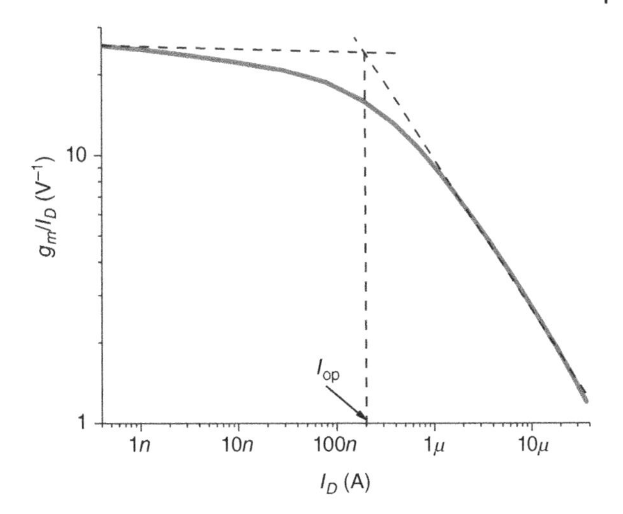

p-channel transistor, the same comes out to be approximately $I_{op} = 120 \times 10^{-9}$ A as shown in Figure 12.5. Putting this value in, we can obtain a range of inversion coefficient IC varying with respect to the variation of I_D as the gate voltage sweeps. Figures 12.6 and 12.7 show the corresponding g_m/I_D vs. IC plot of n-channel and p-channel transistors, respectively. The intersection point of the weak inversion

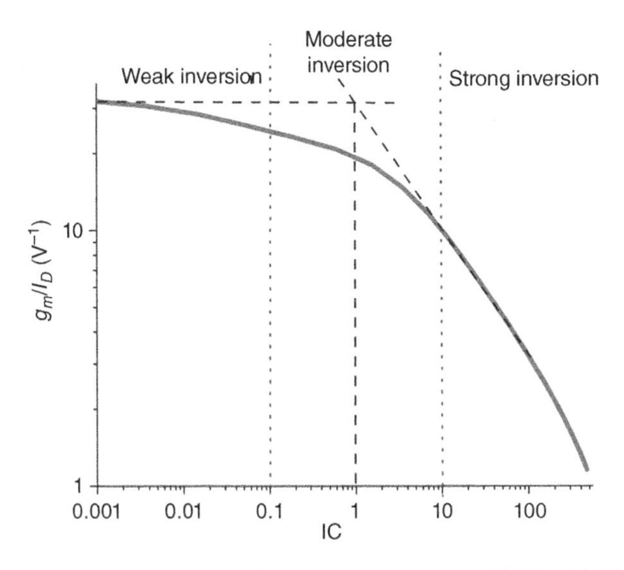

Figure 12.6 g_m/I_D vs. IC plot of an n-channel MOSFET with $W = L = 5\ \mu m$ and $V_{DS} = 1.8$ V.

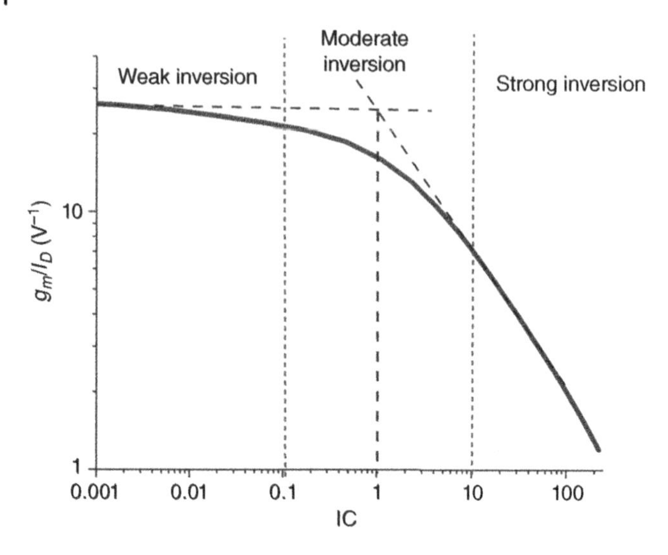

Figure 12.7 g_m/I_D vs. IC plot of a p-channel MOSFET with $W = L = 5\,\mu m$ and $V_{DS} = 1.8$ V.

and strong inversion region asymptotes, as extended toward IC axis, touches the axis at IC = 1. This defines the center of moderate inversion.

Putting the value of technology current, thus obtained in (12.11), we get a range of inversion coefficients for the various values of drain current, corresponding to the dc variation of the gate voltage. Based on the value of inversion coefficient, the region of operations is identified as [6, 7]:

- for weak inversion $IC < 0.1$
- for moderate inversion $0.1 < IC < 10$
- for strong inversion $IC > 10$

12.2.3.2 Sub-threshold Swing Factor η

The plot of the variation of the drain current in log scale, with respect to the gate voltage in weak inversion region, gives a linear curve, as shown in Figure 12.8. The inverse of the slope of the linear portion of the curve is the sub-threshold swing S, which in turn can be defined as the amount of drop in gate voltage required to reduce the drain current by 10 times [2]. By measuring the value of the sub-threshold swing S, the sub-threshold swing factor η for n-channel and p-channel MOS transistors may be measured. These come out to be 1.22 and 1.58 for n-channel and p-channel transistors, respectively, in 0.18 μm digital CMOS technology of SCL. For both of them, η gradually reduces as the level of inversion increases.

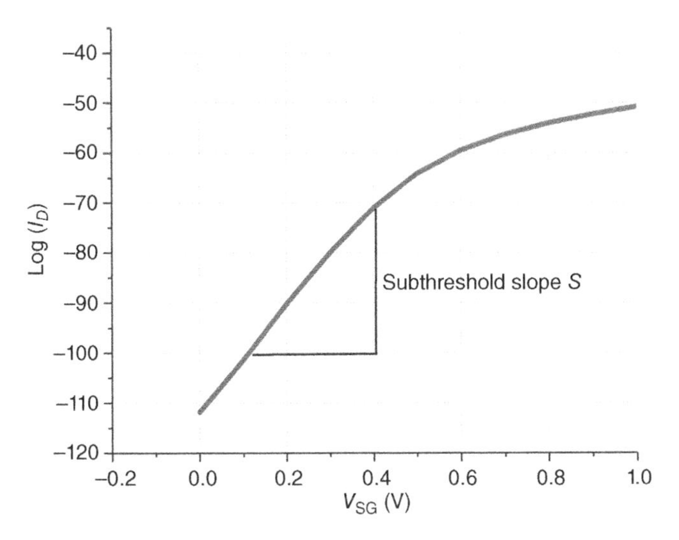

Figure 12.8 Variation of drain current in weak inversion region, with respect to gate voltage for a p-channel MOS transistor with $W = L = 5\ \mu m$.

12.2.4 Small Signal Parameters in Weak Inversion Region

The drain current model is expressed as

$$I_D = I_o \frac{W}{L} \exp\left(\frac{V_{GS} - V_T}{\eta U_T}\right)\left[1 - \underbrace{\exp\left(-\frac{V_{DS}}{U_T}\right)}\right] \tag{12.12}$$

Here, V_{DS} is the drain-to-source voltage.

From the Taylor series expansion of the exponential function, it follows that if the drain-to-source voltage V_{DS} is almost three-to-four times the thermal voltage, U_T, then the expression marked with an underbrace becomes negligible, and thus the drain current, I_D, becomes independent of the drain-to-source voltage V_{DS}. Therefore, the saturation drain voltage is expressed as

$$V_{DSAT} \approx 4U_T \tag{12.13}$$

With this approximation, the current expression in weak inversion saturation region is simplified to

$$I_D = I_o \frac{W}{L} \exp\left(\frac{V_{GS} - V_T}{\eta U_T}\right) \tag{12.14}$$

The overdrive voltage, V_{od}, is expressed as follows

$$V_{od} = V_{GS} - V_T = \eta U_T \ln(IC) \tag{12.15}$$

Since, the transconductance, g_m, is the small signal variation of the drain current, I_D, with respect to the variation of the gate-to-source voltage, V_{GS}, therefore, g_m is expressed as

$$g_m = \frac{\delta I_D}{\delta V_{GS}} \tag{12.16}$$

Differentiating (12.14) with respect to V_{GS}, we obtain

$$g_m = \frac{I_D}{\eta U_T} \tag{12.17}$$

Therefore, the transconductance efficiency factor, g_m/I_D, in weak inversion region is a constant term and is expressed as

$$\frac{g_m}{I_D} = \frac{1}{\eta U_T} \tag{12.18}$$

The standard approach to model the drain-source resistance, at least for drain voltages and currents near the bias point, is to use the concept of Early voltage. The increase in the drain current with variation of the drain-to-source voltage is modeled by multiplying the drain current expression (without any dependency on the drain-to-source voltage) by a factor $\left(1 + V_{DS}/V_A\right)$. It may be noted that the factor V_A accounts for many physical phenomena, such as channel length modulation effect, that due to drain-induced barrier lowering, and a few others [2].

The normalized output conductance g_{ds}/I_D is expressed as

$$\frac{g_{ds}}{I_D} = \frac{1}{V_A + V_{DS}} \approx \frac{1}{V_A} \tag{12.19}$$

V_A is not a constant but depends significantly on channel length L and weakly depends on channel bias condition (inversion level) until velocity saturation sets in [6]. As can be seen in Figure 12.9, that for n-channel MOS transistor, for short channel length, the plot of the Early voltage increases significantly, above the center of moderate inversion (IC = 1). For p-channel MOS transistor this bias dependence is comparatively lesser as seen in Figure 12.10.

It is observed from the two plots that the Early voltage $V_A'(= V_A/L)$, for n-channel and p-channel transistors, are approximately 20 and 30 V/μm, respectively.

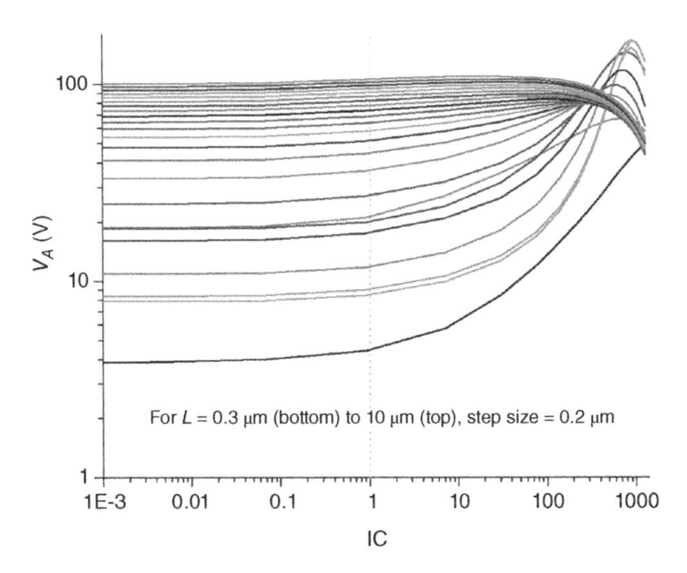

Figure 12.9 Plot of Early voltage of n-channel MOSFET w.r.t inversion coefficient for various values of channel length keeping $W = 5\ \mu m$ and $V_{DS} = 1.8$ V.

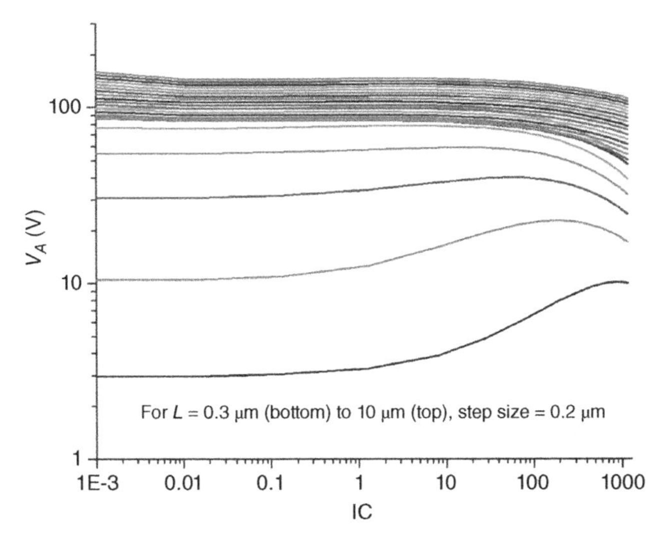

Figure 12.10 Plot of Early voltage of p-channel MOSFET w.r.t inversion coefficient for various values of channel length keeping $W = 5\ \mu m$ and $V_{DS} = 1.8$ V.

12.3 Design Steps for Transistor Sizing Using the IC

Figure 12.11 shows a schematic flow diagram for determining the transistor sizes of an analog amplifier circuit. The various steps are discussed as follows:

- from the supply voltage, V_{DD}, and power budget specifications, determine the drain current (I_D) requirement of the transistor.
- from the specifications of gain bandwidth product (GBW) and load capacitance (C_L), determine the transconductance, g_m.
- hence the transconductance efficiency g_m/I_D is obtained from specifications.
- from the plot of g_m/I_D vs. IC, determine the required level of inversion, i.e. strong, weak, or moderate. For some design problems, it may not be required to compute g_m/I_D to determine the inversion level. In such cases, the inversion level may be identified from the I_D vs. IC plot, after obtaining the drain current from the power budget requirement.
- determine the aspect ratio (W/L), from the following relationship

$$\frac{W}{L} = \frac{I_D}{IC \cdot I_o} \tag{12.20}$$

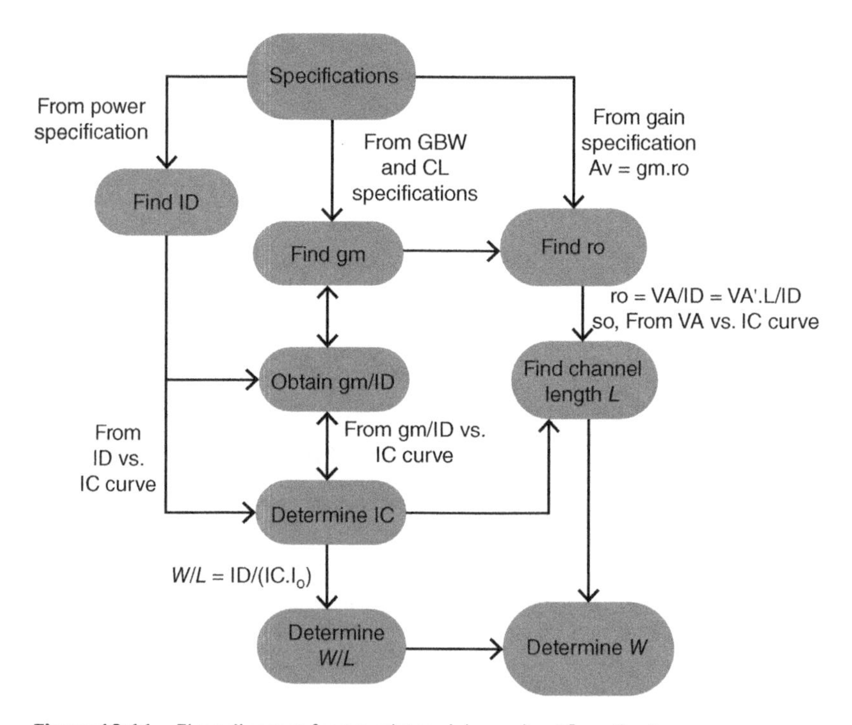

Figure 12.11 Flow diagram for transistor sizing using IC method.

where, I_0 is the technology current of the transistor, whose value is constant for the chosen process technology.

- from gain specification $A_v = g_m \cdot r_o$, obtain the drain resistance r_o using the value of g_m.
- Determine the early voltage V_A from the following relationship

$$r_o = \frac{V_A}{I_D} = \frac{V'_A \cdot L}{I_D} \tag{12.21}$$

- from the plot of V_A vs. IC with channel length as parameter, corresponding to n-channel and p-channel MOS transistors, determine the required channel length, L.
- using the value of the channel length, L, determine the width, W, of the transistor from (12.20).

These design steps are quite generic and applicable to similar other circuits too.

12.4 Design Examples

12.4.1 Design of a Common Source Amplifier

Figure 12.12 shows the schematic design (a) and the layout design (b) of a common source amplifier. The chosen design specifications for designing a common source (CS) amplifier with low-voltage operation are as follows:

- DC gain = 60 dB = 1000
- $V_{DD} = 0.6\,\text{V}$
- Total power consumption $\leq 2.5\,\text{nW}$
- Gain-bandwidth (GBW) product = 10 KHz
- Load capacitance, $C_L = 10\,\text{pF}$

now, total power consumption = $V_{DD} \times I_D$. Therefore, from the above specification, we derive that $I_D \approx 4\,nA$. From the specifications of GBW and C_L, we can find the transconductance, g_m, as follows:

$$g_m = GBW \times C_L = 100\ nS \tag{12.22}$$

We obtain the transconductance efficiency, g_m/I_D, as follows:

$$\frac{g_m}{I_D} = 25\ V^{-1} \tag{12.23}$$

From the g_m/I_D vs. IC plot in Figure 12.6, we can determine the inversion coefficient IC ≈ 0.01.

Now for common-source amplifier DC gain is:

$$|A_v| = g_m(r_{\text{on}}||r_{\text{op}}) \tag{12.24}$$

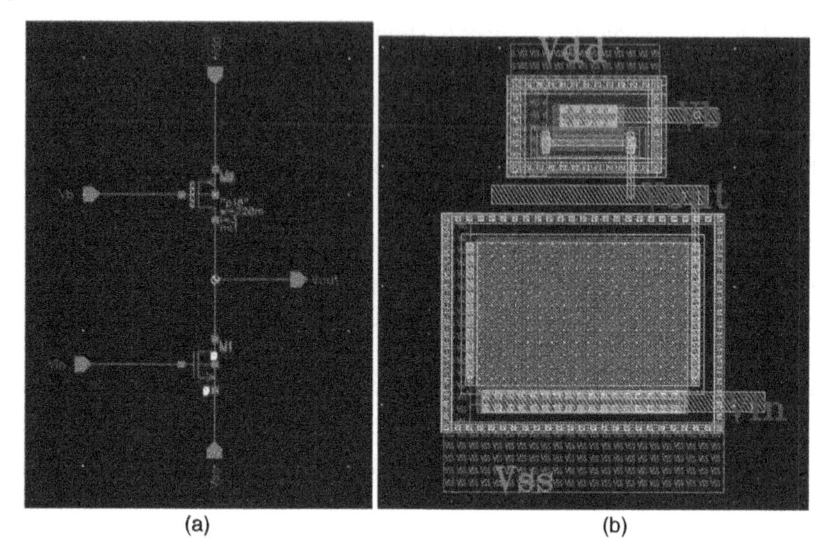

Figure 12.12 Common source amplifier. (a) Schematic design of a common source amplifier and (b) layout design of a common source amplifier.

where, r_{on} and r_{op} are the output resistances of the n-channel and p-channel MOSFET, respectively. Considering them to be equal, $A_v \approx g_m r_{\text{on}}^2$. From the gain specification and the computed values of g_m, the drain resistances are obtained as $r_{\text{on}} = r_{\text{op}} = 20\,G\Omega$. Since for an n-channel MOS transistor, Early voltage $V_A' = 20\,\text{V}/\mu m$; hence, by substituting the values of I_D and r_{on}, we obtain the channel length, L_n, of the input n-channel MOSFET as follows: $L_n \approx 4\,\mu m$. Therefore, width of the transistor, W_n, can be determined as follows:

$$W_n = \frac{L_n \cdot I_D}{\text{IC} \cdot I_{\text{on}}} = 4\,\mu m \tag{12.25}$$

The bias voltage of the n-channel MOS transistor, V_{GSn} is:

$$V_{\text{GSn}} = \eta U_T \ln(\text{IC}) + V_{Tn} = 0.22\ \text{V} \tag{12.26}$$

where V_{Tn} is the threshold voltage of the n-channel MOSFET, and is equal to 0.4 V. Now, for the p-channel load transistor, drain current (I_D) and drain-to-source resistance (r_{op}) are same as those of n-channel transistor. Since,

$$V_{Ap}' \cdot L_p = I_D \cdot r_{\text{op}} \tag{12.27}$$

Therefore,

$$L_p = \frac{I_D}{r_{\text{op}} \cdot V_{A'p}} = 3\,\mu m \tag{12.28}$$

Therefore, width, W_P, of the p-channel transistor is

$$W_p = \frac{L_p \cdot I_D}{IC \cdot I_{\text{op}}} = 6.6\ \mu m \tag{12.29}$$

Gate-to-source voltage, V_{SGp}, of the PMOS transistor is

$$Vp = \eta \cdot U_T \ln(IC) + \left|V_{Tp}\right| = 0.34\ \text{V} \tag{12.30}$$

where, V_{Tp} is the threshold voltage of the p-channel transistor and is equal to $|0.38|$ V. Therefore, bias voltage of the p-channel transistor is $V_b = V_{dd} - V_{\text{SGp}} = (0.6 - 0.34)V = 260\,\text{mV}$.

Simulation result of the common source amplifier shows that both the drain current, output voltage V_{out} and gain of the amplifier circuit in pre-layout and post-layout simulation, closely matches their corresponding analytical result as shown in Figures 12.13–12.15, respectively. A comparative study of the pre-layout, post-layout and analytical results are tabulated in Table 12.1.

12.4.2 Single-Ended Operational Transconductance Amplifier

In this example, we present the complete design of a differential input, single-ended output two-stage operational transconductance amplifier (OTA) using the design methodology discussed earlier. The design specifications are as follows:

- DC gain $(A_0) = 100\,\text{dB} = 10^5$.
- Load capacitor $(C_L) = 1$ pF
- Unity gain bandwidth (UGB) = 50 KHz
- Power consumption $P_{\text{total}} < 70$ nW
- Supply voltage $V_{DD} = 1.8$ V

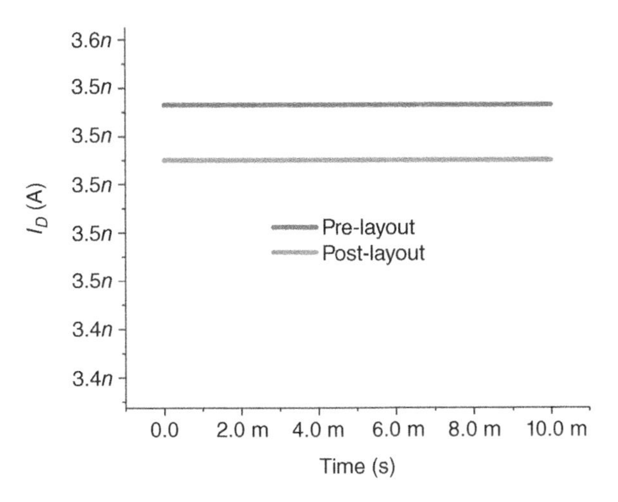

Figure 12.13 Drain current of common source amplifier in pre-layout ans post-layout simulation.

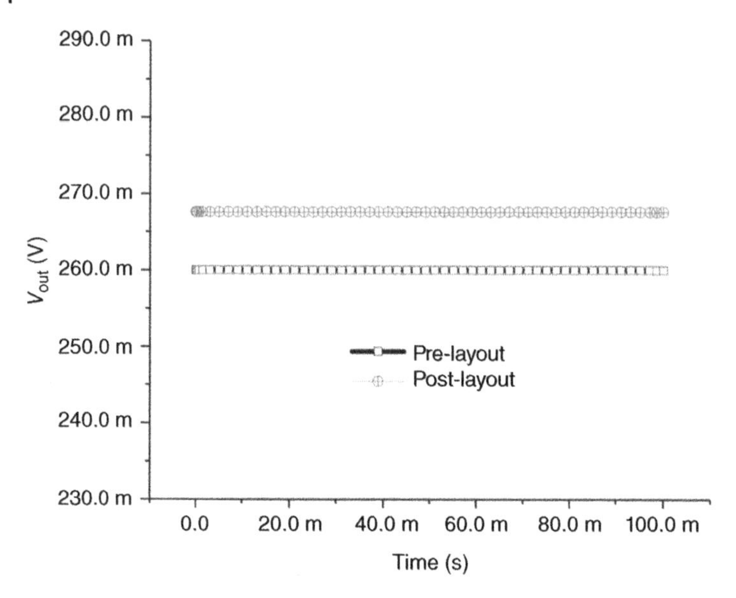

Figure 12.14 Output voltage of common source amplifier in pre-layout and post-layout simulation.

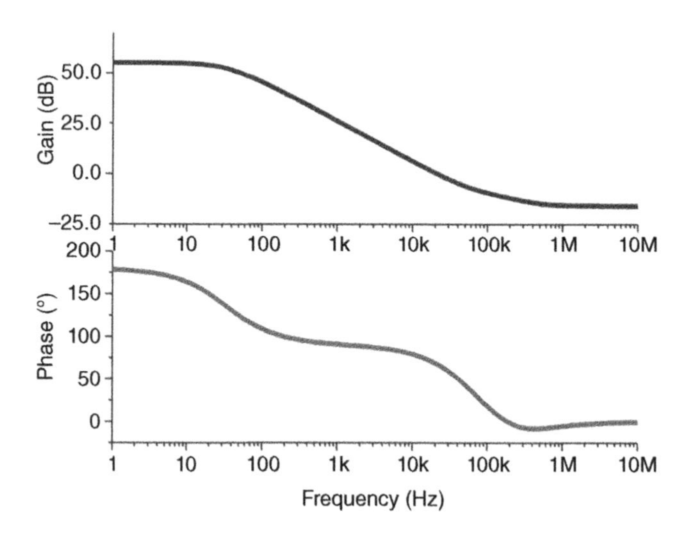

Figure 12.15 Gain of common source amplifier.

Table 12.1 Tabulated study of various performance parameters of the CS stage amplifier in modeling, pre-layout, and post-layout simulation.

Parameter	Model	Pre-layout	Post-layout
Gain (dB)	60	54.5	54.5
I_{ds} (nA)	4	3.53	3.53
V_{out} (mV)	260	260.26	267.28

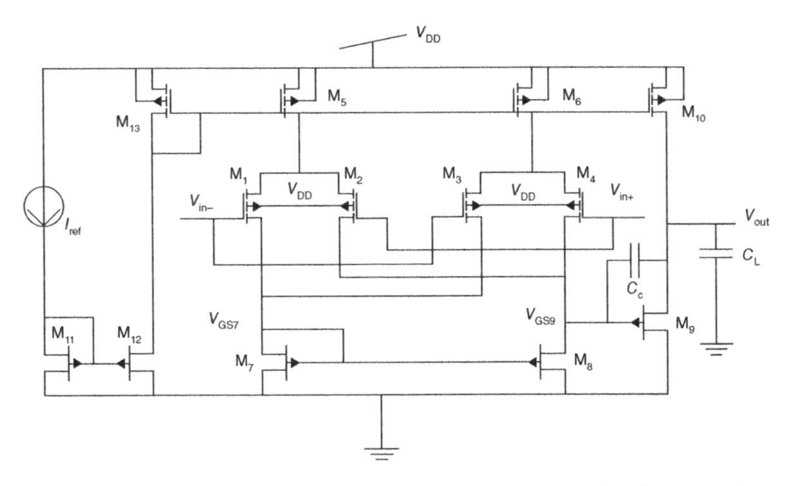

Figure 12.16 Differential-input single-ended output operational transconductance amplifier circuit.

The circuit topology, as considered for the design, is shown in Figure 12.16. Transistor M_1–M_4 are the p-channel input transistors, and transistors M_7–M_8 are the n-channel load transistors for the first gain stage. Transistors M_5 and M_6 provides the necessary bias currents for the first gain stage. Transistor M_9 and M_{10} are the n-channel MOS input and p-channel MOS load transistors of the second gain stage, respectively. The bias currents are generated from a reference current I_{ref} through current mirror circuit comprising of transistors M_{11}–M_{13}. In the present discussion our focus is only on the amplifier section (transistors M_1–M_{10}).

In order to meet the power budget specification, suppose $P_D = 55\,\text{nW}$. Dividing the same by the value of the supply voltage, therefore, the total current flowing through the complete circuit is calculated to be approximately equal to 30 nA. Suppose the first and second gain stages are driven by 5 and 25 nA currents, respectively.

Due to identical structure, the currents I_5 and I_6, flowing through transistors M_5 and M_6, respectively are same. Hence we can say

$$I_5 = I_6 = 2.5\ \text{nA} \tag{12.31}$$

Thus, the current flowing though all the input transistors of the first stage are

$$I_1 = I_2 = I_3 = I_4 = 1.25\ \text{nA} \tag{12.32}$$

From Figure 12.2, we can determine the inversion coefficient, IC_{1-4}, of p-channel input transistors M_1–M_4 as follows:

$$IC_{1-4} \approx 0.008 \tag{12.33}$$

The input transistors are working, thus, in weak inversion mode.

The transconducatance, g_{m1-4}, of the corresponding transistors is obtained as

$$g_{m1-4} = \frac{I_{1-4}}{\eta U_T} = 32\ \text{nS} \tag{12.34}$$

Now the total DC gain $A_0 = A_1 \times A_2$.

Where, A_1 and A_2 are the DC gains of the first and second gain stages, respectively.

Suppose, $A_1 \approx 55\ \text{dB} = 562$, and $A_2 \approx 45\ \text{dB} \approx 180$.

Now considering, the input and the load transistors have equal drain resistances, we get equivalent output impedance, r_{eq1} to be

$$r_{eq1} = \frac{r_{op1-4}}{2} = \frac{r_{on7}}{2} = \frac{r_{on8}}{2} \tag{12.35}$$

here, r_{op1-4} is the drain resistance of the input p-channel MOS transistors M_1-M_4 whereas r_{on7} and r_{on8} are the drain resistances of the n-channel MOS load transistors M_7 and M_8, respectively. Therefore,

$$r_{op1-4} = r_{on7} = r_{on8} = \frac{2 \times A_1}{g_{m1-4}} \approx 36\text{G}\Omega \tag{12.36}$$

Therefore, from V_A vs. IC plot shown in Figure 12.10, we obtain the required channel length, $L_{p1-4} \approx 1.5\ \mu m$, for the p-channel input transistors of the first gain stage.

Therefore, the channel width, W_{p1-4}, for the input transistors is determined as follows:

$$W_{p1-4} = \frac{L_{p1-4} \times I_1}{IC_{1-4} \times I_{op}} = 1.4\ \mu m \tag{12.37}$$

Now, as mentioned earlier, $I_9 = 25 \times 10^{-9}$ A. Therefore, from Figure 12.1, we obtain the inversion coefficient of the n-channel MOS input transistor M_9 of the second gain stage to be

$$IC_9 \approx 0.01 \tag{12.38}$$

Thus, M_9 also operates in weak inversion mode.

Therefore, the transconductance, g_{m9}, of M_9 is

$$g_{m9} = \frac{I_9}{\eta U_T} = 640 \text{ nS} \tag{12.39}$$

Now, suppose the drain resistances, r_{on9} and r_{op10}, of M_9 and M_{10} are equal. Therefore, the equivalent output impedance of the second gain stage is

$$r_{eq2} = (r_{on9} \| r_{op10}) = \frac{r_{on9}}{2} = \frac{r_{op10}}{2} \tag{12.40}$$

Hence,

$$r_{on9} = r_{op10} = \frac{2 \times A_2}{g_{m9}} \approx 500 \text{ M}\Omega \tag{12.41}$$

The Early voltage V_{A9} of transistor M_9 is given by

$$V_{A9} \approx r_{on9} \times I_9 = 12.5 \text{ V} \tag{12.42}$$

and since, channel length, L_{n9}, of transistor M_9 is related to V_{A9} as

$$V_{A9} = V'_{An} \times L_{n9} \tag{12.43}$$

where, V'_{An} is the Early voltage of n-channel MOS transistor per unit μm channel length.

Thus, from the Early voltage vs. IC plot of n-channel MOS transistor, as shown in Figure 12.9, we obtain the L_{n9} as follows:

$$L_{n9} = 0.625\ \mu m \tag{12.44}$$

Using this value of L_{n9}, we can obtain the channel width, W_{n9}, of transistor M_9 from the following relation

$$W_{n9} = \frac{L_{n9} \times I_9}{IC_9 \times I_{on}} \approx 4\ \mu m \tag{12.45}$$

Now, the gate-to-source voltage, V_{GS9}, of transistor M_9, and thus the drain-to-source voltage of transistor M_8 is computed as

$$V_{GS9} = \eta U_T \ln(\text{IC}_9) + V_{Tn} = 0.22 \text{ V} \tag{12.46}$$

Due to having identical structure, we get gate-to-source voltage, V_{GS7}, of the diode-connected transistor M_7 is equal to 0.22 V also.

Thus, the inversion coefficient of transistors M_7 and M_8 is

$$\text{IC}_7 = \text{IC}_8 = \exp\left(\frac{V_{GS7} - V_{Tn}}{\eta U_T}\right) = 0.01 \tag{12.47}$$

Now, using the drain resistance $r_{on7} = r_{on8}$ as obtained earlier and following the logic already implemented for transistor M_9, we get the channel length, $L_{n7} = L_{n8} = 2.25\ \mu m$.

Using the value of $IC_7 = IC_8$, we obtain the channel width, $W_7 = W_8$ of the transistors M_{n7}–M_{n8} as follows:

$$W_{n7} = W_{n8} = \frac{L_{n7} \times I_7}{IC_7 \times I_{on}} = 0.7\,\mu m \tag{12.48}$$

Since, $r_{on9} = r_{op10}$ and $I_9 = I_{10}$, we can say, Early voltage, V_{A10} of transistor M_{10} is also 12.5 V.

Therefore, by the same logic, as implemented for transistor M_9, we can obtain the channel length, L_{p10}, of the transistor M_{10} as follows:

$$L_{p10} \approx 0.4\,\mu m \tag{12.49}$$

Now, from the specification of UGB, we can obtain the required compensation capacitance, C_c, from the following relationship [8]

$$C_c = \frac{g_{m1-4}}{2\pi \text{UGB}} \approx 0.11\ \text{pF} \tag{12.50}$$

Now, since our targeted UGB is 50 KHz. Thus, for stability purpose, the output pole is supposed to be situated far from unity gain frequency. Therefore, suppose, output pole $p_2 = 100$ KHz. We can, therefore, obtain the transconductance, g_{m10} of transistor M_{10} from the following relationship [8]

$$g_{m10} = 2\pi C_L \times p_2 \approx 630\ \text{nS} \tag{12.51}$$

Hence, we get the transconductance efficiency of M_{10} as follows:

$$\frac{g_{m10}}{I_{10}} \approx 25 \tag{12.52}$$

Hence, from g_m/I_D vs. IC plot of PMOS transistor, as shown in Figure 12.7 we obtain the inversion coefficient IC_{10} of $M_{10} \approx 0.006$.

From IC_{10}, we compute the required bias voltage V_{SG10} of M_{10} to be

$$V_{SG10} = \eta U_T \ln(IC_{10}) + |V_{Tp}| \approx 0.18\,\text{V} \tag{12.53}$$

Using the value of r_{op10} already derived earlier and following the same logic implemented for PMOS transistors M_1–M_4 we get width, W_{p10}, and channel length, L_{p10}, of the transistor M_{10} to be 1.16 and 0.416 μm, respectively.

Since transistors M_5 and M_6 share the same bias voltage V_{SG10}, therefore, they have the same inversion coefficient $IC_{5-6} = 0.006$, as transistor M_{10}. Thus transistors M_5 and M_6, too, operate in weak inversion mode.

Considering M_{10}, M_5, and M_6 have the same channel length, using the current ratio logic we can derive the width, $W_5 = W_6 = 0.116\,\mu m$ for the transistors M_5 and M_6. But, since, as per the design rule set by SCL 0.18 μm process technology, the minimum allowable channel width is 0.18 μm, hence we consider $W_5 = W_6 = 0.22\,\mu m$ for reliable circuit operation.

Table 12.2 summarizes the dimensions of all the transistors of the proposed OTA circuit, as computed earlier.

Table 12.2 Table summarizing the dimensions of the transistors in the proposed OTA circuit.

Transistor names	Dimensions W/L ($\mu m/\mu m$)
M_1	1.4/1.5
M_2	1.4/1.5
M_3	1.4/1.5
M_4	1.4/1.5
M_5	0.22/0.416
M_6	0.22/0.416
M_7	0.7/2.25
M_8	0.7/2.25
M_9	4/0.625
M_{10}	1.16/0.4

12.4.2.1 Implementation and Simulation Result

The device dimensions as computed through the methodology described earlier are needed to be tuned slightly to adjust to the technology constraints and better match the overall specifications of the circuit. A comparison between the various node voltages and currents as obtained through SPICE simulation results and calculated through our methodology is tabulated in the Table 12.3. We find quite satisfactory matching between the two results for most of the observable parameters. Some important analyses and corresponding results are summarized below.

Table 12.3 Tabulated report of the derived and simulated results of important branch currents and nodal voltages of the OTA circuit.

Transistors' electrical properties	Derived	Simulated
I_{1-4} (nA)	1.25	1.4
I_{5-6} (nA)	2.5	2.8
I_{9-10} (nA)	25	19
V_{GS7-8} (V)	0.22	0.251
V_{SG5-6} (V)	0.18	0.173
g_{m1-4} (nS)	32	35.9
g_9 (nS)	600	575
$r_{o1-4}(G\Omega)$	36	38
$r_{o9-10}(M\Omega)$	500	355

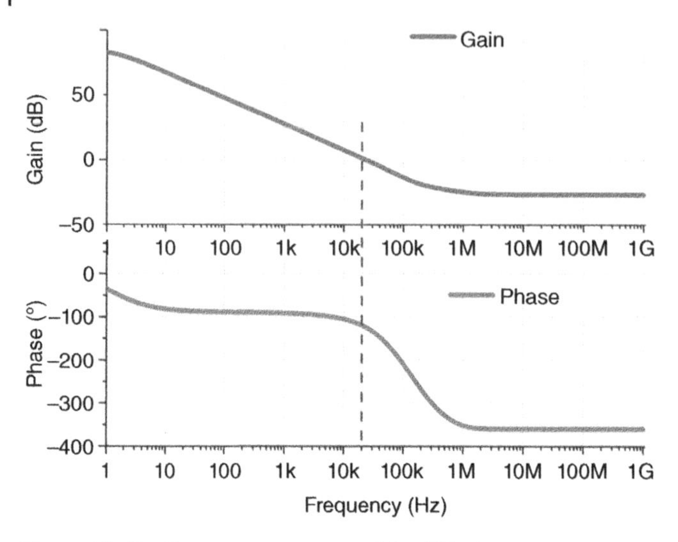

Figure 12.17 Frequency response of the OTA circuit.

Frequency response: In Figure 12.17, the maximum DC gain obtained by simulation is 86.7 dB with cutoff frequency lying at 1.9 Hz. The amplifier shows UGB of 30 KHz with phase margin 61°. The phase crossover frequency takes place nearly at 110 KHz with gain margin 24 dB.

PSRR: The power supply rejection ratio (PSRR) is obtained to be 33 dB over the targeted bandwidth of 15 KHz, as shown in Figure 12.18. Though this data needs to be improved further for better noise performance.

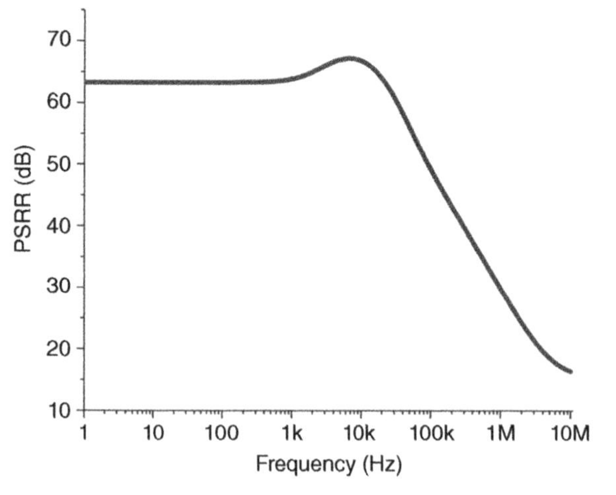

Figure 12.18 PSRR of the OTA circuit.

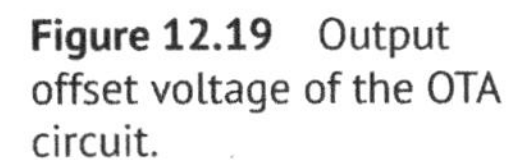

Figure 12.19 Output offset voltage of the OTA circuit.

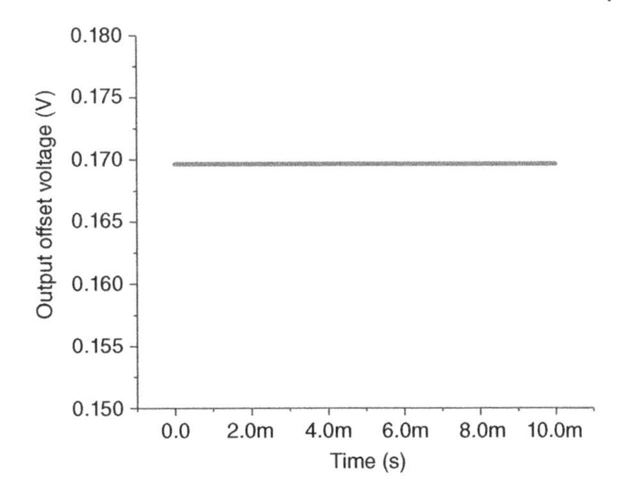

Output offset voltage: With both the input terminals grounded, the output offset voltage is 170 mV as shown in Figure 12.19. Corresponding input referred offset is 8.5 μV.

Input common mode range (ICMR): When a common mode signal V_{icm} is applied to the input terminals and varied from 0 V to the $V_{DD} = 1.8V$, it is observed in Figure 12.20 that, the drain currents I_{1-4} remain constant for the V_{icm} varying upto 1.44 V after which it starts to drop. For $V_{icm} < 280$ mV, the output voltage no longer follows the input. Hence, the common mode range is 280 mV to 1.44 V.

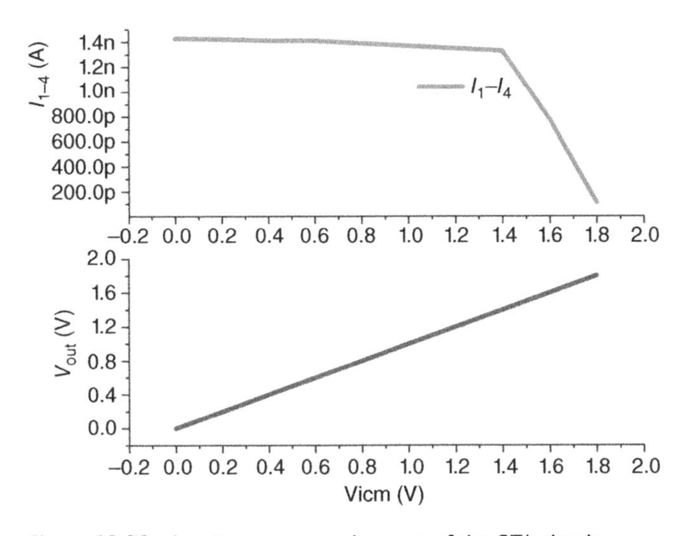

Figure 12.20 Input common mode range of the OTA circuit.

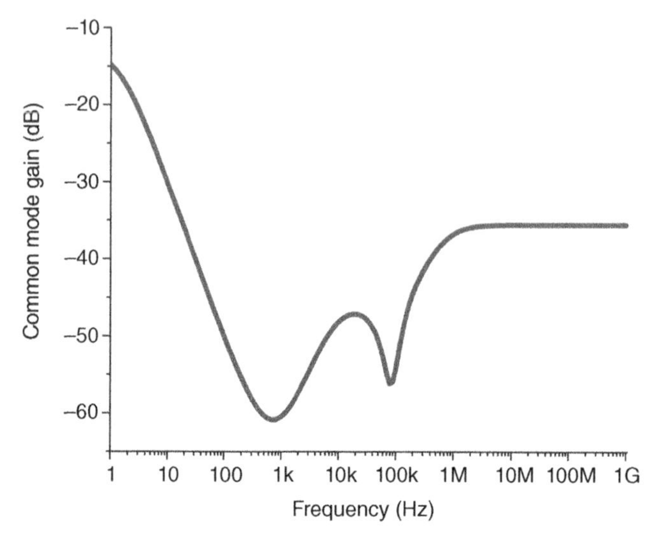

Figure 12.21 Common mode gain of the OTA circuit.

Common mode gain A_{cm}**:** At input frequency 50 Hz, the common mode signal is obtained to be −44 dB, as can be seen in Figure 12.21. Thus the resulting common mode rejection ratio (CMRR) is 130.7 dB.

The complete simulated results are tabulated in Table 12.4.

Table 12.4 Table of simulation results of the OTA circuit.

Parameter	Value
DC gain (dB)	86.7
UGB (KHz)	30
Phase margin (°)	61
Gain margin (dB)	24
Phase cross over frequency (KHz)	110
Input common mode range (V)	0.28–1.44
Common mode rejection ratio (CMRR) (dB)	130
Power supply rejection ratio PSRR (dB)	63

12.5 Summary

This chapter provides a detailed description of the design of an analog amplifier circuit using 180 nm digital CMOS technology of SCL for ultra-low-power VLSI applications. The trade-off for reduction of power dissipation in an amplifier circuit with respect to the signal-to-noise ratio is discussed, and the motivation for operation in the weak inversion region is drawn from the results derived there. A qualitative review of the theory of weak inversion mode operation of MOS transistor is presented with an appropriate drain current model and associated small signal model parameters. The concept of inversion coefficients associated with levels of inversion in a transistor is introduced. The extraction methodology for few important parameters are discussed. The chapter then presents a detail design methodology for low-power design of analog amplifier circuits based on the inversion coefficient. The methodology is explained in detail with two design examples, being thoroughly worked out. The SPICE simulation results of the two designs are also presented and compared with theoretical results. A good matching between the two sets of results is observed.

References

1 Sansen, W. (2015). Minimum power in analog amplifying blocks: presenting a design procedure. *IEEE Solid-State Circuits Magazine* 7 (4): 83–89.

2 Pandit, S., Mandal, C., and Patra, A. (2014). *Nano-Scale CMOS Analog Circuits: Models and CAD Techniques for High-Level Design*. Boca Raton, FL: CRC Press.

3 Svensson, C. and Wikner, J.J. (2010). Power consumption of analog circuits; a tutorial. *Analog Integrated Circuits and Signal Processing* 65: 171–184.

4 Pandit, S. (2015). *Nanoscale MOSFET: MOS Transistor as Basic Building Block*, 145–172. Berlin, Heidelberg: Springer-Verlag.

5 Pandit, S. (2013). *MOSFET Characterization for VLSI Circuit Simulation*. Boca Raton, FL: CRC Press.

6 Binkley, D.M. *Tradeoffs and Optimization in Analog CMOS Design*. Wiley.

7 Binkley, D.M., Blalock, B.J., and Rochelle, J. (2006). Optimizing drain current, inversion level, and channel length in analog CMOS design. *Analog Integrated Circuits and Signal Processing* 47: 137–163.

8 Allen, P. and Holberg, D. (2002). *CMOS Analog Circuit Design*, 2e. Oxford University Press, USA.

13

Ultra-conductive Junctionless Tunnel Field-effect Transistor-based Biosensor with Negative Capacitance

Palasri Dhar, Soumik Poddar, and Sunipa Roy

Electronics and Communication Engineering Department, Guru Nanak Institute of Technology, Maulana Abul Kalam Azad University of Technology, Kolkata, India

13.1 Introduction

The development of biosensors has revolutionized the way we track and monitor crucial processes and parameters in numerous sectors. The evolvement of these devices has led to improvements in many fields, including biomedicine, food security and processing, environmental tracking, protection, and safety. Biosensors are devices that detect and measure biological or biochemical reactions and are used to check whether a sample contains a particular substance. Biosensors are analytical devices that are composed of a biological recognition element, such as an enzyme or antibody, and a transduction element, such as an electrochemical or optical sensor. These components work together to detect and measure the presence or concentration of a particular analyte of interest. Essentially, biosensors are self-contained devices that can recognize and quantify biological or chemical substances by converting the biological or chemical signals into a measurable signal that can be analyzed. This makes biosensors a powerful tool for a variety of applications, including medical diagnostics, environmental monitoring, and food safety testing. The biological recognition element interacts with the analyte, producing a signal that is transduced into an electrical, optical, or other measurable signal that can be read and interpreted by the biosensor.

Typically, biosensors consist of three primary components, as illustrated in Figure 13.1. Examples of these three major components include a biological sensing element, a transducer to convert biological or chemical signal into a measurable electrical signal that can be interpreted and analyzed, and a signal processing system to analyze and interpret the signal. When a biosensor detects the target analyte, there is a biological interaction between the sensing component, which can be made up of various biological elements such as tissues,

Advanced Nanoscale MOSFET Architectures: Current Trends and Future Perspectives,
First Edition. Edited by Kalyan Biswas and Angsuman Sarkar.

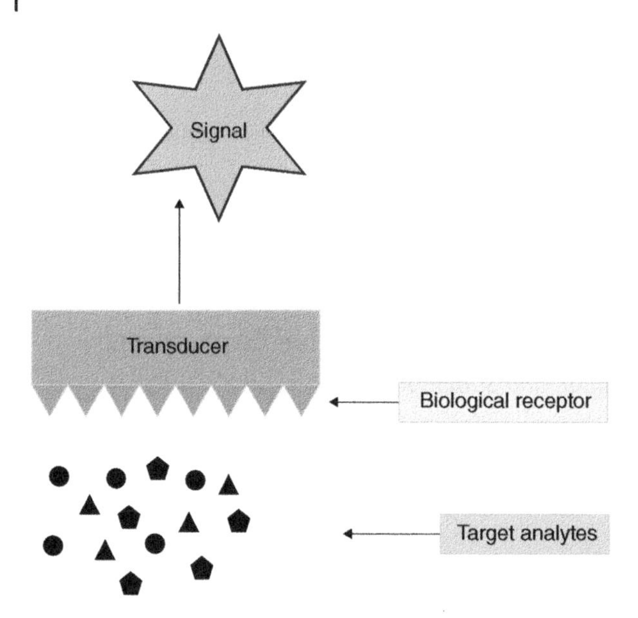

Figure 13.1 Basic diagram of a biosensor.

microbes, cells, cell receptors, peptides, antibodies, or nucleic acids, and the analyte of interest. This interaction generates a signal, which is then converted into an electrical signal that can be measured and quantified by the transducer. After the biosensor's transducer converts the biological or chemical signal into an electrical signal, the signal is typically amplified and processed by a signal processing system. This system is responsible for improving the accuracy and reliability of the biosensor's readings by reducing noise and enhancing the signal. Once the signal has been processed, it is transmitted to a data processor, which generates a measurable output such as a computer display, printout, or color change. The output is designed to provide a clear and understandable result that can be easily interpreted by the user. The data processor is an important component of the biosensor, as it is responsible for converting the electrical signal into a format that is easily understood and used by researchers, clinicians, or other users. By generating a measurable output, the data processor enables the user to quickly and accurately determine the presence and quantity of the analyte of interest, making biosensors a valuable tool for a variety of applications.

Biosensors have been in consideration for a long time. Cremer's observation in 1906 that the electrical potential generated across a glass membrane is directly proportional to the level of acid in an aqueous solution can be identified as the early origins of the biosensor. In 1909, Soren Sorensen developed the concept of pH, which was later utilized by Hughes to create a pH-measuring electrode in 1922.

The title "Father of Biosensors" is commonly attributed to Leland Clark, Jr., who in 1959 developed the first genuine biosensor. This biosensor employed a glucose oxidase electrode to identify the existence of oxygen or hydrogen peroxide within biological samples. Since then, biosensors have undergone significant advancements in terms of sensitivity and selectivity. The emphasis of this chapter is on the ultra-conductive JL-TFET-based biosensor with negative capacitance (NC).

The significance of biosensors in modern society is increasing, concomitant with the requirement for the development of novel methods of bio-molecules sensing. Due to their rapid detection capabilities, low power usage, low cost, label-free biomolecule detection, and CMOS compatibility, FET-based biosensors are a household name in flexible electronics. The sensitivity is highly interlinked with sub-threshold slope (SS), I_{ON}/I_{OFF} current ratio, and new techniques are adopted to improve these parameters. NC offers sub 60 mV/dec SS and tunnel field-effect transistor (TFET) are nominated for high I_{ON}/I_{OFF}. Si – TFET band-to-band tunneling (BTBT) enables SS, i.e. 60 mV/dec SS and reduced I_{OFF}. By hybridizing NC with JL-TFET, more band bending occurs that offers higher I_{ON}, steeper SS without much degradation in I_{OFF}. Also, NC amplifies low gate voltage, for which sensitivity can readily be upgraded. The efficiency of such NC-enabled JL-TFET can be enhanced by dopingless drains and sources that eradicate random dopant fluctuations (RDFs), thermal budget, ambipolar transport, etc. The source may be other than Si to increase the I_{ON}. The JL-TFET may be encompassed with dual drain to restrict tunneling of holes at negative gate bias. Also, the hysteresis due to NC can be controlled by hybrid HfO_2. Such modifications may enhance the bio sensitivity at superb level. To increase the probability of tunneling, a group of III–V based (GaAs/InAs) pockets may be positioned across the silicon film's interfaces in between the source and channel. Keeping in view different types of TFET with dual drain, Ge source, III–V semiconductor-based pocket at source–channel interface, dopingless (charge plasma mediated) drain and source, Zr doped HfO_2 as gate stack materials are highlighted, and the sensing attributes towards biomolecules and breast cancer cell detection are enumerated.

NC has a significant impact on biosensors, as it lowers the SS and increases the current sensitivity in the weak inversion regime. The NC also enables a substantial overdrive of the sensor, which in turn reduces the device's power consumption. This property of NC is advantageous for biosensors, as it enhances their performance and efficiency [1–4].

Conventional Metal–oxide–semiconductor field-effect transistors (MOSFETs) are effective biosensors. Despite their benefits, MOSFET devices face certain negative consequences, such as SCEs including velocity saturation, hot carrier effect, and drain-induced barrier lowering (DIBL), which can have an adverse impact on their low power performance [4–6]. As both static and dynamic power consumption of these devices is increasing exponentially, scaling MOSFETs has become increasingly challenging. TFETs that utilize BTBT have emerged as the most viable option for reducing overall power usage. The reduced sub-threshold swing and low OFF-state leakage current are some of the advantages offered by the TFET, which attracts a lot of interest in biosensing (SS). The sub-threshold fluctuation (SS) of a TFET is also less than 60 mV/dec, which is Boltzmann's limit. The utilization of NC for amplification of low gate voltage would facilitate the low-power operation of the device (Figure 13.2).

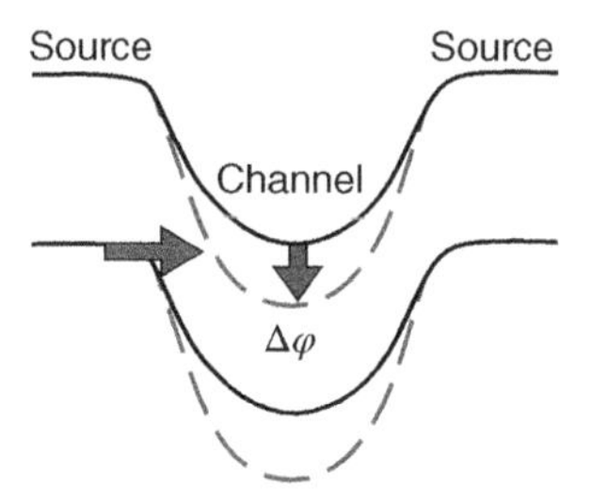

Figure 13.2 Negative capacitance – junctionless tunnel field-effect transistors schematic diagram (TFET).

13.2 Importance of SS and I_{ON}/I_{OFF} in Biosensing

The biosensor's sensitivity is an essential parameter to investigate when analyzing device performance. The sensitivity is determined by the I_{ON} and $I_{OFF.}$ It was discovered that as the biomolecule charges shifts from −ve to +ve in the different proposed JL-FET structures, ON current increased. Figure 13.3 depicts a DM-DGJLT biosensor structure.

It is noted that the I_{OFF} decreases (increases) for negatively (positively) charged biomolecules when compared to neutral biomolecules. The increase (decrease)

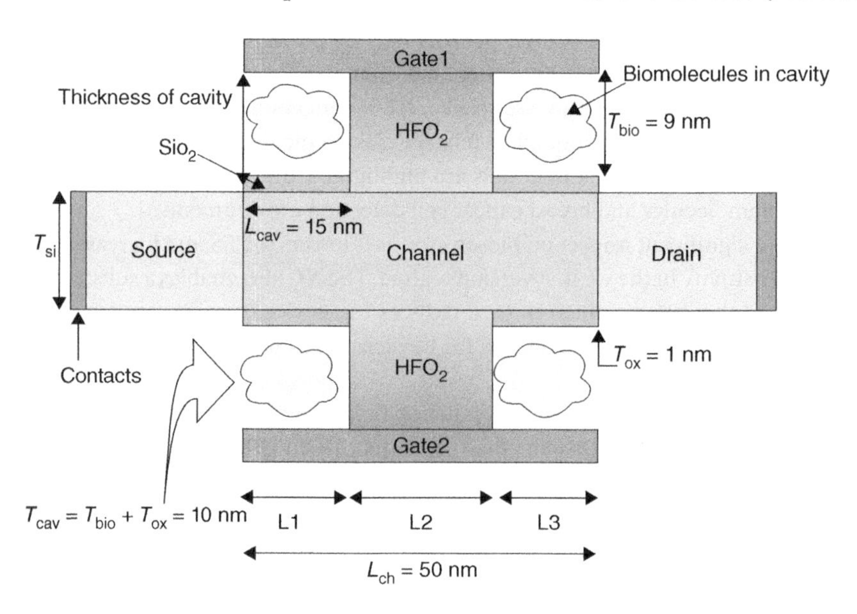

Figure 13.3 High-*k* DM-DGJLT biosensor structure. Source: Adapted from [5, 6].

in negative (positive) charge within the cavity leads to a corresponding increase (decrease) in the source-channel barrier height, which in turn causes a decrement (increment) of the drain current. I_{OFF} is primarily influenced by immobilized bioanalytes, while I_{OS} changes very little. Furthermore, as the k value of bioanalytes increases, so does the device's V_{TH}. As with k, increased gate-to-channel electrostatic interaction raises the source-to-channel barrier height. As a result, V_{TH} rises.

The voltage application effect and the super-linear onset behavior are the typical features and problems for conventional TFETs [5, 6], which can be mitigated in the case of junctionless tunnel field-effect transistors (JL-TFETs). In addition, a narrower work function (W_f) can result in a higher energy band in the off state, leading to a sharper band bending when the device is switched on. This can cause a higher tunneling electric field to mitigate the junction depleted-modulation effect [7–9]. As a result, when W_f decreases, steeper SS and higher I_{ON} are achieved in JL-FET, as given in Figure 13.4a–d. In JL-FET, the longer L_f increases the tunneling area for greater I_{ON}.

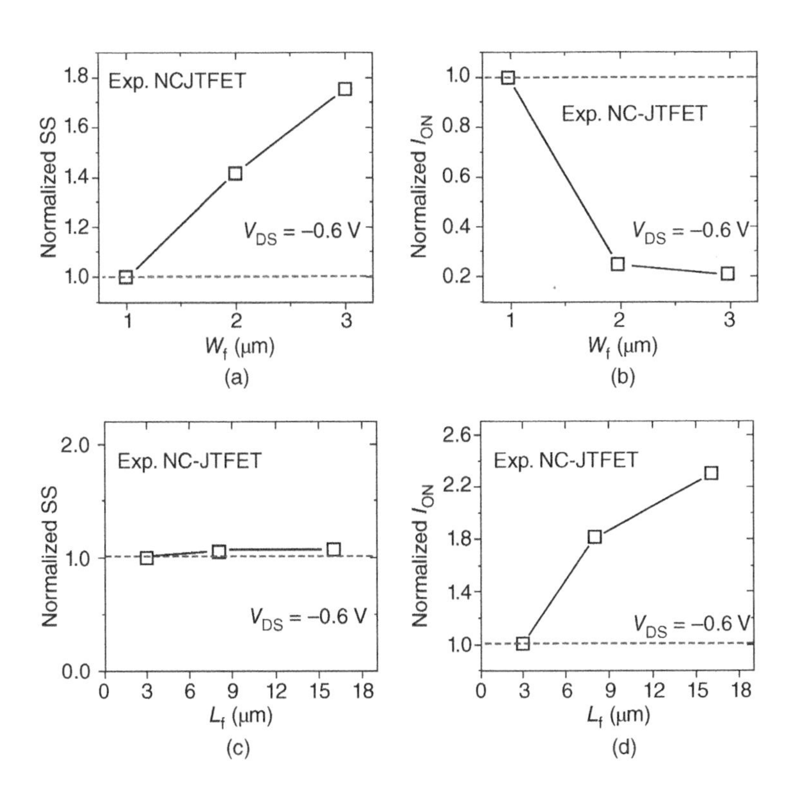

Figure 13.4 Dependence of measured normalized (a) SS and (b) I_{ON} dependence on W_f for junctionless TFET (JL-TFET) devices; Dependence of measured normalized (c) SS and (d) I_{ON} dependence on L_f for JL-TFET devices. Source: Adapted from [10–12].

As the k value of biomolecules increases, the ratio of I_{ON}/I_{OFF} current also increases. When the value of the –ve charge density increases, the I_{ON}/I_{OFF} current ratio decreases significantly. The reason for this decrement in the I_{ON}/I_{OFF} current ratio is due to an increment and decrementing value of electric field at the junction of source and channel. The I_{ON}/I_{OFF} current ratio of negative charge proteins is greater than that of +ve charge proteins. The greater current ratio is there as the –ve charge biomolecules have a lower I_{OFF} current. The I_{OFF} current is reduced in the presence of negatively charged biomolecules, as these molecules repel electrons from the channel surface, leading to an increase in the width of the space charge region and a subsequent decrease in the I_{OFF} current. When the charge of a biomolecule rises, the $I_{ON}/I_{OFF\ current}$ ratio increases. When +ve charge biomolecules are present, they attract electrons to the surface and decrease the space charge region. This leads to an increase in the number of carriers in the channel and reduces the resistance, resulting in an increase in I_{ON} current. Consequently, the I_{ON}/I_{OFF} current ratio decreases. So, we can summarize that negative charge biomolecules have a higher I_{ON}/I_{OFF} current ratio due to the dominance of reduced OFF current, and positive charge biomolecules have a lower I_{ON}/I_{OFF} current ratio due to the dominance of higher I_{ON} current. Higher I_{ON}/I_{OFF} ratios are desirable as they indicate better device performance and lower power dissipation from leakage. A higher I_{ON}/I_{OFF} ratio implies better control over the gate, leading to faster switching.

Increasing the dielectric constant results in an increase in the electric field at the channel and source junction, leading to a reduction in the number of tunneling carriers. On the other hand, as the density of charge grows, the tunneling junction electric field decreases, increasing the number of the tunneling carriers.

$$S_{drain} = \frac{I_{drain}^{bio} - I_{drain}^{air}}{I_{drain}^{air}} \tag{13.1}$$

In Eq. (13.1),

I_{drain}^{air} represents the current without the presence of any biomolecule ($k = 1$) and I_{drain}^{bio} represents the current Drain terminal in the cavity region ($k > 1$).

The improvement in sensitivity of drain current is clearly visible at low values of VDS, where the dielectric constant and charge density varies.

The SS is a critical factor in determining a device's switching behavior. It indicates how effectively the gate voltage controls the channel area. SS is calculated as the change in gate voltage that results in a tenfold change in drain current in the subthreshold region of the device. The device's SS is used to measure its turn-on behavior, and it is expressed in mV/dec. A lower SS value implies better channel control, resulting in a higher ON/OFF current ratio.

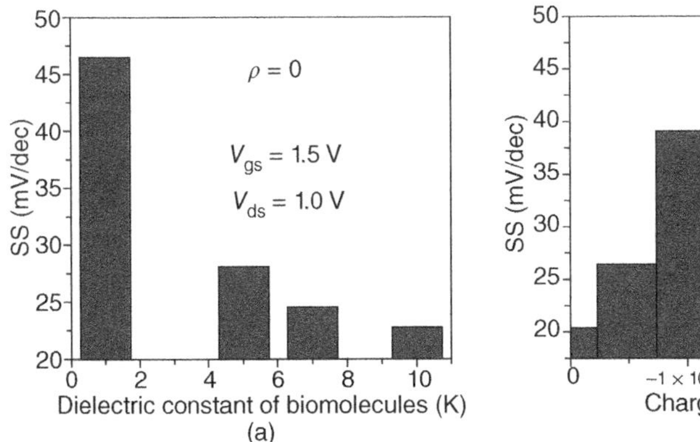

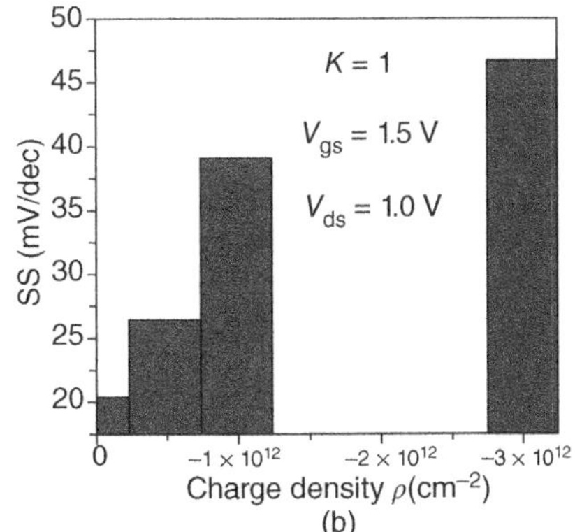

Figure 13.5 (a) Graph depicting the variation of SS in JL-TFET for varying dielectric constant values at $r = 0$. (b) Graph depicting the variation of SS in JL-TFET for varying charge density values at $k = 5$. Source: Chandan and Sharma [23]/John Wiley & Sons.

SS is defined as:

$$SS = \frac{\partial V_{gs}}{\partial \log I_{ds}} \text{mV/dec} \tag{13.2}$$

In Eq. (13.2), it is observed that SS is dependent on its gate voltage (V_{gs}).

In the instance of TFET, however, it is observed that SS is primarily changed due to the mechanism of diffusion current, and SS is also dependent on the mechanism of tunneling current. The sensitivity of the device changes with different dielectric constants and charge densities of biomolecules as shown in Figure 13.5a,b. As the dielectric constant increases, the SS characteristics decrease. This is because the electric field at the source and channel junction under the cavity region increases. A smaller value of SS is preferred for JL-TFET devices, indicating a better detection capacity and response. On the other hand, the SS characteristics increase with decreasing dielectric constant values, which is due to a decrease in the source and channel junction electric field under the cavity region. However, when the charge density increases, the opposite behavior is observed as shown in Figure 13.4a,b.

13.3 Importance of Dopingless Source and Drain in High Conductivity

Dopingless TFET (DL-TFET) was used to produce significant ON-state current and to neutralize fluctuations by random dopants caused by doping. The DL-TFET is distinguished by a low thermal expenditure. Dopingless TFET's are manufactured without using elevated-temperature heating or ion implantation techniques. The sudden junction formed for efficient tunneling is perhaps the most sensitive stage in the TFET manufacturing process. The source region and the drain regions are full of diffused doping atoms. This creates a sudden junction across the intrinsic channel of a doped-TFET. Doping is not a simple job. However, such problems caused by doped TFET have now been resolved by dopingless TFET. This type of structure also reduces BTBT.

Achieving a sharp doping profile is crucial to mitigate the impact of RDFs. In order to enable band-to-band tunneling at the source-channel and drain-channel junction regions, a precise doping profile is crucial, especially due to the presence of RDFs. However, this complexity and potential flaws in the fabrication process can be minimized by reducing the use of metallurgical doping. Therefore, in order to create the source and drain regions in TFETs, a dopingless method has been developed using an appropriate electrode metal work function on each side of the source and drain regions [7–9].

The charge plasma concept is another concept that is utilized to generate the drain and source regions in the dopingless device. As a result, chemical doping is not required in dopingless JL-TFETs.

The dopingless method involves creating the source and drain regions of TFETs without using metallurgical doping, which can be challenging due to random dopant fluctuations and fabrication complexity. Instead, suitable metals with specific work functions, such as Pt (5.93 eV) and Hf (3.9 eV) for the source and drain regions, respectively, are applied to overcome these issues.

Increasing the dielectric constant (k) of biomolecules within the cavity region of a TFET leads to a significant increase in charge carrier concentration in the same region, resulting in higher sensitivity. This improved sensitivity enables more accurate detection of biomolecules within the cavity of aJL-TFET.

Figure 13.6 depicted shows the surface potential contour of a biosensor based on dopingless-JL-TFET for the channel area. The biosensor is designed to detect both non-tumorigenic cell lines (MCF-10A) and breast cancerous cell

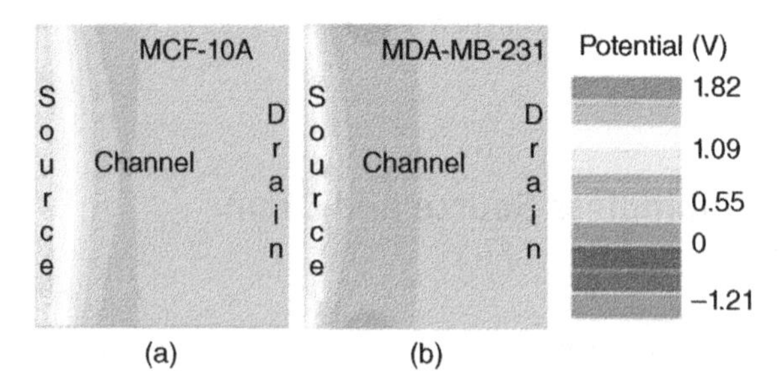

Figure 13.6 Shows the surface potential contour of a biosensor based on dopingless-junctionless tunnel field-effect transistors (TFETs) for both non-tumorigenic cell line (MCF-10A) and breast cancerous cell line (MDA-MB-231). Source: Adapted from [14].

lines (MDA-MB-231). The contour plot indicates the distribution of the surface potential across the biosensor's channel area for both cell lines. The variations in the surface potential distribution help in the detection and differentiation of the two cell lines, making it a promising tool for cancer diagnosis and research.

13.4 Relation of Negative Capacitance with Non-hysteresis and Effect on Biosensing

The hysteresis behavior is a crucial aspect to consider for ferroelectric-based transistors. This behavior is not suitable for logic applications. However, researchers working on biosensing have made a promising discovery. They have found that the Negative capacitance-junctionless tunnel-field effect transistor (NC-JLTFET) they fabricated shows almost no hysteresis behavior at room temperature. This is an exciting development for biosensing applications since hysteresis can cause inaccuracies in the detection process. With this discovery, NC-JLTFETs can be a reliable tool for biosensing applications, improving the accuracy and reliability of detection [15–17].

To gain a better understanding of the NC effect, it is essential to analyze the polarization variation with gate voltage, specifically the hysteresis curve of the MOS-capacitor that uses Si-doped Ferro material as the gate oxide. The hysteresis curve demonstrates the relationship between polarization and the gate voltage. As the voltage is increased or decreased, the polarization changes correspondingly. In a conventional MOS-capacitor, the hysteresis curve follows a typical S-shape, indicating the presence of hysteresis.

However, when using a Si-doped Ferro material as the gate oxide, the hysteresis curve takes on a different shape. In this case, the curve shows a negative slope, which is a characteristic of the NC effect. The negative slope indicates that the

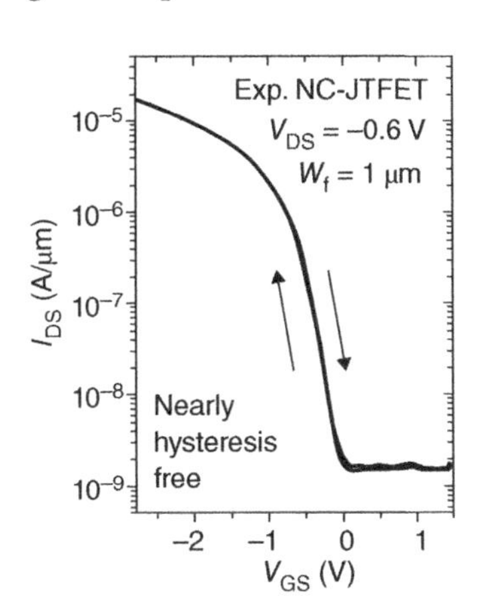

Figure 13.7 Measured hysteresis properties of fabricated negative capacitance-tunnel field-effect transistors ($L_f = 15\,\mu m$).

ferroelectric material's polarization is decreasing as the voltage increases, resulting in an apparent reduction in capacitance. This effect can be utilized to improve the performance of MOS-capacitors and other electronic devices, making them more energy-efficient and faster.

Hafnium(IV) oxide could be used as the ferroelectric substance (HfO_2). Furthermore, in some JL-TFET biosensors, the intrinsic voltage amplification behavior is investigated, and the relationship between the gate voltage and the resulting surface potential in the channel region of a JL-TFET is illustrated. The measured hysteresis properties of fabricated NC-TFET with $L_f = 15$ are given in Figure 13.7.

13.5 Variation of Source Material on Biosensing

In order to reduce the potential barrier width between the bands in an electronic device, it is advantageous to use a material with a low energy band gap, such as germanium, in the source border. This allows carriers to move more quickly from the valence band of the source to the conduction band of the channel, resulting in a more efficient device.

In contrast, the drain material, such as silicon, remains constant to maintain a high energy barrier width. This is important in the OFF state of the device, as a high energy barrier width helps to keep the OFF current low. By carefully selecting the appropriate materials for the source and drain regions, it is possible to optimize the device's performance and minimize power consumption. When a −ve gate voltage is applied, to optimize the ON current in an electronic device, germanium can be utilized in the source region. The reason why germanium is commonly used as a channel material in some electronic devices is because it has a narrow energy band gap, which makes it easier for electric charge carriers to move from the valence band of the source to the conduction band of the channel. By using germanium in the source region, the device's performance can be enhanced, resulting in a more efficient device.

Additionally, using dual metals on the electrode of the drain terminal can create a potential barrier that restricts hole tunneling. This is important in maintaining a low OFF current in the device, as hole tunneling can cause current leakage. By carefully designing the electrode of the drain terminal and selecting appropriate materials, it is possible to create a potential barrier that restricts hole tunneling while still allowing for efficient current flow in the ON state. The use of germanium in the source region and dual metals on the electrode of the drain terminal can lead to significant improvements in the performance of electronic devices. Specifically, it can result in a large I_{ON}/I_{OFF} current ratio, a smaller subthreshold swing (SS) point, and a standard SS.

The I_{ON}/I_{OFF} current ratio refer to the ratio of current in the ON state to the current in the OFF state. By utilizing appropriate materials and design techniques, it is possible to create a device that has a large I_{ON}/I_{OFF} current ratio, which is desirable in many applications.

The SS is a measure of how quickly the device can transition from the OFF state to the ON state. By reducing the point subthreshold swing, it is possible to achieve a faster transition, resulting in more efficient device operation. Additionally, achieving a standard SS is important to ensure that the device can function consistently and predictably in different operating conditions.

Overall, the use of germanium in the source region and dual metals on the electrode of the drain terminal can lead to significant improvements in the performance of electronic devices, making them more efficient and reliable in a variety of applications [13].

The proposed device design is based on the principle of charge plasma, obviating the need for chemical doping or ion implantation procedures. An important advantage of the suggested device is that it can be manufactured using a low thermal budget process, making it an attractive option.

The suggested device offers greater reliability because the difference in the source material raises the possibility of barriers at drain/channel interfaces. The JL-TFET utilizes an oxide stack consisting of a lower-k silicon dioxide (SiO_2) and a higher-k dielectric material, such as tantalum dioxide (TaO_2), titanium dioxide (TiO_2), and hafnium dioxide (HfO_2), to enhance its sub-threshold behavior.

Incorporating low bandgap materials towards the source side is an approach that tackles the primary restrictions of TFET, specifically the ambipolar conduction and ON-state current.

13.6 Importance of Dual Gate and Ferroelectricity on Biosensing

The addition of a second gate terminal on the opposite side of the body enhances the flexibility of TFETs. Double gate terminal devices are particularly advantageous because they provide superior control over short-channel effects (SCEs), which can be problematic in sub-50 nm nodes. In addition, double gate terminal devices reduce subthreshold leakage current and increase the overall ON-current, as they form two conductive channels. This makes them well-suited for circuits designed in the sub-50 nm regime. Isolating the gates also aids in reducing power consumption in JL-TFETs. Moreover, the presence of two types of gates above the layer of Si channel allows for the conversion of the highly doped n+–n+–n+ substrate region in the drain, channel, and source to a p+–i–n+ substrate region without the need for any doping.

The operation of a dual metal gate JL-TFET is based on the difference in work function between the gate material and the channel region. This work function difference leads to the depletion of charge carriers in the OFF state, which obstructs the flow of current and widens the depletion region. The application of a positive voltage to the gate of an n-type dual metal gate JL-TFET permits the ingress of carriers into the channel region, causing a reduction in the depletion region and the onset of conduction. The nanogap cavity can be occupied by air or another biomolecule having a dielectric constant greater than 1, resulting in a variation in the gate capacitance, and as a result, the channel potential and performance metrics change. The sensitivity of the biosensors can be determined based on the performance parameters. Thus, when a biomolecule enters the nanogap cavity, the performance measure changes, demonstrating the effectiveness of biosensors in detecting biomolecules [18].

The dual metal gate JL-TFET provides a solution to the issue of ambipolar behavior. The proposed device's reliability is improved by placing two metals with different work functions on the drain electrodes, thus potential barrier is created at the interfaces between the drain and channel. Incorporating dual metals on the drain electrode mitigates the adverse effects of hot carriers that can negatively impact the device's reliability [19]. The dual material gate method, with the higher work function gate metal (M2) positioned towards the drain side and the lower work function metal (M1) towards the source side, is an effective approach for enhancing the drive characteristics. Figure 13.8 shows 2-D schematic of dopingless negative capacitance ferroelectric TFET.

In order to obtain a low SS in a MOSFET gate stack, a ferroelectric dielectric may be employed in place of the conventional insulator (Table 13.1). The utilization of this type of insulator results in an increase in gate bias, which leads to a reduction in the SS. The permittivity of the ferroelectric layer is a critical characteristic that can be altered by adjusting the applied gate bias. To incorporate the

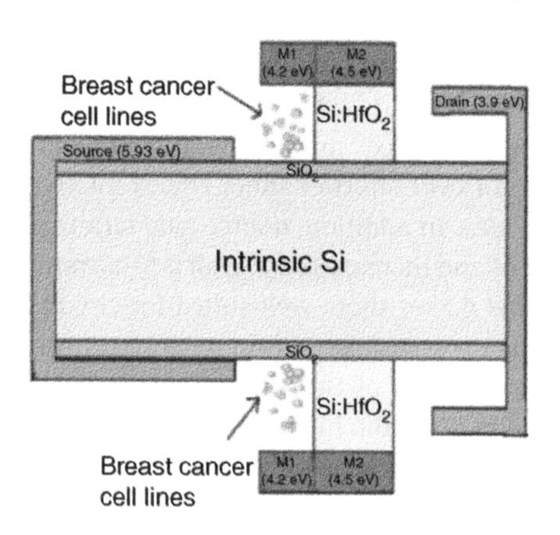

Figure 13.8 2-D schematic of dopingless dual gate negative capacitance ferroelectric TFET. Source: Adapted from [20].

Table 13.1 Device simulations parameters (Replacing the conventional insulator of a MOSFET integrated gate structure with a ferroelectric dielectric).

Parameters	TFET
Channel length (L_g)	100 nm
Source doping (N_A)	2×10^{19} cm^{-3}
Drain doping (N_D)	2×10^{18} cm^{-3}
Device layer thickness (t_{Si})	30 nm
Gate-oxide thickness (t_{ox})	3 nm
Buffer thickness (t_{ins})	—
Fe layer thickness (t_{Fe})	—
Gate work-function	4.7 ev

NC effect of the ferroelectric dielectric layer, it can also be employed as a substrate. When the thickness of the ferroelectric layer is increased, both the SS and threshold voltage also increase. When a gate voltage is applied, the energy band between the source and channel contracts, leading to the occurrence of tunneling current. The presence of trapped charges in the ferroelectric material causes the channel to form and the tunneling mechanism to initiate at a lower voltage. A thinner ferroelectric film has a greater NC, this, in turn, leads to an increase in the number of trapped charges within the film. Consequently, tunneling can begin at a reduced threshold voltage due to trapped charges. The ferroelectric material serves as a buffer, reducing both V_{Th} and SS. During transitions from OFF to ON, the ferroelectric capacitance captures charges from the channel, which helps to hasten the turn-OFF speed.

The intrinsic amplification of the electric potential applied to the gate terminal is achieved by the ferroelectric material. Ferroelectric charge plasma tunnel also increases sensitivity, which helps in optimum biosensing. The drain current increases with the transconductance. In 2010, Lattanzio and his research group proposed the use of a ferroelectric material in the integrated structure of a gate to achieve a NC effect and intrinsic voltage amplification. The incorporation of a ferroelectric material in the design of a tunnel field-effect transistor (TFET) can lead to increased drain current and transconductance, even at high temperatures beyond the Curie temperature of the material. This is achieved through the use of a ferroelectric tunnel FET (FE-TFET), which can take advantage of the unique properties of ferroelectric materials to improve its performance characteristics. By utilizing ferroelectric materials in this way, it is possible to enhance the operation of TFETs, enabling them to function more effectively under demanding conditions.

The FE gate stack utilizes the P(VDE-TrFE) principle to achieve a steep SS. When the NC effect is introduced, the electric field across the tunnel junction is significantly increased. Incorporating a ferroelectric insulator into the gate-stack, along with a conventional oxide, creates a NC effect. This effect acts as a step-up voltage transformer, leading to improved ON-current and a steeper SS in the device. By utilizing this approach, the ferroelectric insulator can help enhance the performance of the transistor beyond what would be achievable with only a conventional oxide in the gate-stack.

The attainment of extremely low power consumption, particularly in standby mode, can be expressed as achieving ultra-low power, which is a critical technological requirement for devices used in Internet of Things (IoT) applications. Negative capacitance FETs (NCFETs) with ferroelectric (FE) gates have shown advantages in terms of SS for low supply voltages of less than 60 mV/dec, but as the transport is now also dependent on thermionic emission, standby leakage current cannot be decreased [21–23]. A ferroelectric dielectric $Hf_{0.5}Zr_{0.5}O_2$ (HZO) film, for example, exhibits substantial improvements in I_{ON} without I_{OFF} degradation, SS, and nearly non-hysteresis behavior (Figure 13.9).

The uniform tunneling rate, known as G_{BTBT}, in a TFET is achieved due to the double-gate structure, which balances the tunneling speeds for both interfaces. This balanced tunneling rate is crucial for the optimal operation of the TFET. The tunneling rate can be expressed as equation:

$$G_{BTBT} = AE^{\sigma} \exp\left(-\frac{B}{E}\right) \tag{13.3}$$

In Eq. (13.3),

A is a constant and is dependent on the electrons' effective mass,
B is the constant of probability of tunneling,
σ is the constant of transition.

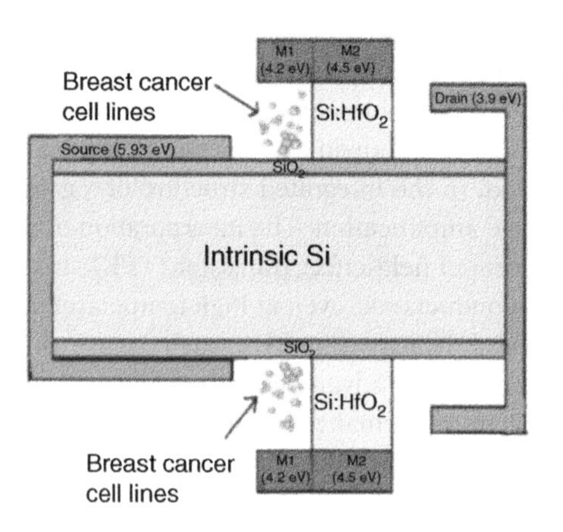

Figure 13.9 Two-dimensional diagram of a dopingless negative capacitance ferroelectric TFET [24]/IEEE.

The immobilization procedure begins with an elevation in the cell lines' dielectric constant within the cavity region. This increase in dielectric constant is the first step in the immobilization process, which involves fixing the cells in a specific location to enable further analysis or manipulation. By increasing the cell lines' dielectric constant, it becomes possible to manipulate the electrical properties of the cells, thereby facilitating immobilization. This results in a downward bending of the energy bands, as illustrated in Figure 13.10a. The consequent displacement of both bands causes a decrease in the barrier width towards the source-channel junction and an increase in electron tunneling. The shift in the source valence band and the channel region conduction band is dependent on variations in flatband voltage. It is worth emphasizing that when the dielectric constant values in the cavity region increase (from $K = 4.5–32$), there is a corresponding increase in the change in electric potential that occurs at the junction between the source and channel of a semiconductor device on its

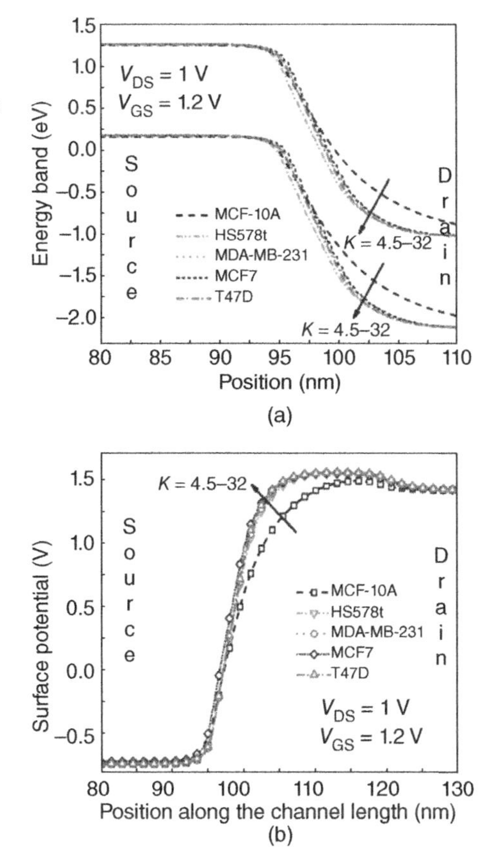

Figure 13.10 (a) depicts the energy band diagram and (b) depicts the surface potential of dopingless TFET-based biosensor as a function of device length when the dielectric constant of various breast cancerous cell lines is varied. Source: Adapted from [25].

surface, as illustrated in Figure 13.10b [26]. The alteration in surface potential can be credited to two primary factors: the sharper gradient of the junction profile and the presence of a binding site below the gate electrode. The combined effect of these factors results in more efficient control of the current flow in the device, making it possible to achieve better performance characteristics.

13.7 Effect of Dual Material Gate on Biosensing

An effective approach for enhancing the adaptability of the TFET is to introduce a second gate terminal on the opposite side of the body. Devices with double gate terminals provide improved management of SCEs, a challenge in nodes smaller than 50 nm. Additionally, they decrease subthreshold leakage current and increase the ON-current, as the formation of two conductive channels makes them more appropriate for circuits designed in the sub-50 nm range. Creating isolated gates also helps reduce power consumption in the JL-TFET. By having two kinds of gates above the channel layer of silicon, the highly doped n+–n+–n+ substrate region in the drain, channel, and source can be transformed into a p+–i–n+ substrate region without doping. The performance of the JFET-based biosensor has been improved using a dual material gate, where a metal with a high work function is placed near the source end to minimize leakage current, and a metal with a low work function is placed near the drain to keep the ON-state current constant. The two metalwork functions cause the threshold voltage near the source side to be greater than the threshold voltage near the drain side, which accelerates charge carriers in the conduit. In DMG FET structures, the step profile in the surface potential helps reduce SCEs by screening the drain potential variations. Asymmetric etching or asymmetric lift-off can be used to create the gate [27].

By placing two metals with different work functions on the drain electrode, a potential barrier is created at the drain/channel interfaces, which improves the device's reliability by reducing hot carrier effects [28]. Additionally, a dual material gate was utilized to improve the biosensor's performance. A metal with a high work function was placed near the source end to minimize leakage current, while a metal with a low work function was placed near the drain to maintain a constant ON-state current. Device performance was improved by using varying work functions for the control-gate and P-gate, resulting in a higher I_{ON}/I_{OFF} ratio and lower SS [29].

Dual metal gate JL-FET transistors are effective in solving the problem of ambipolar behavior, which is commonly observed in single-material gate JL-FET transistors. A potential barrier can be created at the drain/channel interfaces in the dual metal gate structure by placing two metals with different work functions on the drain electrode. This helps to improve the device's reliability by reducing hot carrier effects. The use of dual metals on the drain electrode also helps to reduce SS and improve the I_{ON}/I_{OFF} ratio. Overall, the dual metal gate JL-FET transistor is a promising solution for achieving better device performance and reliability [30–33].

References

1. Mascini, M., Marrazzaet, G., and Chianella, I. (2019). Disposable DNA electrochemical sensor for hybridization detection. *Biosensors and Bioelectronics* 14 (1): 43–51.
2. Sarkar, A., Das, A.K., De, S., and Sarkar, C.K. (2012). Effect of gate engineering in double-gate MOSFETs for analog/RF applications. *Microelectronics Journal* 43 (11): 873–882.
3. Pal, A. and Sarkar, A. (2014). Analytical study of dual material surrounding gate MOSFET to suppress short-channel effects (SCEs). *Engineering Science and Technology, an International Journal* 17 (4): 205–212.
4. Miller, M.M., Sheehan, P.E., Edelstein, R.L. et al. (2021). A DNA array sensor utilizing magnetic microbeads and magnetoelectronic detection. *Journal of Magnetism and Magnetic Materials* 225 (2): 138–144.
5. Souteyrand, E., Cloarec, J.P., Bessueille, F. et al. (2010). Control of immobilization and hybridization on DNA chips by fluorescence spectroscopy. *Journal of Fluorescence* 10 (3): 247–253.
6. Scheumann, V., Busse, S., Menges, B., and Mittler, S. (2021). Sensitivity studies for specific binding reactions using the biotin/streptavidin system by evanescent optical methods. *Biosensors and Bioelectronics* 17 (8): 704–710.
7. Chakraborty, A. and Sarkar, A. (2017). Analytical modeling and sensitivity analysis of dielectric-modulated junctionless gate stack surrounding gate MOSFET (JLGSSRG) for application as biosensor. *Journal of Computational Electronics* 16: 556–567. https://doi.org/10.1007/s10825-017-0999-2.
8. Sarkar, A., De, S., Dey, A. et al. (2012). Analog and RF performance investigation of cylindrical surrounding-gate MOSFET with an analytical pseudo-2D model. *Journal of Computational Electronics* 11: 182–195. https://doi.org/10.1007/s10825-012-0396-9.
9. Drummond, T.G., Hill, M.G., and Barton, J.K. (2018). Electrochemical DNA sensors. *Nature Biotechnology* 21 (10): 1192–1199.
10. Biswal, S.M., Baral, B., De, D., and Sarkar, A. (2016). Study of effect of gate-length downscaling on the analog/RF performance and linearity investigation of InAs-based nanowire tunnel FET. *Superlattices and Microstructures* 91: 319–330.

11 Sarkar, A. and Jana, R. (2014). The influence of gate underlap on analog and RF performance of III–V heterostructure double gate MOSFET. *Superlattices and Microstructures* 73: 256–267.

12 Nabaei, V., Chandrawati, R., and Heidari, H. (2018). Magnetic biosensors: modelling and simulation. *Biosensors and Bioelectronics* 103: 69–86.

13 Schaertel, B.J. and Firstenberg eden, R. Biosensors in the food industry: present and future. *Journal of Food Protection* 51 (10): 811–820.

14 Kim, Y.P., Jang, D.Y., Kim, H.S. et al. (2017). Sublithographic vertical gold nanogap for label-free electrical detection of protein-ligand binding. *Journal of Vacuum Science and Technology B* 25 (2): 443–447.

15 Biswas, K., Sarkar, A., and Sarkar, C.K. (2018). Fin shape influence on analog and RF performance of junctionless accumulation-mode bulk FinFETs. *Microsystem Technologies* 24: 2317–2324. https://doi.org/10.1007/s00542-018-3729-1.

16 Biswas, K., Sarkar, A., and Sarkar, C.K. (2015). Impact of barrier thickness on analog, RF and linearity performance of nanoscale DG heterostructure MOSFET. *Superlattices and Microstructures* 86: 95–104. ISSN 0749-6036. https://doi.org/10.1016/j.spmi.2015.06.047.

17 Chen, X., Guo, Z., Yang, G.-M. et al. (2017). Electrical nanogap devices for biosensing. *Materials Today* 13 (11): 28–41.

18 Kwon, Y.S., Nguyen, V.T., and Gu, M.B. (2017). Aptamer-based environmental biosensors for small molecule contaminants. *Current Opinion in Biotechnology* 45: 15–23.

19 Malhotra, B.D., Kumar, S., and Pandey, C.M. (2016). Nanomaterials based biosensors for cancer biomarker detection. *Journal of Physics Conference Series* 704: 1–11.

20 Barbaro, M., Bonfiglio, A., and Raffo, L. (2016). A charge-modulated FET for detection of biomolecular processes: conception, modeling, and simulation. *IEEE Transactions on Electron Devices* 53 (1): 158–166.

21 Choi, Y.K., Kim, C.H., Jung, C., and Park, H.G. (2018). Novel dielectric modulated field-effect transistor for label-free DNA detection. *BioChip Journal* 2 (2): 127–134.

22 Singh, S. and Singh, S. (2022). Dopingless negative capacitance ferroelectric TFET for breast cancer cells detection: design and sensitivity analysis. *IEEE Transactions on Ultrasonics, Ferroelectrics, and Frequency Control* 69 (3): 1120–1129.

23 Chandan, B. and Sharma, D. (2017). Junctionless based dielectric modulated electrically doped tunnel FET based biosensor for label-free detection. *Micro & Nano Letters* 13 (4): 452–456.

24 Venugopal, K., Rather, H.A., Rajagopal, K. et al. (2017). Synthesis of silver nanoparticles (Ag NPs) for anticancer activities (MCF 7 breast and A549 lung cell lines) of the crude extract of *Syzygium aromaticum*. *Journal of Photochemistry and Photobiology B: Biology* 167: 282–289.

25 Hoai, T.T., Yen, P.T., Dao, T.T. et al. (2020). Evaluation of the cytotoxic effect of rutin prenanoemulsion in lung and colon cancer cell lines. *Journal of Nanomaterials* 2020: 1–11.

26 Zhang, Y., Li, M., Gao, X. et al. (2019). Nanotechnology in cancer diagnosis: progress, challenges and opportunities. *Journal of Hematology & Oncology* 12 (1): 1–13.

27 Chen, E.Y. and Liu, W.F. (2018). Biomolecule immobilization and delivery strategies for controlling immune response. *Nanotechnology and Regenerative Medicine* 207–225.

28 Kanungo, S., Chattopadhyay, S., Gupta, P.S., and ET L. (2015). Comparative performance analysis of the dielectrically modulated full-gate and short gate tunnel FET-based biosensors. *IEEE Transactions on Electron Devices* 62 (3): 994–1001.

29 Sarkar, A. and Sarkar, C.K. (2013). RF and analogue performance investigation of DG tunnel FET. *International Journal of Electronics Letters* 1 (4): 210–217. https://doi.org/10.1080/21681724.2013.854158.

30 Biswal, S.M., Baral, B., De, D., and Sarkar, A. (2015). Analytical subthreshold modeling of dual material gate engineered nano-scale junctionless surrounding gate MOSFET considering ECPE. *Superlattices and Microstructures* 82: 103–112. ISSN 0749-6036. https://doi.org/10.1016/j.spmi.2015.02.018.

31 Kundu, A., Das, J.C., De, D. et al. (2023). Design of secure reversible select, cross and variation (RSCV) architecture in quantum computing. In: *2023 IEEE Devices for Integrated Circuit (DevIC)*, Kalyani, India, 243–247. https://doi.org/10.1109/DevIC57758.2023.10134998.

32 Biswas, K., Sarkar, A., and Sarkar, C.K. (2017). Spacer engineering for performance enhancement of junctionless accumulation-mode bulk FinFETs. *IET Circuits, Devices and Systems* 11 (1): 80–88.

33 Sarkar, A., De, S., Chanda, M., and Sarkar, C.K. (2016). *Low Power VLSI Design: Fundamentals*. Walter de Gruyter GmbH & Co KG.

14

Conclusion and Future Perspectives

Kalyan Biswas[1] *and Angsuman Sarkar*[2]

[1] *ECE Department, MCKV Institute of Engineering, Liluah, Howrah, West Bengal, India*
[2] *ECE Department, Kalyani Government Engineering College, Kalyani, Nadia, West Bengal, India*

14.1 Applications

Recent advancements in nanoscale devices have contributed to major developments in electronics and computing, leading to smaller, faster, and more portable systems that can accomplish and store more and more information. These developments have constantly supported new application areas as discussed below.

14.1.1 Opportunities in Big Data

In the era of information technology, nanoscale plays a significant part in taking care of huge information. These devices play a significant role by helping the realization of ultra-dense memory that can be useful to store remarkable amounts of information. Simultaneously, its advancements provide the inspiration to produce very effective algorithms for storing, handling, and processing information without damaging its dependability. Different architectures of personal computers (PC) propelled by the human intellect may also make use of energy more effectively and suffer less from excess heat – one of the very important issues due to the size reduction of electronic devices.

14.1.2 Fight Against Environment Change

One of the main ways to prevent climate change is to create a more efficient way of producing energy or to create products that use less energy. Thanks to advances in low-power nanoscale devices, we already see different low-power devices which consume little power in their circuit applications. Different energy

Advanced Nanoscale MOSFET Architectures: Current Trends and Future Perspectives,
First Edition. Edited by Kalyan Biswas and Angsuman Sarkar.

harvesting methodologies also are invented with the help of nanomaterials and nanoscale devices which are promising. Nanoparticles with proper design can harvest energy from the environment like light, temperature variations, wind, glucose, and numerous other sources with very high efficiency of conversion.

14.1.3 Creation of Graphene

One of the most potential applications of nanoscale devices is in the exploitation of graphene or similar super-materials by using passive nanostructures. As it is known that graphene is a material that is made of a lonely sheet of carbon having thickness of one atom. It is extremely hard and shows the potential of electrical conduction. It is very flexible as well and can be elongated up to 20% of its original length. It can perpetually bend without breaking. So, enormous possibilities are seen in future with respect to the graphene-based technologies like flexible screens in our smartphones and flexible solar cells for energy generation.

14.1.4 Nano Systems

Probably the most exciting aspect of nanoscale materials and devices is that its projections for the future. Nanosystems are the next step in realizing the sci-fi future we all dream of. Nanosystems always include individual nano devices or sensors or systems that can produce products on an atomic scale and work with other nanoparticles and machines to create molecular structures for specific tasks and actions. For example, nanosystems that can produce synthetic materials are capable of developing without any external input. These systems not only use less fossil fuels but also use carbon dioxide in the atmosphere, which is beneficial for protecting the environment and climate. Other possible applications take into account disaster-resistant buildings, self-healing structures, protective gear, etc.

14.1.5 Nanosensors

The small size of nanosensors using nanoscale MOSFETS makes them very attractive for medical applications because many biosensors exist and work at the micro- and nano-scale. In particular, nanoscale sensors are generally smaller than the physical body, allowing nanotechnology to assist in certain tasks such as administering drugs, and measuring and repairing damaged cells. Nano-devices, or nano-engineered materials, are less invasive than current medical devices and can be implanted directly into the body, reducing times for biochemical reactions and increasing precision. Composed of thousands of nanowires (usually made of carbon nanotubes or graphene), nanosensors can measure biomarkers from inside or outside the body using only breath or blood drops.

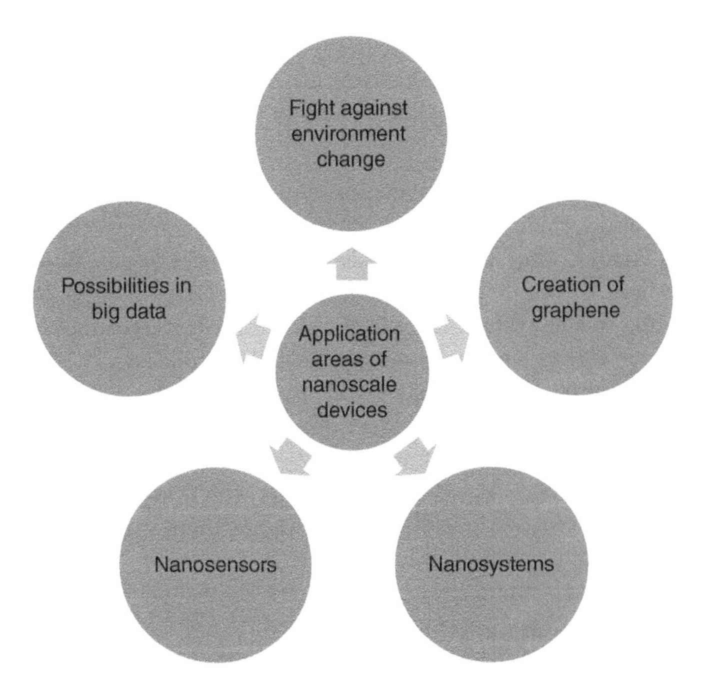

Figure 14.1 Application areas of nanoscale devices.

With further progress, these sensors will be sensitive enough to simultaneously test and identify many different cancers and diseases with a single sensor. In general, it can be said that the future applications of nanotechnology in medicine are very promising, but this does not promise that there will be no problems. More and more research and development studies should be carried out in the medical field and it should be ensured that the use of this technology does not harm patients. A summary of application areas is shown in Figure 14.1.

14.2 Some Recent Developments

A broad field of nano-device engineering is targeted at wearable devices. These devices include utilization of flexible electronics, rings, wristbands, smart glasses as well as connectable, wearable, and embedded contact lenses. The trend of wearable sensors is an emergent one having endless possibilities for performance improvement, enhanced security, and improved health. Nanotechnologies and nanoscale devices such as emerging MOSFEST technologies [1–16] are fundamentals to those technology breakthroughs as we witness and experience the change. According to recent reports, the global market for nanotechnology is expected to reach a size of US$112.8 Billion by 2030 (https://www.reportlinker.com/p0326269/Global-Nanotechnology-Industry.html). Along with TCAD modeling use of Artificial Intelligence has become a trend in semiconductor device modeling [17–23].

- Researchers at Stanford University, Massachusetts Institute of Technology, and many other leading researchers have found a way to use carbon nanotubes to build integrated circuits. They have developed methodologies to eliminate metallic nanotubes and keep only the semiconducting nanotubes in order to make the circuit work. In a published report in Nature Electronics, researchers have shown the way to fabricate carbon nanotube field-effect transistor (CNT-FET) in mass production on 200-mm wafers which is the industry standard in semiconductor chip design and development. Researchers from MIT have revealed that the commercial fabrication of CNTFET devices is possible with existing facilities and the available equipment systems which are normally used to fabricate the silicon-based transistors. This development can sustain the backbone of the current computing industry.
- Researchers from IMEC and Nantero have developed and demonstrated a memory chip using carbon nanotubes. This memory is called NRAM which stands for nanotube-based non-volatile random access memory and is anticipated to replace dense flash memory chips. The researchers of MIT have developed a new technology that can reduce defects and increase overall control in the CNFET fabrication process using traditional silicon techniques present in foundries. A 16-bit microprocessor having over 14 000 CNFETs are already has been demonstrated to perform similarly to conventional microprocessors.
- A research group belonging to the "Institute for Materials and Systems for Sustainability (IMaSS)" of Nagoya University, Japan has produced a nanosheet device in collaboration with NIMS. They have reported the maximum energy storage performance of the device. These results were published in the journal "Nano Letters" [24].
- Researchers from Beckman Institute for Advanced Science and Technology [25] have found a new system of material for an electronic switch that can change current at the nanoscale under the influence of external stimuli. The material has a unique structure for this molecular switch which is formed by locking a linear molecular backbone into a ladder-type structure. A new research has found that the ladder-like molecular structure greatly improves the stability of the product, making it perfect for single-molecule applications in electronics. This kind of development delivers a noteworthy step in forward direction toward the growth of functional molecular electronic devices. This device meets more or less all requirements for a single-molecule device i.e. stability

in the environment, can be switched on and off several times, is electrically conductive, and there are many molecular states that can be used.

- Chemists from Massachusetts Institute of Technology have estimated the energy transfer of photosynthetic light-harvesting proteins. At the time of photosynthetic cells absorbing light from the sun, energy packets known as photons jump between a chain of "light-harvesting" proteins till they reach the centre of photosynthetic reaction. There, cells are capable of converting the energy into electrons, which in due course power the production of sugar molecules [26].
- Material inspired by molecular gastronomy is developed by Queen Mary University of London to create smart wearables which surpass similar devices in terms of strain sensitivity. They have incorporated graphene material into seaweed to create nanocomposite microcapsules for sustainable and highly tunable electronics. When assembled into a network, microcapsules can record muscle tone, respiration, pulse rate, and blood pressure with very high precision and in real-time [27].
- A team of researchers from Pohang University of Science & Technology (POSTECH) developed a highly stretchable and high-performance organic polymer semiconductor. They have claimed that using this approach, it is possible to preserve up to 96% of the electrical performance parameters for the polymer semiconductor, even if it is stretched up to 80%. At the same time, the developed semiconductor material revealed considerably enhanced stretchability and stability in comparison to conventional semiconductors [28]. This development will enable integration of nanoscale MOSFET devices along with wearable electronic devices for further development.
- Researchers from Lund University have revealed how to make new configurable transistors and control them at a new, more accurate level. To realize this possibility, ferroelectric materials are employed in order to grasp this potential. These are exceptional materials that may alter their inner polarisation when put under the influence of an electric field. This recent work has studied new ferroelectric memories in the form of transistors with tunnel barriers to produce novel circuit architectures [29].
- A team of University of Minnesota has produced an energy-efficient and tunable superconducting diode. It is an encouraging component for the future of electronic devices that could be helpful to scale up quantum computers and improve the artificial intelligence systems. They have fabricated the device with three Josephson junctions, which are created by sandwiching a piece of non-superconducting material between superconductors. The researchers have demonstrated the connection of superconductors with different layers of semiconductors. This distinctive design helps the scientists to control the behavior of the device using voltage [30].

14.3 Future Perspectives

Nanoscale technology has enabled the quick advancements of device engineering that feature more compact and more efficient computing devices. It finds applications in all areas starting from wearable devices to other smart devices. In the first generation of nanoscale devices researchers have observed the advancements by enhancing the material properties by incorporating "passive nanostructures" It is expected that "active nanostructures" will be used in the second generation.

For the next few years, 2D materials and Graphene will persist as a very important domain of research in science and engineering. The available data of huge quantity and the expected extraordinary performance of nanoscale devices leave slight doubt about the future potential of 2D materials for their electronics, photonics, and sensing applications.

At device level, most frontline manufacturers of semiconductor devices are now moving from FinFETs structures to stacked nanosheet FET designs for entering the most advanced complementary metal–oxide–semiconductor (CMOS) technology nodes. Though these nanosheet device architectures are still based on Si-based channels but variations of such nanosheet designs are in the level of research and development for the upcoming technology nodes. These new designs involve forming more tightly spaced n-to-p nanosheets known as fork sheets or mixing p- and n-type nanosheets on top of each other. Additional scaling of the length of channel requires reduction of the channel thickness by a similar scale which can confirm sufficient control to overturn "short channel effects (SCEs)". However, if the thickness of the silicon wafer is reduced to the desired level, the resulting charge scattering at the interface increases, which causes a significant reduction in the mobility of carriers in the channel. In this situation, 2D materials are expected to be the decisive version for nanosheets, as these materials are "self-passivated in the third dimension and charge carrier mobility is not strongly affected from surface scattering" [31].

2D materials are well-suited for gas sensing and bio sensing due to their intrinsically large surface-to-volume ratio and flexible functionalization. For this reason, any molecule or charged particles in the neighborhood of few 2D layered materials can alter their conductivity. These materials also exhibit exceptional mechanical properties that make them enable to find applications in the area of Micro-electro-mechanical systems (MEMS)/nano-electro-mechanical systems (NEMS).

To realize energy-efficient hardware design, neuromorphic computing targets to deliver devices for brain-inspired computing and architectures for applications in the domain of artificial intelligence. At the level of the device model, neuromorphic computing requires the integration of memory technology with logic to permit "Computing-In-Memory" and "memristive device" characteristics that simulate synapses and neurons. The bottleneck of 2D material-based devices and their applications in electronics is the readiness levels of required manufacturing technology.

With the progress of technology, the energy consumption of CMOS devices has increased continuously due to the reduction of supply voltage. The high demand for portable devices needs additional provision of energy consumption and longer battery life. Different device technologies as well as circuit designs have always appeared to be of high performance and low leakage during standby. Variable and multiple threshold CMOS devices along with double-gated SOI technology with effect of back-gate going to create a visible effect in electronic field development. Triple-well CMOS technology may help to achieve the dynamic control of threshold voltage. All this know-hows are going to diminish leakage power when the device is in the idle state which will help to develop new technologies.

A broad area of nanotechnology-based device engineering focuses on wearable devices. Flexible electronics, rings, wristbands, smart glasses along with contact lenses that are wearable and embedded are major examples. The direction of wearable sensors is a developing area with immeasurable possibilities for security, performance, and health.

14.4 Conclusion

With the advancements of Nano technologies significant advancements in device engineering are realized. Advanced nano-devices help in achieving more compact electronic devices and capable computers. Device engineering through microprocessors and semiconductor chips used in electronic circuitry helps to develop new generation smartphones. These new generation smartphones have more computing capabilities than the computers that is only possible due to increase in nanoscale device performance. These advanced nanoscale devices are applied in electronic devices like computers, laptops, mobiles, televisions, and electronic sensors which is utilized in most transportation systems such as e-vehicles, airplanes, trains, and other autonomous vehicles. It is also essential parts of satellites that constantly monitor the Earth for environmental, and security purposes. Nanoscale devices shrinks the device size to a convenient limit for the home application as well as increases the performance and its intelligence capabilities. A current status of developments in the field of nano-electronics and its applications are summarized. Most of the recent developments related to the nanoscale electronic devices in recent time are examined and discussed in this book.

References

1 Sarkar, A. and Deyasi, A. (2022). *Low-dimensional Nanoelectronic Devices: Theoretical Analysis and Cutting-edge Research*. CRC Press.

2 De Sarkar, S. and Sarkar, C.K. (2013). *Asymmetric Halo and Symmetric SHDMG & DHDMGn-MOSFETs Characteristic Parameter Modeling*, vol. 26, no. 1, 41–55. IJNM, Wiley.

3 Sarkar, A., De, S., Dey, A., and Sarkar, C.K. (2012). A new analytical subthreshold model of SRG MOSFET with analogue performance investigation. *International Journal of Electronics* 99 (2): 267–283.

4 Sarkar, A., De, S., Dey, A., and Sarkar, C.K. (2012). Analog and RF performance investigation of cylindrical surrounding-gate MOSFET with an analytical pseudo-2D model. *Journal of Computational Electronics* 11: 182–195.

5 Sarkar, A. (2014). Study of RF performance of surrounding gate MOSFET with gate overlap and underlap. *Advances in Natural Sciences: Nanoscience and Nanotechnology* 5 (3): 035006.

6 Baral, B., Das, A.K., De, D., and Sarkar, A. (2016). An analytical model of triple-material double-gate metal–oxide–semiconductor field-effect transistor to suppress short-channel effects. *International Journal of Numerical Modelling: Electronic Networks, Devices and Fields* 29 (1): 47–62.

7 Chakraborty, A. and Sarkar, A. (2015). Investigation of analog/RF performance of staggered heterojunctions based nanowire tunneling field-effect transistors. *Superlattices and Microstructures* 80: 125–135.

8 Biswas, K., Sarkar, A., and Sarkar, C.K. (2015). Impact of barrier thickness on analog, RF and linearity performance of nanoscale DG heterostructure MOSFET. *Superlattices and Microstructures* 86: 95–104.

9 Biswas, K., Sarkar, A., and Sarkar, C.K. (2016). Impact of Fin width scaling on RF/Analog performance of junctionless accumulation-mode bulk FinFET. *ACM Journal on Emerging Technologies in Computing Systems (JETC)* 12 (4): 1–12.

10 Sarkar, A., De, S., Chanda, M., and Sarkar, C.K. (2016). *Low Power VLSI Design: Fundamentals*. Walter de Gruyter GmbH & Co KG.

11 Biswas, K., Sarkar, A., and Sarkar, C.K. (2017). Spacer engineering for performance enhancement of junctionless accumulation-mode bulk FinFETs. *IET Circuits, Devices and Systems* 11 (1): 80–88.

12 Chakraborty, A. and Sarkar, A. (2017). Analytical modeling and sensitivity analysis of dielectric-modulated junctionless gate stack surrounding gate MOSFET (JLGSSRG) for application as biosensor. *Journal of Computational Electronics* 16: 556–567.

13 Biswal, S.M., Baral, B., De, D., and Sarkar, A. (2019). Simulation and comparative study on analog/RF and linearity performance of III–V

semiconductor-based staggered heterojunction and InAs nanowire (NW) tunnel FET. *Microsystem Technologies* 25: 1855–1861.

14 Biswas, K., Sarkar, A., and Sarkar, C.K. (2018). Fin shape influence on analog and RF performance of junctionless accumulation-mode bulk FinFETs. *Microsystem Technologies* 24: 2317–2324.

15 Deyasi, A. and Sarkar, A. (2018). Analytical computation of electrical parameters in GAAQWT and CNTFET with identical configuration using NEGF method. *International Journal of Electronics* 105 (12): 2144–2159.

16 Basak, A. and Sarkar, A. (2020). Drain current modelling of asymmetric junctionless dual material double gate MOSFET with high K gate stack for analog and RF performance. *Silicon* 1–12.

17 Sarkar, C.K. (2013). *Technology Computer Aided Design: Simulation for VLSI MOSFET*. CRC Press.

18 Sarkar, A., De, S., and Sarkar, C.K. *VLSI Design and EDA Tools*, 1e. SciTech Publications.

19 Sinha, S., Biswas, K., Purkayastha, T. et al. (2016). On the electronic properties of guanine functionalized zigzag single-walled carbon-nanotube. *Journal of Nanoengineering and Nanomanufacturing* 6 (1): 3–8.

20 Biswas, K., Ghoshhajra, R., and Sarkar, A. (2022). High electron mobility transistor: physics-based TCAD simulation and performance analysis. In: *HEMT Technology and Applications*, 155–179. Singapore: Springer Nature.

21 Ghoshhajra, R., Biswas, K., and Sarkar, A. (2022). Device performance prediction of nanoscale junctionless FinFET using MISO artificial neural network. *Silicon* 1–10.

22 Ghoshhajra, R., Biswas, K., Sultana, M., and Sarkar, A. (2023). Ensemble Learning strategy in modeling of future generation nanoscale devices using machine learning. *2023 IEEE Devices for Integrated Circuit (DevIC)*, Kalyani, India, pp. 546–550, https://doi.org/10.1109/DevIC57758.2023.10134917.

23 Bishop, M.D., Hills, G., Srimani, T. et al. (2020). Fabrication of carbon nanotube field-effect transistors in commercial silicon manufacturing facilities. *Nature Electronics* 3: 492–501. https://doi.org/10.1038/s41928-020-0419-7.

24 Kim, H.-J., Morita, S., Byun, K.-N. et al. (2023). Ultrahigh energy storage in 2D high-κ perovskites. *Nano Letters* 23 (9): 3788. https://doi.org/10.1021/acs.nanolett.3c00079.

25 Li, J., Peng, B.-J., Li, S. et al. (2023). Ladder-type conjugated molecules as robust multi-state single-molecule switches. *Chemistry* https://doi.org/10.1016/j.chempr.2023.05.001.

26 Wang, D., Fiebig, O.C., Harris, D. et al. (2023). Elucidating interprotein energy transfer dynamics within the antenna network from purple bacteria. *Proceedings of the National Academy of Sciences* 120 (28): https://doi.org/10.1073/pnas.2220477120.

27 Aljarid, A.K.A., Dong, M., Yi, H. et al. (2023). Smart skins based on assembled piezoresistive networks of sustainable graphene microcapsules for high precision health diagnostics. *Advanced Functional Materials* 33 (41): 2303837. https://doi.org/10.1002/adfm.202303837.

28 Kim, S.H., Chung, S., Kim, M. et al. (2023). Designing a length-modulated azide photocrosslinker to improve the stretchability of semiconducting polymers. *Advanced Functional Materials* 33 (23): https://doi.org/10.1002/adfm.202370142.

29 Zhu, Z., Persson, A.E.O. & Wernersson, LE. Reconfigurable signal modulation in a ferroelectric tunnel field-effect transistor. *Nature Communications* 14, 2530 (2023). https://doi.org/10.1038/s41467-023-38242-w

30 Gupta, M., Graziano, G.V., Pendharkar, M. et al. (2023). Gate-tunable superconducting diode effect in a three-terminal Josephson device. *Nature Communications* 14 (1): 3078. https://doi.org/10.1038/s41467-023-38856-0.

31 Lemme, M.C., Akinwande, D., Huyghebaert, C. et al. (2022). 2D materials for future heterogeneous electronics. *Nature Communications* 13: 1392. https://doi.org/10.1038/s41467-022-29001-4.

Index

Advanced Nanoscale MOSFET Architectures: Current Trends and Future Perspectives, First Edition. Edited by Kalyan Biswas and Angsuman Sarkar.

Printed and bound by CPI Group (UK) Ltd, Croydon, CR0 4YY
16/04/2025
14658586-0003